Texts in Computing

Volume 10

Foundations of Logic
and
Theory of Computation

Second Edition

CW00606452

Texts in Computing Series Editor
Ian Mackie   mackie@lix.polytechnique.fr

Foundations of Logic and Theory of Computation

Second Edition

Amílcar Sernadas

and

Cristina Sernadas

ISBN 978-1-904987-88-8

College Publications
Scientific Director: Dov Gabbay
Managing Director: Jane Spurr
Department of Computer Science
King's College London, Strand, London WC2R 2LS, UK

http://www.collegepublications.co.uk

Original cover design by Richard Fraser
Cover produced by Laraine Welch
Printed by Lightning Source, Milton Keynes, UK

Preface to the Second Edition

Four years after the initial publication of the book and, so, three generations of students later on, the authors found worthwhile to collect in this second edition the many improvements of the text that were meanwhile prepared. The typos and errors found in the first edition were corrected. The material towards Kleene's normalization theorem was completely rewritten. Several proofs were improved, some with more details or hints to the reader, others streamlined. Some guiding comments were added here and there. Finally, Part IV was enriched with more exercises, most selected from written exams.

The authors are grateful to their students and colleagues who pointed out some of typos and errors in the first edition. The still remaining or freshly introduced mistakes and omissions are of course the sole responsibility of the authors.

Lisbon, August 1, 2012.

Amílcar Sernadas
Cristina Sernadas

Department of Mathematics
Instituto Superior Técnico, Universidade Técnica de Lisboa

Security and Quantum Information Group
Instituto de Telecomunicações, Lisboa

Preface

The main objective of the book is to provide a self-contained introduction to mathematical logic and computability theory for students of Mathematics or Computer Science. It starts with the basics of computability theory, presents the language, semantics, Hilbert calculus and Gentzen calculus of first-order logic, develops the notion of first-order theory, proves the incompleteness theorems about arithmetic and concludes with a decidable theory of arithmetic.

The material is organized around the failures and successes of Hilbert's programme for the formalization of Mathematics. It is widely known that the programme failed with Gödel's incompleteness theorems and related negative results about arithmetic. Unfortunately, the positive outcomes of the programme are less well known, even among mathematicians. The book covers key successes, like Gödel's proof of the completeness of first-order logic, Gentzen's proof of its consistency by purely symbolic means, and the decidability of a couple of useful theories, including Presburger arithmetic. It also tries to convey the message that Hilbert's programme made a significant contribution to the advent of the computer as it is nowadays understood and, thus, to the latest industrial revolution.

Part I starts with Hilbert's programme and moves on to computability. Part II presents first-order logic, including Gödel's completeness theorem and Gentzen's consistency theorem. Part III is focused on arithmetic, representability of computable maps, Gödel's incompleteness theorems and decidability of Presburger arithmetic. Part IV provides detailed answers to selected exercises.

The book can be used in several ways at late undergraduate level or early graduate level. An undergraduate course would concentrate on Parts I and II, leaving out Chapter 9, and sketching the way to the 1st incompleteness theorem. A more advanced course might skip early material already known to the students and concentrate on the positive and negative results of Hilbert's programme, thus covering Chapter 9 and Part III in full. Other paths through the book are possible as depicted in Figure 1.

The text is an outgrowth of the translation[1] into English of the 2006 Portuguese draft. It is the result of many years of experience of teaching logic and computability to undergraduate and graduate students.

The authors would like to express their deepest gratitude to their students who reacted to early drafts of the book, several colleagues (especially Luís Cruz-Filipe, Paulo Mateus, Jaime Ramos and João Rasga) who helped in the debugging stage, and the anonymous referee for many suggestions. The remaining mistakes and omissions are of course the sole responsibility of the authors.

The excellent working environment provided by the Department of Mathematics of Instituto Superior Técnico (IST) and the Security and Quantum Information Group (SQIG) of Instituto de Telecomunicações is also acknowledged by the authors.

Lisbon, March 14, 2008.

Amílcar Sernadas
Cristina Sernadas

Department of Mathematics
Instituto Superior Técnico, Universidade Técnica de Lisboa

Security and Quantum Information Group
Instituto de Telecomunicações, Lisboa

---

[1]By our colleague Luís Cruz-Filipe.

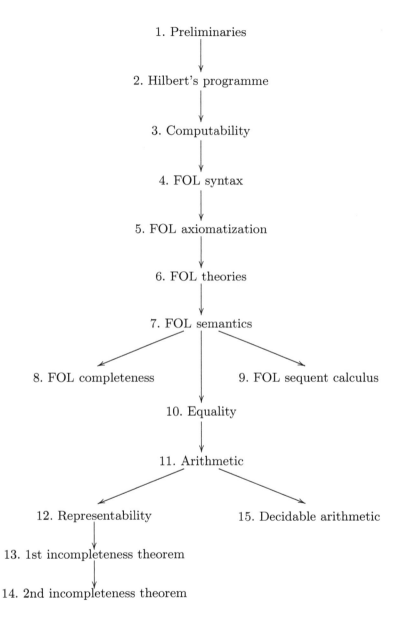

Figure 1: Chapter dependence graph.

x

Contents

Chapter 1

Preliminaries

As it is becoming usual, herein $\mathbb{N}$ denotes the set of natural numbers (that is, the set of non negative integers) and $\mathbb{N}^+$ denotes the set of positive integers. Furthermore, $\mathbb{Z}$ denotes the set of integers and $\mathbb{Q}$ denotes the set of rational numbers.

Given a set D, the set of all subsets of D is denoted by $\wp D$, and, the set of all finite subsets of D by $\wp_{\text{fin}} D$.

Following an European tradition, albeit not popular in the anglo-saxon literature, $\#D$ stands for the cardinality of D.

1.1   Functions versus maps

It is convenient to adopt a notation that distinguishes (possibly partial) functions from total functions, or maps. Therefore,

$$f : C_1 \to C_2$$

should be read as stating that f is a total function or *map* from C_1 into C_2, while

$$f : C_1 \rightharpoonup C_2$$

as stating that f is a (possibly partial) *function* from the source set C_1 into the target set C_2. In the first case, $\text{dom}\, f = C_1$ (the domain of f is C_1), while in the second case it is only known that $\text{dom}\, f \subseteq C_1$.

The set of all maps from C to D is denoted by D^C. Furthermore, one may write

$$(C \to D)$$

for D^C when declaring the domain and the target of a map. For instance, one may write $f : (A \to B) \to (C \to D)$ for $f : B^A \to D^C$.

1

1.2   Lambda notation

It is frequently convenient to use the lambda notation proposed by Alonzo
Church in [10] for describing functions. One writes

$$\lambda x_1, \ldots, x_n . e$$

to denote the n-ary function that assigns to each tuple $x_1, \ldots, x_n$ the value of
expression e that may depend on $x_1, \ldots, x_n$.

   For instance, $\lambda x, y . x + y$ denotes the map that assigns to each pair x, y
the value $x + y$ and should not be confused with the expression $x + y$ that
denotes a numerical value. Clearly, $\sin = \lambda x . \sin(x) = \lambda y . \sin(y)$. As a
further illustration, consider the difference between $\lambda x . x^2$ and $\lambda x, y . x^2$.

   Sometimes, the lambda notation is used together with the declaration of
the source and target sets of the function being defined. For example,

$$\lambda n . \{m \in \mathbb{N} : m < n\} : \mathbb{N} \to \wp \mathbb{N}$$

is the map that assigns to each natural number n the set of all natural numbers
that are smaller than n.

1.3   Inductive definitions as fixed points

It is very common to define a subset C of a given set D in a inductively way
by stating which elements of D are in C (the basis of the inductive definition)
and by giving mechanisms that provide additional elements of C when applied
to known elements of C (the step rules of the inductive definition).

   As an illustration, consider the following inductive definition of the set E
of even natural numbers as a subset of $\mathbb{N}$:

- $0 \in E$;

- $k + 2 \in E$ whenever $k \in E$.

Clearly, E is the least fixed point containing the singleton set $\{0\}$ of the
following operator:

$$O_E = \lambda A . A \cup \{a + 2 : a \in A\} : \wp \mathbb{N} \to \wp \mathbb{N}.$$

That is,
$$E = \mathrm{lfp}(O_E, \{0\})$$

where, in general, given $O : \wp U \to \wp U$ and $B \subseteq U$ for some universe U at
hand, $\mathrm{lfp}(O, B)$, the least fixed point of O containing B, is such that:

- $O(\mathrm{lfp}(O, B)) = \mathrm{lfp}(O, B)$;

- $B \subseteq \mathrm{lfp}(O, B)$;

- $\mathrm{lfp}(O, B) \subseteq C$ for every C such that $O(C) = C$ and $B \subseteq C$.

It is known[1] that such least fixed point of O containing B exists when O is monotonic.[2] Furthermore, if O is defined using only finitary rules,[3] then[4]

$$\mathrm{lfp}(O, B) = \bigcup_{n \in \mathbb{N}} O^n(B).$$

In particular, the reader should be able to verify that

$$\mathrm{lfp}(O_E, \{0\}) = \bigcup_{n \in \mathbb{N}} O_E^n(\{0\}).$$

These ideas apply to any inductively defined set (provided that only finitary rules are used) and will be put to good use in Chapter 5.

Observe that the inductive definition of E is free in the sense that, by successive application of the step rule from the basis element, no element is ever regenerated, that is, each time a new element is generated that was not previously known to belong to E.

In general, a finitary inductive definition (with set B of basis elements and associated operator O) of a set C is said to be *free* if, for each $c \in C$, $c \in B$ or and only or there is only and only one $n \in \mathbb{N}$ such that $c \in (O^{n+1}(B) \setminus O^n(B))$.

Given a free inductively defined set C_1, it is very frequent to define a map $f : C_1 \to C_2$ by taking advantage of the inductive structure of C_1, as follows:

- the value of f should be stated for each element in the basis;

- for each step rule of the form

$$h(c_1, \ldots, c_n) \in C_1 \text{ whenever } c_1, \ldots, c_n \in C_1$$

the value of $f(h(c_1, \ldots, c_n))$ should be stated as a function of values $f(c_1), \ldots, f(c_n)$.

In this situation, one says that map f was *inductively defined on the structure of set C_1*. Note the need for the freeness assumption in order to guarantee that f is well defined.

For instance, consider the map $f : E \to \mathbb{N}$ such that:

---

[1]Thanks to the Knaster–Tarski theorem.

[2]That is, $O(A) \subseteq O(A')$ whenever $A \subseteq A'$.

[3]A finitary rule generates a new element from each finite set of elements already known to belong to the set being inductively defined, while an infinitary rule builds new elements from infinite sets of known elements.

[4]Recall that $O^0(B) = B$ and $O^{n+1}(B) = O(O^n(B))$.

1. $f(0) = 0$;

2. $f(k + 2) = f(k) + 1$ for each $k \in E$.

Given the inductive nature of E, clauses 1 and 2 above are sufficient to determine the value of f when applied to any element of E. The reader should be able to verify that $f(k) = k/2$ for every $k \in E$.

1.4 Proofs by induction

The reader is probably familiar with proofs by induction on the natural numbers, using the the following simple principle:

Induction over $\mathbb{N}$
From

$$\begin{cases} \text{(Basis)} & \alpha(0) \\ \text{(Step)} & \text{for every } n \in \mathbb{N}: \\ & \text{if } \alpha(n) \\ & \text{then } \alpha(n+1), \end{cases}$$

conclude

$$\text{for every } n \in \mathbb{N}: \alpha(n).$$

When proving the step, it is necessary to derive $\alpha(n+1)$ from $\alpha(n)$, the so called *induction hypothesis* (IH).

A variant of the above principle is frequently useful when it is necessary to invoke the induction hypothesis over any $k < n + 1$:

Complete induction over $\mathbb{N}$
From

$$\begin{cases} \text{(Basis)} & \alpha(0) \\ \text{(Step)} & \text{for every } n \in \mathbb{N}: \\ & \text{if for every } k < n + 1: \alpha(k) \\ & \text{then } \alpha(n+1), \end{cases}$$

conclude

$$\text{for every } n \in \mathbb{N}: \alpha(n).$$

Furthermore, it is left to the interested reader to verify that complete induction over $\mathbb{N}$ can be stated more elegantly as follows:

Complete induction over $\mathbb{N}$

From

$$\text{(Step)} \quad \text{for every } n \in \mathbb{N} :$$
$$\text{if for every } k < n : \alpha(k)$$
$$\text{then } \alpha(n),$$

conclude

$$\text{for every } n \in \mathbb{N} : \alpha(n).$$

Proofs by induction are not restricted to proofs about natural numbers. The most general principle applies to any well-founded preorder.

Recall that a binary relation $\preccurlyeq$ in a set C is a *preorder* if it is reflexive and transitive. Such a relation is said to be *well-founded* if every non empty subset D of C has a minimal element. That is, the relation is well-founded if for every non empty $D \subseteq C$ there is $e \in D$ such that there is no $d \in D$ such that $d \prec e$, where relation $\prec$ is the strict order induced by $\preccurlyeq$ as follows

$$a_1 \prec a_2 \ \text{if} \ \begin{cases} a_1 \preccurlyeq a_2 \\ a_1 \not\approx a_2, \end{cases}$$

with relation $\approx$ being the equivalence relation induced by $\preccurlyeq$ as follows

$$a_1 \approx a_2 \ \text{if} \ \begin{cases} a_1 \preccurlyeq a_2 \\ a_2 \preccurlyeq a_1. \end{cases}$$

Note that $\approx$ is indeed an equivalence relation, since it is reflexive, symmetric and transitive, as the reader can easily check.

Structural induction over a well-founded preorder $\preccurlyeq$ **in set** C

From

$$\text{(Step)} \quad \text{for every } c \in C :$$
$$\text{if for every } d \prec c : \alpha(d)$$
$$\text{then } \alpha(c),$$

conclude

$$\text{for every } c \in C : \alpha(c).$$

Clearly, the induction principles over $\mathbb{N}$ can be proved from structural induction which is left as an exercise to the interested reader.

As an illustration of structural induction over a well-founded preorder, see below the proof of the following fact:

$$\text{for every } e \in E : e \equiv 0 \text{ mod } 2.$$

Given $n_1, n_2, n \in \mathbb{Z}$, recall that n_1 and n_2 are said to be *congruent modulo n*, written

$$n_1 \equiv n_2 \operatorname{mod} n,$$

if their difference $n_1 - n_2$ is a multiple of n.

The proof is made using the well-founded preorder

$$\preccurlyeq \subset E^2$$

induced by the inductive definition of E as follows. The rule $k+2 \in E$ whenever $k \in E$ imposes that $k \prec k + 2$ for each $k \in E$. The envisaged preorder $\preccurlyeq$ is obtained by transitive and reflexive closure of this binary relation $\prec$. This preorder is indeed well-founded, as the reader will easily verify. Remark that on free inductive definitions like this one, the induced preorder $\preccurlyeq$ is in fact a partial order because it is also antisymmetric.

Using the induction hypothesis

$$\text{for every } d \prec e : d \equiv 0 \text{ mod } 2$$

one has to prove that

$$e \equiv 0 \text{ mod } 2$$

holds. There are two cases to consider:

- either $e = 0$, in which case $0 \equiv 0$ mod 2 since 0 is multiple of 2;

- or $e \neq 0$, in which case $e - 2 \prec e$ and, by IH, $e - 2 \equiv 0$ mod 2, from where it can be concluded that

$$(e - 2) + 2 \equiv 0 \text{ mod } 2$$

and, thus, $e \equiv 0$ mod 2.

Finally, the thesis results by structural induction on $\preccurlyeq$.

This idea can be and will be heavily used for applying structural induction to the well-founded preorder induced by any inductive definition of a set. In such cases, one says that the proof is made by *induction on the structure* of each element.

Observe that, for each step rule of the form

$$h(c_1, \ldots, c_n) \in C \text{ whenever } c_1, \ldots, c_n \in C$$

in the inductive definition of C, the induced partial order $\preccurlyeq$ is such that:

$$c_i \prec h(c_1, \ldots, c_n) \text{ for } i = 1, \ldots, n.$$

Structural induction can also and will be frequently applied for proving that a property holds for every element of a set C for which a map $f : C \to \mathbb{N}$ is known. In that case, the proof is said to be carried out by *induction on f*. Indeed, map f induces the following well-founded preorder over C:

$$c_1 \preccurlyeq_f c_2 \text{ if } f(c_1) \le f(c_2).$$

This technique is widely adopted by logicians for proving properties about terms, formulas or derivations, using maps like depth and length.

1.5 Programming in *Mathematica*

It is assumed that the reader has, at least, some rudimentary skills in computer programming, preferably using a high level imperative language. In this book, procedures are frequently written in *Mathematica* [52, 8], a very high level contemporary programming language (with symbolic facilities and a large number of built-in functions) that is very easy to learn at the low sophistication level required here.

This section provides a very quick introduction to *Mathematica* for those who do not know it. As a first illustration of the basic constructs of the language, consider the following three alternative programs for computing the sum of the natural numbers less than or equal to the parameter k:

- Imperative:

```
Function[k,
    j = 0;
    s = 0;
    While[j ≤ k
        s = s + j;
        j = j + 1
    ];
    s
]
```

Observe that the value returned by the function is the value of the last evaluated expression, hence the need to conclude the procedure with the evaluation of s.

- Recursive:

$$f = \mathsf{Function}[k,$$
$$\mathsf{If}[k == 0,$$
$$0,$$
$$k + f[k-1]$$
$$]$$
$$]$$

Note that the application of a function h to an argument x is written $h[x]$ in *Mathematica*.

- Functional:

$$\mathsf{Function}[k,$$
$$\mathsf{Apply}[\mathsf{Plus}, \mathsf{Table}[i, \{i, k\}]]$$
$$]$$

Expression $\mathsf{Table}[h[i], \{i, m\}]$ denotes the list $\{h[1], \ldots, h[m]\}$, also represented in *Mathematica* by the expression $\mathsf{List}[h[1], \ldots, h[m]]$.

The primitive *Mathematica* function Apply replaces the head of the second argument (List in the case above) by the first argument (Plus in the case above). Thus, the result above is $\mathsf{Plus}[h[1], \ldots, h[m]]$ as desired.

By the way, the empty list is denoted in *Mathematica* by $\{\}$ or by $\mathsf{List}[]$, and $w[[i]]$ denotes the i-th element of a list w, while $a[[i, j]]$ and $a[[i]][[j]]$ both denote element a_{ij} of a matrix a (represented in *Mathematica* as a list of lists, that is, as a list of rows).

More generally, if a is an expression, then $a[[i, j]]$ and $a[[i]][[j]]$ both denote the j-th argument of the i-th argument of the expression. For instance,

$$h[c, g[x, y]][[2, 1]] \text{ is } x.$$

Furthermore, $a[[0]]$ is the head of expression a. So,

$$h[c, g[x, y]][[0]] \text{ is } h$$

and

$$h[c, g[x, y]][[2, 0]] \text{ is } g.$$

However,

$$h[c, g[x, y]][[1, 0]] \text{ is } \mathsf{Symbol},$$

while

$$h[c[], g[x, y]][[1, 0]] \text{ is } c.$$

For this reason, later on, when manipulating terms and formulas, atomic terms (variables and constants) are represented explicitly as functions with no arguments. Thus, for instance, variable x is represented in the programs of Chapter 4 by $x[\,]$.

Other primitive functions of *Mathematica* used in this book that operate on lists have obvious meanings, like Length, First, Rest and Append. Function Map is used to apply the first argument (a function) to each element of the second argument (a list). Function Take is used herein only as follows: Take$[w, n]$ is the list of the first n elements of list w. All these functions also operate on expressions. For instance, Length applied to $f[a, b, c]$ returns 3.

By now the reader will have guessed that all primitive *Mathematica* functions start with a capital letter. For this reason, it is recommended to use lower case letters in the names of objects introduced by the programmer.

Strings of characters are represented as usual in most programming languages. For instance, "*abc*" stands for the string abc of length 3. As expected, many of the programs in this book involve strings. However, only a few primitive *Mathematica* functions dealing with strings are needed, namely:

- StringLength that obviously returns the length of the string argument.

- StringTake that works like Take but for strings.

- ToString that when applied to a *Mathematica* expression returns the corresponding string of characters.

- SyntaxQ that when applied to a string of characters checks if it can be parsed as a well formed *Mathematica* expression.

- ToExpression that parses the string argument, returning the corresponding *Mathematica* expression when successful.

String concatenation can be achieved in *Mathematica* using the binary operator <>. For instance, "*abc*" <> "*xy*" results in the string "*abcxy*".

Concerning characters, only two primitive functions of *Mathematica* are used:

- FromCharacterCode returns the character with the code given as argument.

- ToCharacterCode does the reverse.

With respect to numerical expressions, remark that, like in the every day language of mathematicians, $x\,y$ stands for $x \times y$ in *Mathematica*. Otherwise, *Mathematica* is similar to other high level programming languages.

Boolean expressions follow also the usual syntax. Primitive constants, like True and False are self explanatory, as well as the functional connectives Not, And and Or. The latter two have the usual infix counterparts && and ||. Moreover, the usual logical connectives can also be used: $\neg$, $\wedge$ and $\vee$.

Primitive infix predicates include $==$, $===$, $!=$ (also written $\neq$), $=!=$, $\leq$ (also written $<=$), etc. The hard equality ($===$) and its negation ($=!=$) always evaluate to true or false, while the soft equality ($==$) and its negation ($!=$) are lazily evaluated. The difference is especially important when working at the symbolic level. For instance, if no values were given before to x and y, $x == y$ evaluates to itself while $x === y$ evaluates to false. Primitive functional predicates in *Mathematica* have names ending with a Q, for instance EvenQ and SyntaxQ.

Function FactorInteger, when applied to a natural number greater than 1, returns the list of pairs {prime number, power} in the prime factorization of the argument, sorted by the first element of the pairs. For instance,

$$\text{FactorInteger}[40] \text{ is } \{\{2,\,3\},\,\{5,\,1\}\}.$$

Finally, primitive function TimeConstrained, when applied to a *Mathematica* program p, a natural number t and a value v, runs p up to t units of time, returning v if the computation is not finished until then. This primitive function is used in the following definition of the function stceval (*success and time constrained evaluation*) which will be put to good use in Chapter 3:

stceval$[e_, t_, v_] := $ Check[TimeConstrained$[e, t, v], v]$;
SetAttributes[stceval, HoldFirst]

According to this definition,

$$\text{stceval}[e, t, v]$$

returns the value of expression e if the evaluation ends with no error within t units of time; otherwise, it returns v.

Part I

Formalizing Mathematics

Chapter 2

Hilbert's Programme

The great mathematician David Hilbert believed that Mathematics should be developed axiomatically, in a manner analogous to what he had achieved for Geometry [26]. One of his main worries was the issue of the consistency of Mathematics. In particular, the consistency of his axiomatization of Geometry was established by presenting a model based on the reals. However, the consistency of Calculus was itself at stake, since it relied on models based upon sets. Recall that, at the time, Set Theory faced the problems caused by Bertrand Russell's paradox (see in [51] his original letter to Gottlob Frege).

At that stage, Hilbert concluded that one should prove the consistency of an axiomatization without resorting to other theories, by a direct method relying on pure mechanizable symbol manipulation techniques.[1] This is the second of the famous list of twenty-three problems proposed by Hilbert at the International Congress of Mathematicians in 1900. Hilbert raised another important question related to his idea of formalizing mathematics: is it the case that every mathematical problem is decidable? In the years that followed, Hilbert contributed to the development of the first-order predicate calculus (see a translation in [28]), where he intended to develop axiomatizations for different fragments of Mathematics and proofs of their consistency and decidability.

The programme of formalizing Mathematics, started in this way by Hilbert, suffered a major blow in 1931, due to the famous theorems proved by Kurt Gödel about the incompleteness of the arithmetic of natural numbers (see his original article in [20]).

Notice that the arithmetic of natural numbers is a very simple fragment of Mathematics (at least in comparison with other parts of this science) and, even so, its full formalization was thus shown by Gödel to be impossible.

---

[1]Nowadays, one would say by a computer program.

Notwithstanding the ultimate failure of Hilbert's programme, the work carried out towards the formalization of Mathematics contributed significantly to the advent of computers as they are understood nowadays and, hence, to the latest industrial revolution.

From the strict point of view of the role of formal logic in Mathematics, the aforementioned negative results should be balanced against some positive outcomes of Hilbert's programme:

- Proof by Kurt Gödel in 1929 of the completeness of the first order calculus (in his PhD thesis reproduced and translated in [20]), thus capturing in a symbolic system the properties of quantifiers and connectives which are at the core of the reasoning methods used by mathematicians.

- Proof by Gerhard Gentzen in 1936, using purely symbolic techniques, of the consistency of first-order predicate logic (see [18]).

- Partial formalization in first-order predicate logic of useful fragments of contemporary Mathematics, frequently using Zermelo-Fraenkel set theory that, while avoiding Russell's paradox, captures a significant part of the mathematical knowledge about sets (for an introduction see Chapter 9 of Shoenfield's textbook [42]).

- Proof of the decidability of several useful theories, like those of real closed fields (Alfred Tarski in [47, 50]), algebraically closed fields, Presburger arithmetic (Mojżesz Presburger [37]), term algebras, atomless Boolean algebras, Abelian groups (Wanda Szmielew [44]), dense linear orders without left and right endpoints, and random graphs. Some of these decidable theories are crucial ingredients of computer systems in Robotics, CAD (Computer Aided Design) and other economically significant areas.

Furthermore, the work aiming at formalizing Mathematics is still going on in several directions, whether by relaxing the notion of procedure, whether by considering other fragments of Mathematics that are formalizable in the traditional sense. The latter have been used for extracting algorithms from constructive proofs.

Therefore, it seems fair to rethink Hilbert's programme in a more positive way than it is usually believed. For more details on the positive impact of Hilbert's programme see, for instance, the survey [54].

The notions of "formal language" and "mechanizable procedure" are at the heart of Hilbert's programme of formalizing Mathematics. Their study requires a brief look at what is nowadays known as Computability Theory, another fruitful outcome of the programme. Computability Theory was developed with contributions from Alan Turing, Alonzo Church and Stephen Kleene, among others.

Chapter 3

Computability theory

The basic notions of computability theory used in this book are introduced in
this chapter, namely those of computable function, computable set and computably enumerable set. It is common in the literature to work within the
universe of natural numbers (see, for instance, [38, 6]). Herein, these notions
are defined within any collection of universes built from finite alphabets. A
rigorous notion of Gödelization is used for establishing the bridge between the
two approaches. Several useful results are proved, namely alternative characterizations of computable enumerability.

The concepts of alphabet and formal language are introduced in Section 3.1.
The Church-Turing thesis is discussed in Section 3.2. The latter section ends
with a justification for defining the notion of computable function using *Mathematica*.

The core notions and results of computability theory are developed in Section 3.3. The definitions of computable function and computable set are given
after defining frame, type, universe and formal set. The problem of enumerating universes is discussed in depth before presenting the notion of computably
enumerable set. The algebras of computable sets and of computably enumerable sets are analyzed along the way. In addition, several results about different
characterizations of computably enumerable sets are proved (among them the
projection theorem that plays a key role in Chapter 14). The section ends
with some results concerning the computability and computable enumerability
of the image and inverse image given by computable maps of computable and
computably enumerable sets.

In Section 3.4, the notion of Gödelization is rigorously defined and key
robustness results are established. Section 3.5 concentrates on the notion of
computable function proposed by Kleene. Finally, Sections 3.6 and 3.7 briefly
deal with some more advanced issues in computability theory, like the unde-

cidability of the halting problem, universal functions and enumeration of the set of computable functions, but only to the extent that they are needed in the rest of the book.

3.1 Formal languages

An *alphabet* is defined as a countable, non empty set. A *formal language* (or simply *language*) *over an alphabet* A is a subset of A^*. Recall that A^* stands for the free monoid generated from A, that is, the set of all finite sequences of elements of A endowed with the concatenation operation $\oplus$ whose unit element is the empty sequence ε. In this setting, the elements of the alphabet are known as *symbols* and the elements of the language are known as *words*. The full language A^* is said to be the *linguistic universe* (or, simply, *universe*) over A. The length of a word w is denoted by $|w|$.

Henceforth, $\{1\}^*$ is identified with the set $\mathbb{N}$ of natural numbers, by identifying ε with 0 and concatenation with sum, hence justifying to some extent the unorthodox notation adopted in this book for concatenation.[1] More generally, the sequence of length k is identified with the natural number k (unary representation of the natural numbers).

In the context of this linguistic universe, one might wonder whether the language of even numbers is computable. In other words, is there an "algorithm" that, given an element of $\mathbb{N}$, will answer whether or not it belongs to that language? The answer to this question is clearly positive, but it does raise the problem of rigorously defining the notion of algorithm.

Intuitively, a procedure is a recipe that defines a sequence of basic operations to be executed upon any tuple of data given at the start. An algorithm is a procedure whose execution always terminates, regardless of the tuple of input data, in a finite number of steps. A function is considered to be computable if there is a procedure that calculates its values: if the function is defined for a particular tuple of input data, then execution of the procedure upon that tuple will necessarily finish in a finite number of steps, yielding as result the value of the function on that tuple; otherwise, the execution of the procedure may not terminate, and even if it does no result will be returned. These ideas are made precise in the following sections.

3.2 Church–Turing postulate

Several approaches have been proposed to define formally the notion of computable function, for instance:

---

[1]Most authors write $w\,w'$ for the concatenation of w and w'.

- Alan Turing's machines [48, 25, 49];

- Alonzo Church's lambda-definable functions [10];

- Stephen Kleene's recursive functions [32];

- Emil Post's machines [36];

- universal register machine (an abstraction of present-day computers) [41].

It has been shown that these formal notions of computability are equivalent, leading Kleene to state the following postulate that became known as *Church–Turing Thesis*:

> Any function that can be accepted as computable is formally computable by some Turing machine.

That is, any formal notion of computability identifies only Turing-computable functions.

Notice that the Church–Turing Thesis cannot be proved. Only on a case-by-case basis can it be shown that a particular formalization of computability only identifies as computable precisely the Turing-computable functions.

It is common in the textbooks on Mathematical Logic to formalize the notion of computable function using Kleene's recursive functions. Herein, the programming language *Mathematica* [52, 8] is adopted instead. *Mathematica* is a very high level contemporary programming language (with symbolic facilities and a large number of built-in functions) which is very easy to learn at the level required in this book. This approach has the advantage of capitalizing on the reader's experience in programming, even if using a different programming language.

However, it is also necessary to work with simpler formalizations of the notion of computability. For the purposes of this book, Kleene's notion (presented in Section 3.5 of this chapter) and Gödel's approach (introduced in Chapter 12) are especially useful.

3.3 Computability in *Mathematica*

The objective of this section is to introduce the notion of computable function using the *Mathematica* programming language and afterward the concepts of computable and computably enumerable set.

Assume that a work *frame* has been chosen, that is, a finite set $\mathcal{A}$ of finite alphabets such that $\{1\} \in \mathcal{A}$ and each $A \in \mathcal{A}$ is endowed with a strict total order. That is, for each $A \in \mathcal{A}$ there is a binary relation, say $<_A$, such that, for any $a_1, a_2, a_3 \in A$:

- (totality) either $a_1 <_A a_2$ or $a_2 <_A a_1$ or $a_1 = a_2$;

- (asymmetry) if $a_1 <_A a_2$, then $a_2 \not<_A a_1$;

- (transitivity) if $a_1 <_A a_2$ and $a_2 <_A a_3$, then $a_1 <_A a_3$.

For example, the usual ordering of the Latin alphabet is a strict total order. Any two letters can be compared (for instance $\mathsf{p} < \mathsf{q}$). Moreover, the ordering satisfies asymmetry (it is not the case that $\mathsf{q} < \mathsf{p}$), as well as transitivity. The usual $<$ relation over the set of integers is another example of a strict total order.

The importance of the assumption that every alphabet has a strict order will become clear when proving Proposition 8 below.

A *type* is a (possibly empty) finite sequence of alphabets in $\mathcal{A}$. A *formal set* (or simply *set*)[2] C *of non empty type* $A_1 \dots A_n \in \mathcal{A}^*$ is a subset of $A_1^* \times \cdots \times A_n^*$. Set $A_1^* \times \cdots \times A_n^*$ is known as the *universe* of type $A_1 \dots A_n$. By convention, the universe of the empty type ε is the set $\{\varepsilon\}$. Hence, there are only two sets of the empty type: the empty set and the universe.

Let C_1 and C_2 be sets within the frame at hand. A function $f : C_1 \rightharpoonup C_2$ is said to be *computable* if there is a procedure p, written in the *Mathematica* programming language,[3] such that:

- given an element c in C_1 for which f is defined, the computation of $p[c]$ terminates within a finite number of steps, returning $f(c)$;

- given an element in C_1 for which f is not defined, the computation of $p[c]$ does not return a value, producing an error message or never terminating.

In this situation, function f is said to be *computed* by procedure p. If f is a map (that is, a total function), then p is an algorithm, since its execution should always terminate, regardless of the input. Observe also that nothing is imposed on the behavior of procedure p when provided with an input outside of set C_1.

Exercise 1 What is the cardinality of the set of computable functions from $\mathbb{N}$ to $\mathbb{N}$?

The set of all computable functions (in *Mathematica*) is denoted as $\mathcal{C}$. It is easy to show that every Turing-computable function is in $\mathcal{C}$: it suffices to

---

[2]It is expected that the reader will be able to infer from the context when one is talking about a formal set within a frame or about a set in the informal sense of mathematics at large.

[3]It is imperative to assume that there are no memory constraints. No details are given here on how to encode in *Mathematica* the alphabets of the frame, since it is not necessary to know that encoding in order to develop the theory.

develop an emulator for Turing machines within *Mathematica*. The converse follows from the Church–Turing Thesis.

Proposition 2 Set C is closed under functional composition.

Proof: Let $f : C_1 \rightharpoonup C_2$ and $g : C_2 \rightharpoonup C_3$ be computable functions. Then, the function $g \circ f : C_1 \rightharpoonup C_3$ is computed by the following *Mathematica* procedure:

$$\text{Function}[w,$$
$$g[f[w]]$$
$$]$$

As envisaged, on any given input $w \in C_1$, the execution of the procedure terminates with output $(g \circ f)(w)$ if both $f(w)$ and $g(f(w))$ are defined, and it does not produce a result otherwise. QED

A set C of non empty type $A_1 \ldots A_n$ is said to be *computable* (or *computably decidable*, or yet, simply, *decidable*) if its characteristic map $\chi_C : A_1^* \times \cdots \times A_n^* \to \mathbb{N}$ is computable.

Intuitively speaking, C is computable if, given $w \in A_1^* \times \cdots \times A_n^*$, it is possible to decide in a finite amount of time whether w belongs to the set or to its complement. That is, an algorithm computing the characteristic map χ_C when executed on input w terminates either with a positive answer, in which case $w \in C$, or a negative result, in which case $w \in (A_1^* \times \cdots \times A_n^*) \setminus C$ (that can also be written as $w \in C^c$ when the type is clear from the context).

Both sets of the empty type are computable because their characteristic maps are computable. More generally:

Proposition 3 Every finite set is computable.

Proof: There are two cases to consider:
(i) If $C = \emptyset$ then its characteristic map is computed by the following algorithm:

$$\text{Function}[w,$$
$$0]$$

(ii) Otherwise, let $C = \{c_1, \ldots, c_k\}$ for some $k \in \mathbb{N}^+$. Then, the following algorithm computes its characteristic map:

$$\text{Function}[w,$$
$$\text{Which}[$$
$$w == c_1, \ 1,$$
$$\ldots$$
$$w == c_k, \ 1,$$
$$\text{True}, \ 0$$
$$]$$
$$]$$

QED

Exercise 4 Let $f : C_1 \rightharpoonup C_2$. Show that if set C_1 is finite, then f is computable.

Exercise 5 Show that every universe is computable.

An *enumeration* of a set C is a surjective (onto) map $f : \mathbb{N} \to C$. Observe that enumerations are not required to be injective. In particular, every map from $\mathbb{N}$ to a singleton set C is an enumeration of C.

The objective now is to construct, for each non empty type $A_1 \ldots A_n$, a computable and injective enumeration of the universe $A_1^* \times \cdots \times A_n^*$, capitalizing on the strict order of each alphabet. The construction is carried out by induction on the length n of the type. To this end, we show that (i) there is such a universe enumeration for any type of length 1, and (ii) if there is such a universe enumeration for any type of length n, then there is such a universe enumeration for any type of length $n + 1$.

Proposition 6 The strict total order of each alphabet $A \in \mathcal{A}$ induces a computable injective enumeration of A^*.

Proof: If the alphabet is a singleton set $A = \{a\}$, take

$$\lambda k . k_{A^*} = \lambda k . \begin{cases} \varepsilon & \text{if } k = 0 \\ (k-1)_{A^*} \oplus a & \text{otherwise} \end{cases} : \mathbb{N} \to A^*$$

as the envisaged computable injective enumeration.

Otherwise, consider the function $S_A : A \rightharpoonup A$ such that $S_A(a)$ is the least of all elements greater than a, whenever these exist. Notice that, by identifying elements of A with words of length one, A itself can be seen as having type A^*. Hence, S_A is computable because A is finite. Let $\lambda k . k_A : \mathbb{N} \rightharpoonup A$ be the computable function that for each k returns the kth element of A. More precisely, 0_A is the least element for the strict order in A and $(k+1)_A$ is $S_A(k_A)$. Let also $\text{ord}_A : A \to \mathbb{N}$ be the inverse of $\lambda k . k_A$ and $\top_A = \#A - 1$.

For each $m \in \mathbb{N}$, S_A is computably extended to the set of words of length m in lexicographic order. Finally, it is extended to A^* by requiring that, for each m, the last word of length m be followed by the first word of length $m + 1$.

More specifically, let $\overline{S}_{A^*} : \{0, \ldots, \top_A\}^* \to \{0, \ldots, \top_A\}^*$ be the map such that the word $\overline{S}_{A^*}(w)$ is obtained, resorting to arithmetic in base $\#A$, as follows:[4]

1. build the word w' by prepending $\top_A$ to w;

---
[4]Following a suggestion by our colleague Paulo Mateus.

2. compute $w'' = w' +_{\#A} 1$;

3. finally, obtain $\overline{S}_{A^*}(w)$ from word w'' by removing its first element.

Then,

$$S_{A^*} = \lambda w \,.\, (\overline{S}_{A^*}(\text{ord}_A^*(w)))_A^* : A^* \to A^*,$$

where ord_A^* and $\lambda w \,.\, w_A^*$ are the point wise extensions to sequences of maps ord_A and $\lambda k \,.\, k_A$, respectively.

Finally, the desired enumeration $\lambda k.k_{A^*}$ is the map that for each k returns the kth element of A^* as follows: 0_{A^*} is the empty sequence and $(k+1)_{A^*}$ is $S_{A^*}(k_{A^*})$. It can easily be checked that this map is injective, computable and an enumeration of A^*.   QED

Clearly, when $A = \{1\}$ and thus $A^* = \mathbb{N}$, the enumeration built in the proof above is the identity map.

Proposition 7 If there exists a computable injective enumeration of $A_1^* \times \cdots \times A_n^*$, then there exists a computable injective enumeration of $A_1^* \times \cdots \times A_n^* \times A_{n+1}^*$.

J	0	1	2	3	...
0	0	1	3	6	...
1	2	4	7	11	...
2	5	8	12	17	...
3	9	13	18	24	...
...	...	...	...	...	...

Figure 3.1: Bijection J between $\mathbb{N}^2$ and $\mathbb{N}$.

Proof: The result is easy to establish using the fact (see Figure 3.1) that there is a computable bijection[5]

$$J : \lambda i, j \,.\, i + \frac{1}{2}((i+j)(i+j+1)) : \mathbb{N} \times \mathbb{N} \to \mathbb{N}$$

whose inverse

$$\text{zigzag} : \mathbb{N} \to \mathbb{N} \times \mathbb{N}$$

---

[5]Adapting Georg Cantor's idea for showing that $\mathbb{Q}$ is countable.

is also computable (finding an algorithm for computing zigzag is left as an exercise to the reader). Indeed, given computable injective enumerations g and h of $A_1^* \times \cdots \times A_n^*$ and A_{n+1}^* respectively (the former by hypothesis and the latter thanks to Proposition 6), an injective enumeration of $A_1^* \times \cdots \times A_n^* \times A_{n+1}^*$ can be computed as follows[6]:

$$\text{Function}[k,$$
$$g[\text{zigzag}[k][[1]]] \oplus h[\text{zigzag}[k][[2]]]$$
$$]$$

QED

Proposition 8 For each non empty type $A_1 \ldots A_n$ there is a computable injective enumeration of $A_1^* \times \cdots \times A_n^*$.

Proof: The result follows by induction on n. The base is proved in Proposition 6 and the step in Proposition 7.   QED

In the following, for each non empty type $A_1 \ldots A_n \in \mathcal{A}^*$,

$$\lambda k . k_{A_1^* \times \cdots \times A_n^*}$$

will denote the *standard enumeration of the universe* of type $A_1 \ldots A_n$ built in the proof of the previous result. The unique enumeration of the universe of type ε is the map $\lambda k . \varepsilon$, in the sequel also referred to as the standard enumeration of the empty type universe, notwithstanding the fact that it is not injective.

It is frequently necessary to use power types. Given an alphabet A in $\mathcal{A}$, for each $n \in \mathbb{N}$, the map

$$\lambda k . k_{(A^*)^n}$$

refers to the standard enumeration of the universe $(A^*)^n$ of type

$$\overbrace{A \ldots A}^{n \text{ times}}.$$

Clearly, for $n = 0$, this standard enumeration is the map $\lambda k . \varepsilon$.

Observe that the following results are robust, in the sense that they do not depend upon the standard enumeration chosen for each universe.

A set C is said to be *computably enumerable* (or *computably semi-decidable*, or yet, simply, *semi-decidable*) if either it is empty or it allows a computable enumeration.

Intuitively speaking, C is computably enumerable if, given $w \in C$, it is possible to ascertain this fact in a finite amount of time. Since in this scenario

---

[6]Using $\oplus$ also for concatenating tuples (lists in *Mathematica*).

$C \neq \emptyset$, make use of an algorithm for computing an enumeration f of C as follows: check if $f(i) = w$ for consecutive values of i starting from 0. If indeed $w \in C$, then, after a finite number of attempts, a k will be found such that $f(k) = w$. Otherwise, this verification procedure never ends.

Observe that every universe is computably enumerable thanks to Proposition 8 and the computability of $\lambda k . \varepsilon$. More generally:

Proposition 9 Every computable set is computably enumerable.

Proof:

(i) If the set is of empty type, then it is computable and it is either the empty set or the singleton set $\{\varepsilon\}$. In the former case, it is computably enumerable by definition. In the latter case, it is enumerated by $\lambda k . \varepsilon$.

(ii) Let C be a computable set of non empty type $A_1 \ldots A_n$. If C is empty, then it is computably enumerable by definition. Otherwise, let $c \in C$. Then, the following algorithm computes an enumeration of set C:

$$\begin{aligned} \mathsf{Function}[k, \\ \mathsf{If}[k_{A_1^* \times \cdots \times A_n^*} \in C, \\ k_{A_1^* \times \cdots \times A_n^*}, \\ c \\ ] \\ ] \end{aligned}$$

As envisaged, the computed function is a surjective map from $\mathbb{N}$ to C.   QED

The converse of the previous result does not hold in general, as it will be seen in Section 3.6, where an example is provided of a computably enumerable set that is not computable.

The following proposition gives a sufficient condition for the *characteristic function* of a set (coinciding with the characteristic map for elements of the set but undefined elsewhere) to be computable.

Proposition 10 Let C be a set of non empty type $A_1 \ldots A_n$. If C is computably enumerable, then the function $h : A_1^* \times \cdots \times A_n^* \rightharpoonup \mathbb{N}$ such that

$$h(w) = \begin{cases} 1 & \text{if } w \in C \\ \text{undefined} & \text{otherwise} \end{cases}$$

is computable.

Proof: If C is empty, the thesis follows since the function $\lambda w . \text{undefined}$ (whose domain is the empty set) is easy to compute, for instance as follows:

```
Function[w,
    While[True,
        Null
    ]
]
```

Otherwise, the desired function is computed by the following procedure, where f is a computable enumeration of C:

```
Function[w,
    k = 0;
    While[f[k] != w,
        k = k + 1
    ];
    1
]
```

The execution of the procedure on input w terminates (with value 1) when there is a k such that $f(k)$ is w. This only happens when w is in C. On the other hand, each element of C is captured because there is always such a k. Non termination occurs whenever $w \notin C$ and so the function computed by the procedure is undefined for such a w.   QED

The reader is invited to verify that the previous result carries over to the sets of the empty type.

The following theorem provides a very important and useful characterization of computably enumerable sets that will play a key role in Chapter 14.

Proposition 11 (Projection theorem)
Let $A_1 \ldots A_n$ and $B_1 \ldots B_{n'}$ be non empty types. Then, a set C of type $A_1 \ldots A_n$ is computably enumerable if and only if there exists a computable set R of type $A_1 \ldots A_n B_1 \ldots B_{n'}$ such that:

$$c \in C \iff \exists d \in B_1^* \times \cdots \times B_{n'}^* : c \oplus d \in R.$$

Proof: ($\leftarrow$):
Suppose that R is a computable set of type $A_1 \ldots A_n B_1 \ldots B_{n'}$ and

$$C = \{c \in A_1^* \times \cdots \times A_n^* : \exists d \in B_1^* \times \cdots \times B_{n'}^* : c \oplus d \in R\}.$$

The goal is to show that C is computably enumerable. If C is empty then it is computably enumerable. Otherwise, let c be an element of C. Then, the following algorithm computes an enumeration of C:

```
Function[k,
    If[k_{A_1^* × ··· × A_n^* × B_1^* × ··· × B_{n'}^*} ∈ R,
        Take[k_{A_1^* × ··· × A_n^* × B_1^* × ··· × B_{n'}^*}, n],
        c
    ]
]
```

$(\rightarrow)$:

Let C be a computably enumerable set of type $A_1 \ldots A_n$. If $C = \emptyset$, take R to be the empty set of type $A_1 \ldots A_n B_1 \ldots B_{n'}$, which is clearly computable and satisfies the required property. Otherwise, suppose that C is enumerated by the computable map $f : \mathbb{N} \to C$. Take

$$R = \{c \oplus k_{B_1^* × ··· × B_{n'}^*} : f(k) = c\}$$

which is of the desired type and satisfies

$$c \in C \;\Leftrightarrow\; \exists d \in B_1^* \times \cdots \times B_{n'}^* : c \oplus d \in R$$

because f is an enumeration of C. It remains to verify that R is computable. For this purpose, consider the following algorithm for computing its characteristic map:

```
Function[{w_1, ..., w_n, v_1, ..., v_{n'}},
    If[f[ord[v_1, ..., v_{n'}]] == {w_1, ..., w_n},
        1,
        0
    ]
]
```

where ord is the inverse of $\lambda k \,.\, k_{B_1^* × ··· × B_{n'}^*}$ that can be computed as follows:

```
Function[{v_1, ..., v_{n'}},
    k = 0;
    While[k_{B_1^* × ··· × B_{n'}^*} != {v_1, ..., v_{n'}},
        k = k + 1
    ];
    k
]
```

                                                                    QED

The following result provides another important characterization of computably enumerable sets using computable functions.

Proposition 12 A set is computably enumerable if and only if it is the domain of some computable function with computable source.

Proof: The case where the set is of empty type is left as an exercise. Otherwise:

($\rightarrow$): Immediate consequence of Proposition 10.

($\leftarrow$): Let $C = \operatorname{dom} h$ for some computable function $h : C_1 \rightharpoonup C_2$ with computable set C_1 and set C_2 of types $A_1 \ldots A_n$ and $B_1 \ldots B_{n'}$, respectively. If C is empty, then it is computably enumerable. Otherwise, let $c \in C$. Furthermore, choose some $e \notin C_2$, e.g. the tuple $(1, \ldots, 1) \in \{1\}^{n'+1}$. Then, the following algorithm computes an enumeration of C:

```
Function[k,
    i = k_{ℕ×ℕ}[[1]];
    j = k_{ℕ×ℕ}[[2]];
    If[i_{A_1*×⋯×A_n*} ∉ C_1 ∨ stceval[h[i_{A_1*×⋯×A_n*}], j, e] == e,
        c,
        i_{A_1*×⋯×A_n*}
    ]
]
```

Let f be the map computed by this algorithm. We have to check the following two requirements:

(a) For every $w \in C$, there is $k \in \mathbb{N}$ such that $f(k) = w$.

Assume that $w \in C$. Since $C \subseteq A_1^* \times \cdots \times A_n^*$, there is $i \in \mathbb{N}$ such that $i_{A_1^* \times \cdots \times A_n^*} = w$. From the hypothesis $C = \operatorname{dom} h \subseteq C_1$, it follows that $i_{A_1^* \times \cdots \times A_n^*} \in C_1$ and there is $j \in \mathbb{N}$ such that the execution on input $i_{A_1^* \times \cdots \times A_n^*}$ of the procedure used for computing h terminates after j steps. In short, there exist i and j such that the condition

$$i_{A_1^* \times \cdots \times A_n^*} \notin C_1 \vee \mathsf{stceval}[\mathsf{Function}[x, 1][h[i_{A_1^* \times \cdots \times A_n^*}]], j, e] == e$$

is false. Let k be such that $k_{\mathbb{N} \times \mathbb{N}} = (i, j)$. Then, the execution on input k of the proposed algorithm for computing f terminates with the value $i_{A_1^* \times \cdots \times A_n^*} = w$ as required.

(b) For every $k \in \mathbb{N}$, $f(k) \in C$.

Having in mind a proof by *reductio ad absurdum*, assume that $f(k) \notin C$. Then, since otherwise the execution on k of the proposed algorithm would return c that belongs to C, the condition

$$i_{A_1^* \times \cdots \times A_n^*} \notin C_1 \vee \mathsf{stceval}[\mathsf{Function}[x, 1][h[i_{A_1^* \times \cdots \times A_n^*}]], j, e] == e$$

must be false for $i = k_{\mathbb{N} \times \mathbb{N}}[[1]]$ and $j = k_{\mathbb{N} \times \mathbb{N}}[[2]]$. This means that $i_{A_1^* \times \cdots \times A_n^*} \in C_1$ and the execution on input $i_{A_1^* \times \cdots \times A_n^*}$ of the procedure used for computing h terminates within j steps. Hence, $i_{A_1^* \times \cdots \times A_n^*} \in \mathrm{dom}\, h = C$. Moreover, in this situation $f(k) = i_{A_1^* \times \cdots \times A_n^*}$ and, so, $f(k) \in C$, in contradiction with the assumption $f(k) \notin C$.   QED

According to the previous result, the computably enumerable sets are the domains of computable functions with computable sources. In fact, the interested reader should be able to show that it is possible to relax the assumption on the source. The reader may also wonder if similar results hold for the ranges of computable functions. The next result gives a positive answer to this question concerning functions with computable sources.

Proposition 13 A set is computably enumerable if and only if it is the range of some computable function with computable source.

Proof:

($\rightarrow$) If the set is empty, then it is the range of the function $\lambda k \,.\, \mathrm{undefined}$, which is computable. Otherwise, it allows a computable enumeration; hence, the set is the range of this enumeration.

($\leftarrow$) Let C_1 be a computable set, $f : C_1 \rightharpoonup C_2$ a computable function and

$$C = \mathrm{range}\, f = f(\mathrm{dom}\, f).$$

If C is empty, then it is computably enumerable. Otherwise, $\mathrm{dom}\, f$ is also non empty. Therefore, thanks to the previous proposition, $\mathrm{dom}\, f$ allows a computable enumeration h. Then, $f|_{\mathrm{dom}\, f} \circ h$ is a computable enumeration of C.   QED

Closure properties (namely, under intersection, union, Cartesian product and complementation) of the classes of computably enumerable sets and computable sets are now investigated, leaving to the reader some of the proofs.[7]

Proposition 14 The class of computably enumerable sets of a given type is closed under the binary operations of intersection and union.

Proof: Herein, only binary union of sets of a given non empty type is considered, leaving the rest as an exercise. First observe that if C_1 and C_2 are sets of type $A_1 \ldots A_n$, that is, if C_1 and C_2 are subsets of $A_1^* \times \cdots \times A_n^*$, then their union is also a subset of $A_1^* \times \cdots \times A_n^*$, in other words, their union is also of type

---

[7]Throughout the text, the reader should try to prove the results stated without proof before looking for them in Part IV.

$A_1 \ldots A_n$. It remains to show that $C_1 \cup C_2$ is computably enumerable when both C_1 and C_2 are computably enumerable. There are two cases to consider:

(a) Either C_1 or C_2 is empty:

Without loss of generality, assume that $C_1 = \emptyset$. Then, $C_1 \cup C_2 = C_2$ and, thus, $C_1 \cup C_2$ is computably enumerable, since, by hypothesis, C_2 is computably enumerable.

(b) Both C_1 and C_2 are non empty:

Let $f_i : \mathbb{N} \to C_i$ be a computable enumeration of C_i for $i = 1, 2$. Consider the following algorithm

$$\text{Function}[k,$$
$$\text{If}[\text{EvenQ}(k),$$
$$f_1[\tfrac{k}{2}],$$
$$f_2[\tfrac{k-1}{2}]$$
$$]$$
$$]$$

It is left to the reader the verification of the fact that this algorithm computes an enumeration of $C_1 \cup C_2$.   QED

As expected, Proposition 14 can be extended by induction to finite intersections and finite unions.

Proposition 15 Let I be a finite set, T a type, and C_i a computably enumerable set of type T for each $i \in I$. Then,

$$\bigcap_{i \in I} C_i \quad \text{and} \quad \bigcup_{i \in I} C_i$$

are computably enumerable.

Proof: The case where $T = \varepsilon$ is left as an exercise to the reader. Otherwise, the proof is carried out by induction on the cardinal k of set I:

$k = 0$:

In this case

$$\left(\bigcap_{i \in I} C_i\right) = \left(\bigcap_{i \in \emptyset} C_i\right) = A_1^* \times \cdots \times A_n^* \quad \text{and} \quad \left(\bigcup_{i \in I} C_i\right) = \left(\bigcup_{i \in \emptyset} C_i\right) = \emptyset.$$

Therefore,

$$\bigcap_{i \in I} C_i \quad \text{and} \quad \bigcup_{i \in I} C_i$$

are both computably enumerable.

$k > 0$:

Choose $j \in I$. Then, by the induction hypothesis,

$$\bigcap_{i \in (I \setminus \{j\})} C_i \quad \text{and} \quad \bigcup_{i \in (I \setminus \{j\})} C_i$$

are computably enumerable. So, by Proposition 14,

$$\bigcap_{i \in I} C_i = \left(\bigcap_{i \in I \setminus \{j\}} C_i \right) \cap C_j$$

and

$$\bigcup_{i \in I} C_i = \left(\bigcup_{i \in I \setminus \{j\}} C_i \right) \cup C_j$$

are also computably enumerable. QED

The reader is invited to provide counterexamples showing that the thesis of the previous proposition does not necessarily follow without the hypothesis of the finite cardinality of I. At the end of this chapter a sufficient condition is established for the computable enumerability of the union of an infinite collection of computably enumerable sets (Proposition 33).

Proposition 16 The class of computably enumerable sets is closed under finite Cartesian product.

Observe that the class of computably enumerable sets is not closed under complementation. An example will be provided later on. But the class of computable sets is closed under complementation, as well as under finite intersection, finite union and finite Cartesian product.

Proposition 17 The class of computable sets is closed under the unary operation of complement.

Proposition 18 The class of computable sets of a given type is closed under finite intersection and finite union. The class of computable sets is closed under finite Cartesian product.

The reader should be able to provide counterexamples showing that the intersection and the union of an infinite collection of computable sets are not necessarily computable.

The following two results provide useful sufficient conditions for a set to be computable.

Proposition 19 Let C be a computable set and $D \subseteq C$. If D and $C \setminus D$ are computably enumerable, then they are both computable.

Proof: The thesis follows trivially if the type at hand is the empty type. Let C be of non empty type $A_1 \ldots A_n$ and $D \subseteq C$. If either D or $C \setminus D$ is empty, then the result is again trivial. Otherwise, let $f_D : \mathbb{N} \to D$ and $f_{C \setminus D} : \mathbb{N} \to C \setminus D$ be computable enumerations of D and $C \setminus D$, respectively. Then, the following algorithm computes the characteristic map $\chi_D : A_1^* \times \cdots \times A_n^* \to \mathbb{N}$:

```
Function[w,
    If[w ∉ C,
        0,
        k = 0;
        While[f_D[k] != w ∧ f_{C\D}[k] != w,
            k = k + 1
        ];
        If[f_D[k] == w, 1, 0]
    ]
]
```

Presenting an algorithm to compute $\chi_{C \setminus D}$ is left as an exercise.   QED

The following result is a direct consequence of Proposition 19 and the fact that the universe of every type is computable.

Proposition 20 (Post's theorem)
If a set and its complement are computably enumerable, then both are computable.

As it will be seen later (in Section 3.6), there are computably enumerable sets that are not computable. Let C be such a set. Then, thanks to Post's theorem, C^c cannot be computably enumerable, providing a counter-example for showing that computable enumerability is not closed under complementation.

The following results address the back and forward preservation by computable maps of computable enumerability and computability of sets, but only as far as needed in the sequel. The reader is invited to check if they can be extended to computable functions.

Proposition 21 Let $h : C_1 \to C_2$ be a computable map and $D \subseteq C_1$. If D is computably enumerable, then $h(D)$ is also computably enumerable.

Proof: If D is empty, then $h(D)$ is also empty, and therefore computably enumerable. Otherwise, let $f : \mathbb{N} \to D$ be a computable enumeration of D. Then, $\check{h} \circ f$ is a computable enumeration of $h(D)$, where $\check{h} = \lambda d \, . \, h(d) : D \to h(D)$.   QED

Proposition 22 Let $h : A_1^* \times \cdots \times A_n^* \to C_2$ be a computable map and $D \subseteq C_2$. If D is computably enumerable, then $h^{-1}(D)$ is also computably enumerable.

Proof: Let C_2 be of type $B_1 \ldots B_{n'}$, leaving the case of the empty type to the reader. Since D is computably enumerable, by Proposition 11 there is a computable set R of type $B_1 \ldots B_{n'} E$ such that $d \in D \Leftrightarrow \exists e \in E^* : d \oplus e \in R$. Take

$$Q = \{c \oplus e : h(c) \oplus e \in R\}$$

of type $A_1 \ldots A_n E$. Clearly,

$$c \in h^{-1}(D) \Leftrightarrow \exists e \in E^* : c \oplus e \in Q.$$

Furthermore, Q is computable. Thus, again by Proposition 11, $h^{-1}(D)$ is computably enumerable. QED

The proof above illustrates the use of the projection theorem. The reader is invited to provide an alternative proof using the characteristic function of D.

Proposition 23 Let $h : A_1^* \times \cdots \times A_n^* \to C_2$ be a computable map and $D \subseteq C_2$. If D is computable, then $h^{-1}(D)$ is also computable.

Proof: It is enough to observe that $\chi_{h^{-1}(D)} = \lambda w . \chi_D(h(w))$. QED

On the other hand, the image of a computable set through a computable map is not necessarily computable. In fact, in Section 3.6 an example is provided of a computably enumerable set that is not computable. Let C be such a set (necessarily non empty). Let $f : \mathbb{N} \to C$ be one of its computable enumerations. Clearly, such an f provides the desired counter-example.

3.4 Gödelizations

In his proof of incompleteness of arithmetic, Kurt Gödel needed to represent the formulas and derivations of arithmetic within itself, for which purpose he introduced a notion of coding. In tribute to his work, these codings are nowadays known as Gödelizations.

In the context of a frame $\mathcal{A}$, a *Gödelization* of non empty type $A_1 \ldots A_n$ is a computable injective map

$$g : A_1^* \times \cdots \times A_n^* \to \mathbb{N}$$

such that:

- $g(A_1^* \times \cdots \times A_n^*)$ is a computable set;

- $g^{-1} : g(A_1^* \times \cdots \times A_n^*) \to A_1^* \times \cdots \times A_n^*$ is a computable map.

Maps g and g^{-1} are known as the encoding and decoding maps, respectively, of the Gödelization.

Observe that a computable injective enumeration of $A_1^* \times \cdots \times A_n^*$ induces a Gödelization of the non empty type $A_1 \ldots A_n$. Just take the inverse of the enumeration as g. Clearly, in this case $g(A_1^* \times \cdots \times A_n^*) = \mathbb{N}$.

The following results show that Gödelizations make it possible to work always in the universe of natural numbers.

Proposition 24 (Gödelization theorem)
Let g be a Gödelization and C be a set, both of non empty type $A_1 \ldots A_n$. Then:

1. C is computably enumerable if and only if $g(C)$ is computably enumerable;

2. C is computable if and only if $g(C)$ is computable.

Proof:

1. ($\to$)
Immediate consequence of Proposition 21.

1. ($\leftarrow$)
Immediate consequence of Proposition 22, since $C = g^{-1}(g(C))$ due to injectivity of g.

2. ($\to$)
If C is computable, then C and C^c are both computably enumerable. Hence, $g(C)$ and $g(C^c)$ are also computably enumerable, thanks to Proposition 21. Observe that

$$g(C^c) = g(A_1^* \times \cdots \times A_n^*) \setminus g(C)$$

because g is injective. Hence, since $g(A_1^* \times \cdots \times A_n^*)$ is computable, Proposition 19 applies and it follows that $g(C)$ is computable.

2. ($\leftarrow$)
Immediate consequence of Proposition 23, since $C = g^{-1}(g(C))$.   QED

The computability requirement on g^{-1} was not needed so far. It is however relevant to prove the following result.

Proposition 25 Let g_1 and g_2 be Gödelizations and C_1 and C_2 be sets of non empty types $A_1 \ldots A_n$ and $B_1 \ldots B_{n'}$, respectively. Then, $f : C_1 \rightharpoonup C_2$ is computable if and only if

$$\lambda w \, . \, g_2(f(g_1^{-1}(w))) : g_1(C_1) \rightharpoonup g_2(C_2)$$

is computable.

For this reason, Computability Theory has traditionally been developed in the realm of natural numbers. In such an approach, when one needs to study computability within other universes (for example, in the universe of the formulas of first-order logic) one begins by introducing a Gödelization and afterward work only within the image of the universe in the set of natural numbers. In this book, a more general approach was preferred for two reasons. First, the notion of Gödelization, still necessary to encode arithmetic reasoning within arithmetic itself, can only be rigorously defined in a multi-universe approach to Computability Theory. Second, the adopted high-level programming language facilitates the direct definition of procedures on the universes at hand, without going through their encodings into the universe of natural numbers, with significant advantages in clarity of exposition and understandability.

3.5 Computability à la Kleene

Although the notion of procedure adopted above (using the programming language *Mathematica*) is fully capable of addressing the problems of decidability of first-order logic and its theories, it is convenient to work also with simpler formalizations of the notion of procedure. For instance, it is much easier to prove the representability of computable maps within arithmetic when using a more concise formalization of the notion of procedure.

The formalization proposed by Kleene is especially simple. From a set of basic functions, the universe of computable functions is built using aggregation, composition, recursion and minimization (iterative search for zeros of functions).

In the setting of the minimal frame $\mathcal{A} = \{\{1\}\}$, the set $\mathcal{R}$ of *(possibly partial) recursive functions* is defined inductively as follows:[8]

- for each $k \in \mathbb{N}$, the constant[9]

$$k :\to \mathbb{N}$$

  is in $\mathcal{R}$;

- the unary map
$$Z = \lambda k . 0 : \mathbb{N} \to \mathbb{N}$$

  is in $\mathcal{R}$;

---

[8]Kleene's original work was carried out only in the universe of natural numbers. This single universe is enough for the use of (possibly partial) recursive functions that will be made throughout this book.

[9]Recall that a constant is a 0-ary map, that is, a map with no arguments. More precisely, its domain is the universe of the empty type. It is traditional to write $f :\to C$ for $f : \{\varepsilon\} \to C$, and, in such case, to write f for $f(\varepsilon)$.

- the unary map
$$S = \lambda k \, . \, k + 1 : \mathbb{N} \to \mathbb{N}$$

  is in $\mathcal{R}$;

- for each positive natural number n and each $n' \in \{1, \ldots, n\}$, the n-ary map
$$\mathsf{P}^n_{n'} = \lambda k_1, \ldots, k_n \, . \, k_{n'} : \mathbb{N}^n \to \mathbb{N}$$

  is in $\mathcal{R}$;

- for each pair (n, n') of natural numbers with n' positive, if
$$f_1 : \mathbb{N}^n \rightharpoonup \mathbb{N}, \, \ldots, \, f_{n'} : \mathbb{N}^n \rightharpoonup \mathbb{N}$$

  are all in $\mathcal{R}$, then the function (obtained by *aggregation*)

$$\langle f_1, \ldots, f_{n'} \rangle = \\ \lambda k_1, \ldots, k_n \, . \, (f_1(k_1, \ldots, k_n), \ldots, f_{n'}(k_1, \ldots, k_n)) : \mathbb{N}^n \rightharpoonup \mathbb{N}^{n'}$$

  is in $\mathcal{R}$;

- for each triple (n, n', n'') of natural numbers with n' and n'' both positive, if
$$f_1 : \mathbb{N}^n \rightharpoonup \mathbb{N}^{n'} \text{ and } f_2 : \mathbb{N}^{n'} \rightharpoonup \mathbb{N}^{n''}$$

  are both in $\mathcal{R}$, then the function (obtained by *composition*)

$$f_2 \circ f_1 = \lambda k_1, \ldots, k_n \, . \, f_2(f_1(k_1, \ldots, k_n)) : \mathbb{N}^n \rightharpoonup \mathbb{N}^{n''}$$

  is in $\mathcal{R}$;

- for each natural number n, if both $f_0 : \mathbb{N}^n \rightharpoonup \mathbb{N}$ and $f_1 : \mathbb{N}^{n+2} \rightharpoonup \mathbb{N}$ are in $\mathcal{R}$, then the function (obtained by *primitive recursion*, or, simply, recursion)
$$\mathsf{rec}(f_0, f_1) : \mathbb{N}^{n+1} \rightharpoonup \mathbb{N}$$

    such that
$$\mathsf{rec}(f_0, f_1)(k_1, \ldots, k_n, 0) = f_0(k_1, \ldots, k_n)$$

    and
$$\mathsf{rec}(f_0, f_1)(k_1, \ldots, k_n, k + 1) = \\ f_1(k_1, \ldots, k_n, k, \mathsf{rec}(f_0, f_1)(k_1, \ldots, k_n, k))$$

  is in $\mathcal{R}$;

- for each natural number n, if $f : \mathbb{N}^{n+1} \rightharpoonup \mathbb{N}$ is in $\mathcal{R}$, then, denoting the set

$$\{k : (k_1, \ldots, k_n, j) \in \operatorname{dom} f \text{ for } j < k \ \& \ f(k_1, \ldots, k_n, k) = 0\}$$

by $W^f_{k_1,\ldots,k_n}$, the function (obtained by *minimization*)

$$\min(f) : \mathbb{N}^n \rightharpoonup \mathbb{N}$$

such that,

$$\min(f)(k_1, \ldots, k_n) = \begin{cases} \text{undefined} & \text{if } W^f_{k_1,\ldots,k_n} = \emptyset \\ \text{minimum of } W^f_{k_1,\ldots,k_n} & \text{otherwise} \end{cases}$$

is in $\mathcal{R}$.

It is worthwhile to identify the domain of the functions obtained by the different constructions:

- $(k_1, \ldots, k_n) \in \operatorname{dom} \langle f_1, \ldots, f_{n'} \rangle$ if and only if $(k_1, \ldots, k_n) \in \operatorname{dom} f_k$ for every $k = 1, \ldots, n'$;

- $(k_1, \ldots, k_n) \in \operatorname{dom}(f_2 \circ f_1)$ if and only if $(k_1, \ldots, k_n) \in \operatorname{dom} f_1$ and, furthermore, $f_1(k_1, \ldots, k_n) \in \operatorname{dom} f_2$;

- $(k_1, \ldots, k_n, m) \in \operatorname{dom} \operatorname{rec}(f_0, f_1)$ is inductively defined as follows:

  - $(k_1, \ldots, k_n, 0) \in \operatorname{dom} \operatorname{rec}(f_0, f_1)$ if and only if $(k_1, \ldots, k_n) \in \operatorname{dom} f_0$;
  - $(k_1, \ldots, k_n, k+1) \in \operatorname{dom} \operatorname{rec}(f_0, f_1)$ if and only if

    $$\begin{cases} (k_1, \ldots, k_n, k) \in \operatorname{dom} \operatorname{rec}(f_0, f_1) \\ (k_1, \ldots, k_n, k, \operatorname{rec}(f_0, f_1)(k_1, \ldots, k_n, k)) \in \operatorname{dom} f_1; \end{cases}$$

- $(k_1, \ldots, k_n) \in \operatorname{dom} \min(f)$ if and only if $W^f_{k_1,\ldots,k_n} \neq \emptyset$, that is, if and only if there is $k \in \mathbb{N}$ such that $f(k_1, \ldots, k_n, k) = 0$ and $(k_1, \ldots, k_n, j)$ is in $\operatorname{dom} f$ for every $j < k$.

Exercise 26 Show that $\mathcal{R} \subseteq \mathcal{C}$. Hint: In order to use structural induction, verify that the basic recursive functions are computable and that $\mathcal{C}$ is closed under aggregation, composition, recursion and minimization.

Proving that $\mathcal{C} \subseteq \mathcal{R}$ (in the context of the universe of natural numbers) would be very tedious. For this reason, the Church–Turing thesis is invoked instead as follows. It has already been verified that Kleene computability is equivalent to Turing computability (see for example Chapter 18 of [16]). Thus, $\mathcal{C} \subseteq \mathcal{R}$ because, by the Church–Turing thesis, every function in $\mathcal{C}$ is Turing computable.

In order to illustrate the notion proposed by Kleene, observe that the map

$$\mathsf{add} = \lambda\, k_1, k_2\,.\, k_1 + k_2 : \mathbb{N}^2 \to \mathbb{N}$$

is in $\mathcal{R}$, since it is computed by the following algorithm:

$$\mathsf{rec}(\mathsf{P}_1^1, \mathsf{S} \circ \mathsf{P}_3^3)$$

(written in what one may consider to be the *Kleene* programming language).

Indeed, induction on the second argument shows that

$$\mathsf{rec}(\mathsf{P}_1^1, \mathsf{S} \circ \mathsf{P}_3^3)(k_1, k_2) = k_1 + k_2.$$

Basis ($k_2 = 0$):

$$\mathsf{rec}(\mathsf{P}_1^1, \mathsf{S} \circ \mathsf{P}_3^3)(k_1, 0) = \mathsf{P}_1^1(k_1) = k_1.$$

Step ($k_2 = k + 1$):

$$\begin{aligned}
\mathsf{rec}(\mathsf{P}_1^1, \mathsf{S} \circ \mathsf{P}_3^3)(k_1, k + 1) &= (\mathsf{S} \circ \mathsf{P}_3^3)(k_1, k, \mathsf{rec}(\mathsf{P}_1^1, \mathsf{S} \circ \mathsf{P}_3^3)(k_1, k)) \\
&= \mathsf{S}(\mathsf{rec}(\mathsf{P}_1^1, \mathsf{S} \circ \mathsf{P}_3^3)(k_1, k))
\end{aligned}$$

which, by the induction hypothesis, is equal to

$$\mathsf{S}(k_1 + k) = (k_1 + k) + 1 = k_1 + (k + 1) = k_1 + k_2.$$

Exercise 27 Show that the map

$$\mathsf{neq} = \lambda\, k_1, k_2\,.\, \begin{cases} 1 & \text{if } k_1 \neq k_2 \\ 0 & \text{otherwise} \end{cases} : \mathbb{N}^2 \to \mathbb{N}$$

is in $\mathcal{R}$.

Exercise 28 Show that the function (partial subtraction)

$$\lambda\, k_1, k_2\,.\, \begin{cases} k_1 - k_2 & \text{if } k_1 \geq k_2 \\ \text{undefined} & \text{otherwise} \end{cases} : \mathbb{N}^2 \rightharpoonup \mathbb{N}$$

is in $\mathcal{R}$ by checking that it is computed by the following procedure:

$$\mathsf{min}(\mathsf{neq} \circ \langle \mathsf{P}_1^3, \mathsf{add} \circ \langle \mathsf{P}_2^3, \mathsf{P}_3^3 \rangle \rangle)\,.$$

As expected, for each natural number n, a subset of $\mathbb{N}^n$ is said to be *recursive* if the corresponding characteristic map is recursive, and it is said to be *recursively enumerable* if either it is empty or it allows a recursive enumeration.

3.6 Undecidability of the halting problem

In order to give an example of a computably enumerable set that is not computable, consider the problem of termination of the execution of *Mathematica* procedures.

Let A_M be the finite (total and strictly ordered)[10] alphabet of the programming language *Mathematica*. In other words, A_M is the set of characters of this language. In order to illustrate the available operations in *Mathematica* over characters and strings, observe that A_M^* is enumerated by the map computed by the following algorithm:[11]

```
Function[k,
    w = Map[
              Function[v, FromCharacterCode[v[[2]] − 1]],
              FactorInteger[k]
          ];
    s = "";
    While[w != {},
        s = s <> First[w];
        w = Rest[w]
    ];
    s
]
```

Let $\mathcal{P}$ be the set of all *Mathematica* programs, used herein to compute functions of type $A_M^* \rightharpoonup A_M^*$. Observe that $\mathcal{P} \subset A_M^*$. Therefore, $\mathcal{P}$ is a set of type A_M.

For each $p \in \mathcal{P}$ and each $w \in A_M^*$, let $p(w)\!\downarrow$ state that execution of program p with input $w \in A_M^*$ terminates with result taken in A_M^*. Consider the following set of type A_M:

$$\Delta = \{p \in \mathcal{P} : p(p)\!\downarrow\}.$$

Proposition 29 Set Δ is computably enumerable.

Proof: Set Δ is the domain of the function from A_M^* to A_M^* computed by the following procedure:

```
Function[s,
    ToExpression[s][s]
]
```

---

[10]The ordering is as follows: $a < b$ if ToCharacterCode["a"] < ToCharacterCode["b"]. The enumeration $\lambda k . k_{A_M}$ is the map FromCharacterCode.

[11]This enumeration of A_M^* is not injective. Finding an algorithm that computes an injective enumeration of A_M^* is left as an exercise to the reader. Hint: Recall the proof of Proposition 6.

Therefore, thanks to Proposition 12, Δ is computably enumerable.   QED

Proposition 30 Set Δ is not computable.

Proof: The proof is by *reductio ad absurdum*. Suppose that χ_Δ is computable. Then, the function

$$f = \lambda p \, . \begin{cases} \text{undefined} & \text{if } p(p)\!\downarrow \\ 0 & \text{otherwise} \end{cases} : \mathcal{P} \rightharpoonup \mathbb{N}$$

is computable, since it is computed by the following procedure P_f:

```
Function[p,
    k = 0;
    While[k + χ_Δ(p) != 0,
        k = k + 1
    ];
    k
]
```

Moreover, in particular,

$$f(P_f) = \begin{cases} \text{undefined} & \text{if } P_f(P_f)\!\downarrow \\ 0 & \text{otherwise.} \end{cases}$$

Therefore:[12]

- if $P_f \in \operatorname{dom} f$ then $P_f(P_f)\!\downarrow$ and, so, $f(P_f) = $ undefined, that is, $P_f \notin \operatorname{dom} f$;

- if $P_f \notin \operatorname{dom} f$ then not $P_f(P_f)\!\downarrow$ and, so, $f(P_f) = 0$, that is, $P_f \in \operatorname{dom} f$.

Clearly, both cases are impossible.   QED

The previous proposition yields as immediate corollary the following result (shown independently by Alonzo Church [10] and Alan Turing [48]):

Proposition 31 (Undecidability of the halting problem)
The set
$$H = \{(p, w) \in \mathcal{P} \times A_M^* : p(w)\!\downarrow\}$$
of type $A_M A_M$ is not computable.

---

[12]Recall that the execution of a program p computing h terminates when given input w if and only if $w \in \operatorname{dom} h$.

3.7 Enumerating the computable functions

Given $p \in \mathcal{P}$, let ϕ_p denote the (possibly partial) function from A_M^* to A_M^* computed by the procedure p.

Observe that there exists a universal *Mathematica* procedure capable of emulating every procedure in $\mathcal{P}$, in the sense that it computes the following *universal function*:

$$u = \lambda\, p, w \,.\, \phi_p(w) : \mathcal{P} \times A_M^* \rightharpoonup A_M^*$$

The universality comes from the fact that u takes any program p and any sequence w and returns the result of executing p with input w. Clearly,

$$\phi_p = \lambda\, w \,.\, u(p, w).$$

Indeed, the *Mathematica* system is capable of computing this universal function using, for instance, the following procedure:

```
Function[{p, w},
    ToExpression[p][w]
]
```

The unexpected simplicity of this universal procedure is consequence of the power of the primitive *Mathematica* function ToExpression that recognizes in the argument (the string of characters p) a proper *Mathematica* expression (a procedure) that can be applied to the argument (w).

Returning to the problem of enumerating the computable functions, observe that there are computable enumerations of set $\mathcal{P}$ of type A_M because $\mathcal{P}$ is non empty and computable. Choose once and for all one such computable enumeration

$$\lambda\, k \,.\, k_{\mathcal{P}}.$$

Whenever $p = i_{\mathcal{P}}$, one says that i is an *index* of program p (according to the chosen enumeration). Each program may have a countable infinity of indexes (it depends on the choice of the enumeration). However, it is useful to introduce the map

$$\mathrm{mind} : \mathcal{P} \rightarrow \mathbb{N}$$

returning the least index of the given program. This map is computed by the following algorithm:

```
Function[p,
    k = 0;
    While[p != kₚ,
        k = k + 1;
    ];
    k
]
```

Note that there is an algorithm to decide whether two programs are the same (as sequences of characters). But, of course, there is no algorithm to decide whether two programs compute the same function.

Let $\mathcal{C}_1^1$ be the set of computable functions from A_M^* to A_M^*. The chosen enumeration of $\mathcal{P}$ induces the enumeration

$$\lambda k . \phi_{k_{\mathcal{P}}} = \lambda k . (\lambda w . u(k_{\mathcal{P}}, w))$$

of $\mathcal{C}_1^1$ that is denoted in the sequel by

$$\lambda k . k_{\mathcal{C}_1^1}.$$

Clearly, a function $f : A_M^* \rightharpoonup A_M^*$ is computable if and only if there exists i such that

$$f = \phi_{i_{\mathcal{P}}} = \lambda w . u(i_{\mathcal{P}}, w) = i_{\mathcal{C}_1^1}.$$

Whenever $f = i_{\mathcal{C}_1^1}$, one says that i is an *index* of f. Clearly, every computable function has a countable infinity of indexes.

Exercise 32 Is the map computing the least index of a function in $\mathcal{C}_1^1$ computable?

The following result illustrates how indexes of computable functions can be used and it introduces a technique used in Chapter 5 when discussing the properties of derivation and, in particular, the semi-decidability of the Hilbert calculus.

For the sake of readability, one says that $f \in \mathcal{C}_1^1$ enumerates a subset S of A_M^* if $S = \text{range } f|_{\mathbb{N}}$. Recall that $\mathbb{N} = \{1\}^* \subset A_M^*$.

Proposition 33 Let $s : \mathbb{N} \to \mathbb{N}$ be a computable map and assume that, for each $i \in \mathbb{N}$, $C_i \subseteq A_M^*$ is enumerated by $s(i)_{\mathcal{C}_1^1}$. Then,

$$\bigcup_{i \in \mathbb{N}} C_i$$

is computably enumerable.

Proof: The following algorithm computes an enumeration of $\bigcup\limits_{i\in\mathbb{N}} C_i$:

$$
\begin{aligned}
&\text{Function}[k, \\
&\quad i = k_{\mathbb{N}\times\mathbb{N}}[[1]]; \\
&\quad j = k_{\mathbb{N}\times\mathbb{N}}[[2]]; \\
&\quad s[i]_{C_1^1}[j] \\
&]
\end{aligned}
$$

Indeed,

$$s(i)_{C_1^1}(j) = u(s(i)_{\mathcal{P}}, j)$$

and, thus, its value can indeed be computed because u is computable.   QED

Notice that the previous result could have been stated intuitively as follows: the set

$$\bigcup_{i\in\mathbb{N}} C_i$$

is computably enumerable if each C_i is computably enumerable and the map

$$\lambda\, i \,.\, C_i : \mathbb{N} \to \wp(A_M^*)$$

is "computable". The problem with this formulation lies in the fact that the notion of computable function has not been extended to higher-order types. This is possible, but not essential, since, as it was shown above, one is able to rephrase the question in terms of basic types by resorting to indexes of functions.

Part II

First-order predicate logic

Chapter 4

Syntax

The first attempts at formalizing Mathematics were made by Gottlob Frege and Charles Pierce who introduced, independently of each other, the notions of variable and quantifier, the essential ingredients of the language of first-order predicate logic besides the propositional connectives. For an historical overview see, for instance, [5].

The main goal of this chapter is to define the languages of first-order predicate logic as understood today, including signatures, alphabets, terms, formulas and substitution. Computability requirements on the signature are imposed in order to ensure the computability of the key syntactic sets and maps, like the set of formulas and the substitution map. The important concept of free term for a variable in a formula is carefully introduced and motivated by the subtle problems that appear when reasoning about instantiation of variables.

Section 4.1 is dedicated to signatures and alphabets. Terms and formulas are presented in Section 4.2. Section 4.3 concentrates on defining what are free and bound occurrences of a variable in a formula. Finally, the notions of free term for a variable in a formula and substitution are presented in Section 4.4.

4.1 Signatures

The objective here is to identify the symbols that can be used to construct formulas. There are two groups of such symbols: those that are defined once and for all (e.g. connectives and variables) and those that can vary depending on the purpose at hand (function and predicate symbols collected in the signature). All these symbols are built in this book using the characters in the alphabet A_M of the *Mathematica* language.

Punctuation symbols (comma and brackets), connectives (like $\neg$ and $\Rightarrow$),

and quantifiers (namely, $\forall$ and $\exists$) are collected in the set R of *reserved symbols*.
For each $k \in \mathbb{N}$, let the string

$$\text{`` } \overbrace{x \ldots x}^{k+1 \text{ times}} \text{ ''}$$

be denoted by x_k. Furthermore, let

$$X = \{x_k : k \in \mathbb{N}\} \subset (A_M \setminus R)^*.$$

The elements of X are called *variables*. Each variable x_k is also denoted by
k_X. Note that X is computable as a subset of A_M^* and that the map $\lambda k \, . \, k_X$
is a computable injective enumeration of X.

A *first-order signature* (or, simply, signature) is a triple

$$\Sigma = (F, P, \tau)$$

such that:

- F and P are computable subsets of $(A_M \setminus R)^*$ disjoint from each other
  and from X;

- $\tau : F \cup P \to \mathbb{N}$ is a computable map.

The elements of F are said to be *function symbols*. The elements of P are said
to be *predicate symbols*. Map τ returns the *arity* of its argument. Let:

- F_n denote $\{f \in F : \tau(f) = n\}$ (set of function symbols of arity n);

- P_n denote $\{p \in P : \tau(p) = n\}$ (set of predicate symbols of arity n).

Note that these two sets are computable. Indeed, observe that $\chi_{F_n}(f) = 1$
if and only if $\tau(f) = n$ and $\chi_F(f) = 1$, and $\chi_{P_n}(p) = 1$ if and only if $\tau(p) = n$
and $\chi_P(p) = 1$. Thus, the computability of χ_{F_n} and χ_{P_n} follows from the
computability of τ, F and P.

The elements of F_0 are also called *constant symbols*. As usual, it is assumed
that $P_0 = \emptyset$ and that $P \neq \emptyset$.

The set

$$A_\Sigma = R \cup X \cup F \cup P \subset A_M^*$$

is the *first-order alphabet* induced by signature Σ. Observe that this set is
computable as well.

The following exercise allows the reader to become familiar with the task of
defining signatures suitable for asserting properties of interesting mathematical
objects.

Exercise 1 Define appropriate signatures to express properties of:

- groups;

- natural numbers;

- sets.

The following exercise provides a possible encoding of the strings over a very general alphabet (a Gödelization in the sense of Section 3.4).

Exercise 2 For each $k, n \in \mathbb{N}$, let the strings

$$\text{`` } \overbrace{f \ldots f}^{k+1 \text{ times}} \text{ '' } <> \text{ `` } \overbrace{a \ldots a}^{n \text{ times}} \text{ ''}$$

and

$$\text{`` } \overbrace{p \ldots p}^{k+1 \text{ times}} \text{ '' } <> \text{ `` } \overbrace{a \ldots a}^{n \text{ times}} \text{ ''}$$

be denoted by f_k^n and p_k^n, respectively. Consider the triple $\Sigma = (F, P, \tau)$ such that:

- $F = \bigcup_{n \in \mathbb{N}} F_n$ with $F_n = \{f_k^n : k \in \mathbb{N}\}$;

- $P = \bigcup_{n \in \mathbb{N}^+} P_n$ with $P_n = \{p_k^n : k \in \mathbb{N}\}$;

- $\tau(f_k^n) = n$;

- $\tau(p_k^n) = n$.

(a) Show that this triple is a first-order signature.

(b) Assuming that p_i stands for the ith prime number and

$$h = \lambda a . \begin{cases} 3 & \text{if } a = \text{``[''} \\ 5 & \text{if } a = \text{``]''} \\ 7 & \text{if } a = \text{``,''} \\ 9 & \text{if } a = \text{``}\neg\text{''} \\ 11 & \text{if } a = \text{``}\Rightarrow\text{''} \\ 13 & \text{if } a = \text{``}\forall\text{''} \\ 7 + 8(k+1) & \text{if } a = x_k \\ 9 + 8(k+1) & \text{if } a = f_k^0 \\ 11 + 8(2^n 3^{k+1}) & \text{if } a = f_k^n \ \& \ n > 0 \\ 13 + 8(2^n 3^{k+1}) & \text{if } a = p_k^n \end{cases} : A_\Sigma \to \mathbb{N},$$

show that the map[1]

$$g = \lambda\, a_1 \ldots a_m \cdot \begin{cases} 0 & \text{if } m = 0 \\ \mathsf{p}_0^{h(a_1)} \times \cdots \times \mathsf{p}_{m-1}^{h(a_m)} & \text{otherwise} \end{cases} : A_\Sigma^* \to \mathbb{N}$$

is a Gödelization of type A_Σ.

4.2   Terms and formulas

Before stating what is a formula in the language of first-order logic, it is necessary first to say what is a term. Indeed, intuitively speaking, terms are used to refer to individuals in the universe of discourse at hand, and formulas are used to express properties of those individuals.

The set T_Σ of *terms* over a signature Σ is inductively defined as follows:

- $x[\,] \in T_\Sigma$ whenever $x \in X$;

- $f[t_1, \ldots, t_n] \in T_\Sigma$ whenever $f \in F_n$, $t_1, \ldots, t_n \in T_\Sigma$ and $n \in \mathbb{N}$.

Note that set T_Σ is not empty, since it contains the set of variables. It also contains the (possibly empty) set of constant symbols. In algebraic terms, the set of terms is the free algebra generated from $X \cup F_0$ using as operations the symbols in $\bigcup_{n \geq 1} F_n$. Recall that on free inductive definitions like this one, the induced preorder $\preceq$ is in fact a partial order because it is also antisymmetric.

In the sequel, $T_\Sigma \subset A_\Sigma^*$ is taken to be a subset of A_M^* as follows:

- $x <> \text{``}[\,]\text{''} \in T_\Sigma$ whenever $x \in X$;

- $f <> \text{``}[\text{''} <> t_1 <> \text{``},\text{''} <> \cdots <> \text{``},\text{''} <> t_n \text{``}]\text{''} \in T_\Sigma$ whenever $f \in F_n$, $t_1, \ldots, t_n \in T_\Sigma$ and $n \in \mathbb{N}$.

The computability requirements imposed on the signature are essential in the proof of the following result.

Proposition 3 Set T_Σ is computable.

Proof: The following algorithm computes the characteristic map χ_{T_Σ}:

---

[1]A variant of the Gödelization adopted in Chapter 3 of [35].

```
Function[w,
    If[SyntaxQ[w] == False,
        0
      ,
        ew = ToExpression[w];
        sh = ToString[ew[[0]]];
        lsa = Map[Function[ea, ToString[ea]], Apply[List, ew]];
        Which[
            sh ∈ X,  If[Length[ew] == 0, 1, 0],
            sh ∈ F,  If[Length[ew] != τ[sh], 0, χ_{T_Σ^*}[lsa]],
            True,  0
        ]
    ]
]
```

where map $\chi_{T_\Sigma^*}$ returns 1 if its argument is a (possibly empty) list of terms and 0 otherwise. This map is computed by the following algorithm:

```
Function[ls,
    If[ls == {},
        1
      ,
        If[χ_{T_Σ}[First[ls]] == 1 && χ_{T_Σ^*}[Rest[ls]] == 1,
            1
          ,
            0
        ]
    ]
]
```

Note that the two algorithms invoke each other mutually. The computation terminates because in each call to χ_{T_Σ} the term provided as argument is simpler and in each call to $\chi_{T_\Sigma^*}$ the list provided as argument is shorter. As expected, the decidability of the sets of variables and function symbols as well as the computability of τ are needed here.   QED

The set L_Σ of *formulas* over Σ, called the *first-order language* over Σ, is inductively defined as follows:

- $p[t_1, \ldots, t_n] \in L_\Sigma$ whenever $p \in P_n$, $t_1, \ldots, t_n \in T_\Sigma$ and $n \in \mathbb{N}^+$;

- $\neg[\varphi] \in L_\Sigma$ whenever $\varphi \in L_\Sigma$ — *negation*;

- $\Rightarrow[\varphi, \psi] \in L_\Sigma$ whenever $\varphi, \psi \in L_\Sigma$ — *implication*;

- $\forall[x, \varphi] \in L_\Sigma$ whenever $x \in X$ and $\varphi \in L_\Sigma$ — *universal quantification.*

Observe that set L_Σ is not empty because T_Σ is not empty and it was assumed that there is at least one predicate symbol. A formula of the form $p[t_1, \ldots, t_n]$ is said to be *atomic* (or *basic*). Denote by B_Σ the set of all atomic formulas. The set of formulas is the free algebra generated from B_Σ using as operations the connectives and the universal quantification for the different variables.

In the sequel, $L_\Sigma \subset A_\Sigma^*$ is taken to be a subset of A_M^*. The task of rewriting the inductive definition of L_Σ using the string manipulation operations available in *Mathematica* is left to the reader.

Proposition 4 Set L_Σ is computable.

Proof: The development of an algorithm computing the characteristic map of L_Σ is left as an exercise.   QED

The *prefix notation* was adopted herein to facilitate the development of procedures and algorithms for manipulating terms and formulas. In this way, T_Σ and L_Σ are both subsets of the language of *Mathematica* expressions. Thus, it is possible to recognize the sequences of characters corresponding to terms and formulas in a very practical way by appealing directly to the recognizer of the *Mathematica* system. It was also assumed that the characters used are available in A_M and that they are not used by the system for a different purpose.

Although prefix notation is used by some authors, *infix notation* (using round brackets) prevails, as follows:

- x in place of $x[\,]$;

- f in place of $f[\,]$;

- $f(t_1, \ldots, t_n)$ in place of $f[t_1, \ldots, t_n]$;

- $p(t_1, \ldots, t_n)$ in place of $p[t_1, \ldots, t_n]$;

- $(\neg\,\varphi)$ in place of $\neg[\varphi]$;

- $(\varphi \Rightarrow \psi)$ in place of $\Rightarrow[\varphi, \psi]$;

- $(\forall x\,\varphi)$ in place of $\forall[x, \varphi]$.

Infix notation will be used hereafter except in the development of procedures and algorithms.

The remaining connectives and the existential quantifier are introduced, as usual, by means of abbreviations:

- $(\varphi \vee \psi)$ for $((\neg\,\varphi) \Rightarrow \psi)$ — disjunction;

- $(\varphi \wedge \psi)$ for $(\neg(\varphi \Rightarrow (\neg\,\psi)))$ — conjunction;

- $(\varphi \Leftrightarrow \psi)$ for $(\neg((\varphi \Rightarrow \psi) \Rightarrow (\neg(\psi \Rightarrow \varphi))))$ — equivalence;

- $(\exists x\,\varphi)$ for $(\neg(\forall x\,(\neg\,\varphi)))$ — existential quantification.

Exercise 5 Write an algorithm that expands the corresponding abbreviations in prefix notation.

Note that one should not speak about the first-order language, but rather about first-order languages, since each signature induces a specific language.

Exercise 6 Use the signatures defined in Exercise 1 in order to express in the corresponding first-order languages interesting assertions about:

- groups;

- natural numbers;

- sets.

4.3 Free and bound variables

In the context of a formula, it is important to distinguish between variables that occur free and variables that are bound to a quantifier. For this purpose, it is convenient to define first the set of variables that occur in a term.

The map $\mathrm{var}_\Sigma : T_\Sigma \to \wp X$ assigning to each term the set of variables occurring in it is inductively defined as follows:

- $\mathrm{var}_\Sigma(x) = \{x\}$;

- $\mathrm{var}_\Sigma(c) = \emptyset$;

- $\mathrm{var}_\Sigma(f(t_1,\ldots,t_n)) = \mathrm{var}_\Sigma(t_1) \cup \cdots \cup \mathrm{var}_\Sigma(t_n)$.

Exercise 7 Show that the map $\lambda\, x, t\,.\, \chi_{\mathrm{var}_\Sigma(t)}(x) : X \times T_\Sigma \to \mathbb{N}$ is computable. Conclude that the set $\mathrm{var}_\Sigma(t)$ is computable for each term t.

Term t is said to be *closed* (or *rigid*) if $\mathrm{var}_\Sigma(t) = \emptyset$. The set of closed terms over Σ is denoted cT_Σ. The reason why these terms are also called rigid will become clear in Chapter 7: their denotation does not depend on the values given to the variables. On the other hand, the more common name of closed terms emphasizes the fact that these terms do not have variables. Some authors refer to closed terms as *ground* terms since substitutions do not affect them. Observe that $cT_\Sigma = \emptyset$ whenever $F_0 = \emptyset$.

Exercise 8 Show that cT_Σ is computable.

The map $\mathrm{fv}_\Sigma : L_\Sigma \to \wp X$ assigning to each formula the set of variables occurring *free* in it is inductively defined as follows:

- $\mathrm{fv}_\Sigma(p(t_1, \ldots, t_n)) = \mathrm{var}_\Sigma(t_1) \cup \cdots \cup \mathrm{var}_\Sigma(t_n)$;

- $\mathrm{fv}_\Sigma((\neg \, \varphi)) = \mathrm{fv}_\Sigma(\varphi)$;

- $\mathrm{fv}_\Sigma((\varphi \Rightarrow \psi)) = \mathrm{fv}_\Sigma(\varphi) \cup \mathrm{fv}_\Sigma(\psi)$;

- $\mathrm{fv}_\Sigma((\forall x \, \varphi)) = \mathrm{fv}_\Sigma(\varphi) \setminus \{x\}$.

As an example, observe that

$$\mathrm{fv}_\Sigma(((\forall x_1 \, p(x_1)) \vee (\exists x_2 \, q(x_1, x_2, x_3)))) = \{x_1, x_3\}.$$

Exercise 9 Show that the map $\lambda \, x, \varphi \, . \, \chi_{\mathrm{fv}_\Sigma(\varphi)}(x) : X \times L_\Sigma \to \mathbb{N}$ is computable. Conclude that the set $\mathrm{fv}_\Sigma(\varphi)$ is computable for each formula φ.

The map $\mathrm{bv}_\Sigma : L_\Sigma \to \wp X$ that assigns to each formula the set of variables occurring *bound* in it is inductively defined as follows:

- $\mathrm{bv}_\Sigma(p(t_1, \ldots, t_n)) = \emptyset$;

- $\mathrm{bv}_\Sigma((\neg \, \varphi)) = \mathrm{bv}_\Sigma(\varphi)$;

- $\mathrm{bv}_\Sigma((\varphi \Rightarrow \psi)) = \mathrm{bv}_\Sigma(\varphi) \cup \mathrm{bv}_\Sigma(\psi)$;

- $\mathrm{bv}_\Sigma((\forall x \, \varphi)) = \{x\} \cup \mathrm{bv}_\Sigma(\varphi)$.

It is straightforward to check that

$$\mathrm{bv}_\Sigma(((\forall x_1 \, p(x_1)) \vee (\exists x_2 \, q(x_1, x_2, x_3)))) = \{x_1, x_2\}$$

and that

$$\mathrm{bv}_\Sigma(((\forall x_2 \, p(x_2)) \vee (\exists x_2 \, (\forall x_4 \, q(x_2, x_3))))) = \{x_2, x_4\}.$$

Observe that the same variable may occur (several times) free and (several times) bound in a given formula. That is, it is not always the case that $\mathrm{fv}_\Sigma(\varphi) \cap \mathrm{bv}_\Sigma(\varphi) = \emptyset$.

Exercise 10 Show that the map $\lambda \, x, \varphi \, . \, \chi_{\mathrm{bv}_\Sigma(\varphi)}(x) : X \times L_\Sigma \to \mathbb{N}$ is computable. Conclude that the set $\mathrm{bv}_\Sigma(\varphi)$ is computable for each formula φ.

In the quantification formula $(\forall x \, \varphi)$ the free occurrences of x in φ (whenever they exist) are said to be *captured by the quantifier*.

Formula φ is said to be *closed* if $\mathrm{fv}_\Sigma(\varphi) = \emptyset$. The set of closed formulas over Σ is denoted cL_Σ. The role of closed formulas will become clear in the subsequent chapters.

Exercise 11 Show that cL_Σ is computable.

4.4 Free terms and substitutions

The main objective herein is to introduce the notion of substitution of a variable by a term in a formula and to identify the situations where such a substitution is appropriate as an inference tool.

The notion of substitution of variables by terms in terms is easy to define and, thus, left to the reader in the following exercise.

Exercise 12 Let $y_1, \ldots, y_m$ be distinct variables and $u_1, \ldots, u_m$ be terms. Define inductively the map

$$\lambda t \,.\, [t]_{u_1, \ldots, u_m}^{y_1, \ldots, y_m} : T_\Sigma \to T_\Sigma$$

assigning to each term t the term obtained by simultaneously and uniformly replacing each variable y_i in t by term u_i, for each $1 \le i \le m$.

Exercise 13

Compute $[f(x_1, x_2, x_3)]_{x_3, h(x_1, x_4, x_3)}^{x_1, x_3}$.

It is now possible to define substitution of variables by terms in formulas. Let $y_1, \ldots, y_m$ be distinct variables and $u_1, \ldots, u_m$ be terms. The map

$$\lambda \varphi \,.\, [\varphi]_{u_1, \ldots, u_m}^{y_1, \ldots, y_m} : L_\Sigma \to L_\Sigma$$

is inductively defined as follows:

- $[p(t_1, \ldots, t_n)]_{u_1, \ldots, u_m}^{y_1, \ldots, y_m}$ is $p([t_1]_{u_1, \ldots, u_m}^{y_1, \ldots, y_m}, \ldots, [t_n]_{u_1, \ldots, u_m}^{y_1, \ldots, y_m})$;

- $[(\neg\,\varphi)]_{u_1, \ldots, u_m}^{y_1, \ldots, y_m}$ is $(\neg [\varphi]_{u_1, \ldots, u_m}^{y_1, \ldots, y_m})$;

- $[(\varphi \Rightarrow \psi)]_{u_1, \ldots, u_m}^{y_1, \ldots, y_m}$ is $([\varphi]_{u_1, \ldots, u_m}^{y_1, \ldots, y_m} \Rightarrow [\psi]_{u_1, \ldots, u_m}^{y_1, \ldots, y_m})$;

- $[(\forall x\, \varphi)]_{u_1, \ldots, u_m}^{y_1, \ldots, y_m}$ is:

  - $(\forall x\, [\varphi]_{u_1, \ldots, u_m}^{y_1, \ldots, y_m})$ whenever $x \notin \{y_1, \ldots, y_m\}$;
  - $(\forall x\, [\varphi]_{u'_1, \ldots, u'_m}^{y_1, \ldots, y_m})$ where u'_j is u_j for $j \ne k$ and u'_k is x whenever x is y_k.

The map so defined assigns to each formula φ the formula obtained by simultaneously and uniformly replacing each free occurrence of variable y_i in φ by term u_i, for each $1 \le i \le m$.

Exercise 14 Compute $[(\forall x\, \varphi)]_t^x$.

Exercise 15 Compute $[(\forall x_1\, p(x_1, x_2, x_3))]_{x_3, h(x_1, x_4, x_3)}^{x_1, x_3}$.

Recall the idea of instance of a quantification: from $(\forall x \, \varphi)$ infer the formula $[\varphi]_t^x$ obtained from φ by replacing every free occurrence of x with term t. The following exercise motivates the fact that not every substitution is desirable with respect to this kind of inference.

Exercise 16 Consider the formula

$$(\forall x_1(\exists x_2(x_1 < x_2)))$$

and its instance

$$(\exists x_2(c < x_2))$$

obtained by replacing x_1 with c. Consider now its instance

$$(\exists x_2(x_2 < x_2))$$

obtained by replacing x_1 with x_2. What is the problem with the latter substitution?

The following notion of free term for a variable in a formula helps in avoiding unwanted substitutions. The set $\rhd_\Sigma \subseteq T_\Sigma \times X \times L_\Sigma$ is inductively defined as follows:

- $(t, x, p(t_1, \ldots, t_n)) \in \rhd_\Sigma$;

- $(t, x, (\neg \varphi)) \in \rhd_\Sigma$ whenever $(t, x, \varphi) \in \rhd_\Sigma$;

- $(t, x, (\varphi \Rightarrow \psi)) \in \rhd_\Sigma$ whenever $(t, x, \varphi) \in \rhd_\Sigma$ and $(t, x, \psi) \in \rhd_\Sigma$;

- $(t, x, (\forall y \, \varphi)) \in \rhd_\Sigma$ whenever

  (i) either y is x

  (ii) or the two following conditions are fulfilled:

$$\begin{cases} \text{if } x \in \mathrm{fv}_\Sigma(\varphi) \text{ then } y \notin \mathrm{var}_\Sigma(t); \\ (t, x, \varphi) \in \rhd_\Sigma. \end{cases}$$

When $(t, x, \varphi) \in \rhd_\Sigma$ it is said that *term t is free for variable x in formula* φ, usually written $t \rhd_\Sigma x : \varphi$. Intuitively, $t \rhd_\Sigma x : \varphi$ means that substituting t for x in φ does not lead to the capture of variables in t by quantifiers in φ.

Exercise 17 State in which conditions a term is free for a variable in formulas using conjunction, disjunction, equivalence and existential quantification.

Exercise 18 Compute $t \rhd_\Sigma x : \varphi$ for each of the following combinations:

term t	variable x	formula φ
c	x_1	$(\exists x_2 (x_1 < x_2))$
x_1	x_2	$((\forall x_1\, p(x_1, x_2)) \lor (\exists x_2 (\forall x_4\, q(x_1, x_2, x_3))))$
x_2	x_3	
x_3	x_4	
x_4	x_5	
x_5		
$f(x_2)$		
$g(x_3, x_2)$		
$g(x_1, x_2)$		

The following exercise shows that sometimes it is easy to guarantee that a term is free for a variable in a formula.

Exercise 19 Show that each of the following assertions is a sufficient condition for t to be free for x in φ:

1. $x \notin \mathrm{fv}_\Sigma(\varphi)$;

2. $\mathrm{var}_\Sigma(t) \cap \mathrm{bv}_\Sigma(\varphi) = \emptyset$;

3. t is x.

The next exercise shows an important interplay between the notion of free term for a variable in a formula and the notion of substitution.

Exercise 20 Show that if $y \notin \mathrm{fv}_\Sigma(\varphi)$ and $y \rhd_\Sigma x : \varphi$, then $x \rhd_\Sigma y : [\varphi]_y^x$.

The following exercises establish computability results about substitution and freeness that are needed later on.

Exercise 21 Show that the map $\lambda t \,.\, [t]_{u_1,\ldots,u_m}^{y_1,\ldots,y_m}$ is computable.

Exercise 22 Show that the map $\lambda \varphi \,.\, [\varphi]_{u_1,\ldots,u_m}^{y_1,\ldots,y_m}$ is computable.

Exercise 23 Show that the set $\rhd_\Sigma$ is computable.

The following exercise establishes a key property of composition of substitutions that will be used in the next chapter.

Exercise 24 Show that

$$[[\varphi]_t^x]_u^y = [[\varphi]_u^y]_{[t]_u^y}^x$$

provided that $x \notin \mathrm{var}_\Sigma(u)$.

Hint: Start by stating and proving the corresponding result for terms.

Chapter 5

Hilbert calculus

The chapter is dedicated to setting-up a first-order predicate calculus, using the so called axiomatization approach usually attributed to David Hilbert.[1] This approach is characterized by the use of axioms and inference rules for deriving formulas from hypotheses.[2]

Two alternative definitions of derivation are introduced and shown to be equivalent by using a fixed point construction. Derivation is shown to be extensive, monotonic, idempotent, compact and, more importantly, semi-decidable (the set of formulas derived from a computably enumerable set of hypotheses is shown to be computably enumerable).[3] A rich collection of metatheorems and admissible rules is provided in order to facilitate the practical use of the calculus and also for theoretical reasons.

Section 5.1 presents the axioms, the inference rules and the two notions of derivation. Section 5.2 is dedicated to proving their equivalence. The main properties of derivation are investigated in Section 5.3, with emphasis on the semi-decidability of the Hilbert calculus. Several metatheorems and admissible rules are stated and proved in Section 5.4, including the metatheorem of deduction.

---

[1] However, in fairness, it should also be attributed to Gottlob Frege.

[2] Another first-order predicate calculus is introduced in Chapter 9, using the approach attributed to Gerhard Gentzen. These two calculi are quite different but they share with all calculi the essential characteristic of providing the means for reasoning at a purely symbolic level, that is, by syntactic manipulation of terms and formulas, without taking into account their meaning. Furthermore, the two calculi are equivalent in a significant sense to be made clear in Chapter 9.

[3] The fact that is not decidable in general is proved much later, in Chapter 13.

5.1 Axioms and rules

A Hilbert calculus includes a set of axioms and a set of inference rules. An axiom is a formula.[4] An inference rule is a pair composed by a set of formulas (the premises of the rule) and a formula (the conclusion of the rule).[5] A finitary Hilbert calculus is a Hilbert calculus where the set of premises is a finite set for every inference rule.

The objective of a Hilbert calculus is to provide a purely symbolic mechanism for ascertaining that a formula is derivable from a set of hypotheses, by making use of the hypotheses at hand, the axioms and the inference rules.

Let $\Gamma \subseteq L_\Sigma$. The set $\Gamma^{\vdash_\Sigma}$ of formulas *derivable* from Γ is defined inductively as follows:

- $\Gamma^{\vdash_\Sigma}$ contains the *hypotheses*:

    - $\Gamma \subseteq \Gamma^{\vdash_\Sigma}$.

- $\Gamma^{\vdash_\Sigma}$ contains the *axioms*:

    1. $(\varphi \Rightarrow (\psi \Rightarrow \varphi)) \in \Gamma^{\vdash_\Sigma}$,
    2. $((\varphi \Rightarrow (\psi \Rightarrow \delta)) \Rightarrow ((\varphi \Rightarrow \psi) \Rightarrow (\varphi \Rightarrow \delta))) \in \Gamma^{\vdash_\Sigma}$,
    3. $(((\neg\,\varphi) \Rightarrow (\neg\psi)) \Rightarrow (\psi \Rightarrow \varphi)) \in \Gamma^{\vdash_\Sigma}$,
    4. $((\forall x\, \varphi) \Rightarrow [\varphi]_t^x) \in \Gamma^{\vdash_\Sigma}$ whenever $t \triangleright_\Sigma x : \varphi$,
    5. $((\forall x\, (\varphi \Rightarrow \psi)) \Rightarrow (\varphi \Rightarrow (\forall x\, \psi))) \in \Gamma^{\vdash_\Sigma}$ whenever $x \notin \mathrm{fv}_\Sigma(\varphi)$,

    for each $\varphi, \psi, \delta \in L_\Sigma$, $x \in X$ and $t \in T_\Sigma$.

- $\Gamma^{\vdash_\Sigma}$ is closed under the *inference rules*:

    Modus ponens (MP) also called *detachment rule*

    $$\psi \in \Gamma^{\vdash_\Sigma} \text{ whenever } \varphi, (\varphi \Rightarrow \psi) \in \Gamma^{\vdash_\Sigma},$$

    Generalization (Gen)

    $$(\forall x\, \varphi) \in \Gamma^{\vdash_\Sigma} \text{ whenever } \varphi \in \Gamma^{\vdash_\Sigma},$$

    for each $\varphi, \psi \in L_\Sigma$ and $x \in X$.

---

[4]Accepted as true.
[5]Accepted as true whenever the premises are accepted as true.

The set of all axioms is denoted Ax_Σ and, for $j = 1, \ldots, 5$, the set of axioms of type j is denoted $\mathrm{Ax}j_\Sigma$. For instance,

$$\mathrm{Ax1}_\Sigma = \{(\varphi \Rightarrow (\psi \Rightarrow \varphi)) : \varphi, \psi \in L_\Sigma\}$$

and

$$\mathrm{Ax4}_\Sigma = \{((\forall x\, \varphi) \Rightarrow [\varphi]_t^x) : x \in X, t \in T_\Sigma, \varphi \in L_\Sigma, t \rhd_\Sigma x : \varphi\}.$$

Usually one writes

$$\Gamma \vdash_\Sigma \varphi$$

instead of $\varphi \in \Gamma^{\vdash_\Sigma}$ for stating that φ is derivable from Γ. In this way, *derivation* can be seen as a binary relation defined over the set of all subsets of formulas and the set of formulas, that is, over $\wp L_\Sigma \times L_\Sigma$. One may also write

$$\gamma_1, \ldots, \gamma_n \vdash_\Sigma \varphi$$

for $\{\gamma_1, \ldots, \gamma_n\} \vdash_\Sigma \varphi$.

A *theorem* is a formula derivable from the empty set. It is usual to write

$$\vdash_\Sigma \varphi$$

instead of $\varphi \in \emptyset^{\vdash_\Sigma}$ for stating that φ is a theorem.

As a first example, it can be seen that $(\eta \Rightarrow \eta)$ is a theorem (regardless of the formula η in consideration). It is necessary to show that $(\eta \Rightarrow \eta) \in \emptyset^{\vdash_\Sigma}$:

1. $(\eta \Rightarrow (\eta \Rightarrow \eta)) \in \emptyset^{\vdash_\Sigma}$ because this formula is an axiom of type 1;[6]

2. $(\eta \Rightarrow ((\eta \Rightarrow \eta) \Rightarrow \eta)) \in \emptyset^{\vdash_\Sigma}$ because this formula is an axiom of type 1;[7]

3. $((\eta \Rightarrow ((\eta \Rightarrow \eta) \Rightarrow \eta)) \Rightarrow ((\eta \Rightarrow (\eta \Rightarrow \eta)) \Rightarrow (\eta \Rightarrow \eta))) \in \emptyset^{\vdash_\Sigma}$ because this formula is an axiom of type 2;[8]

4. $((\eta \Rightarrow (\eta \Rightarrow \eta)) \Rightarrow (\eta \Rightarrow \eta)) \in \emptyset^{\vdash_\Sigma}$ because this formula is obtained by MP from the formulas at steps 2 and 3;

5. $(\eta \Rightarrow \eta) \in \emptyset^{\vdash_\Sigma}$ because this formula is obtained by MP from the formulas at steps 1 and 4.

It is usual to present the reasoning above in the compact form of a derivation sequence as follows:

---

[6]Taking η for φ and also for ψ.

[7]Taking η for φ and $(\eta \Rightarrow \eta)$ for ψ.

[8]Taking η for φ, $(\eta \Rightarrow \eta)$ for ψ and η for δ.

$$
\begin{array}{lll}
1 & (\eta \Rightarrow (\eta \Rightarrow \eta)) & \text{Ax1} \\
2 & (\eta \Rightarrow ((\eta \Rightarrow \eta) \Rightarrow \eta)) & \text{Ax1} \\
3 & ((\eta \Rightarrow ((\eta \Rightarrow \eta) \Rightarrow \eta)) \Rightarrow ((\eta \Rightarrow (\eta \Rightarrow \eta)) \Rightarrow (\eta \Rightarrow \eta)))) & \text{Ax2} \\
4 & ((\eta \Rightarrow (\eta \Rightarrow \eta)) \Rightarrow (\eta \Rightarrow \eta)) & \text{MP 2,3} \\
5 & (\eta \Rightarrow \eta) & \text{MP 1,4}
\end{array}
$$

In general, by a *derivation sequence* for $\varphi \in L_\Sigma$ from $\Gamma \subseteq L_\Sigma$ one means a sequence

$$(\psi_1, J_1) \ldots (\psi_n, J_n)$$

such that:

- each $\psi_i \in L_\Sigma$;

- each J_i is the *justification* for ψ_i:

  - if J_i is Hyp, then ψ_i is an element of Γ;

  - if J_i is Axj, then ψ_i is an axiom of type j;

  - if J_i is Gen k, then $k < i$ and ψ_i is $(\forall x\, \psi_k)$ for some $x \in X$;

  - if J_i is MP k_1, k_2, then $k_1, k_2 < i$ and $\{\psi_{k_1}, \psi_{k_2}\} = \{\alpha, (\alpha \Rightarrow \psi_i)\}$ for some $\alpha \in L_\Sigma$;

- ψ_n is φ.

Exercise 1 Provide derivation sequences for:

- *Principle of quantifier exchange* (PQE): $(\forall x (\forall y\, \varphi)) \vdash_\Sigma (\forall y (\forall x\, \varphi))$.

- *Hypothetical syllogism* (HS): $(\varphi_1 \Rightarrow \varphi_2), (\varphi_2 \Rightarrow \varphi_3) \vdash_\Sigma (\varphi_1 \Rightarrow \varphi_3)$.

The *universal closure* of a formula φ such that $\mathrm{fv}_\Sigma(\varphi) = \{x_{i_1}, \ldots, x_{i_n}\}$ with $i_1 < \cdots < i_n$ is the closed formula $(\forall \varphi) = (\forall x_{i_1} (\ldots (\forall x_{i_n}\, \varphi) \ldots))$.

Exercise 2 Show that

$$\Gamma \vdash_\Sigma \varphi \text{ if and only if } \Gamma \vdash_\Sigma (\forall x\, \varphi) \text{ if and only if } \Gamma \vdash_\Sigma (\forall \varphi).$$

Exercise 3 Provide derivation sequences for:

1. $(\forall x (\varphi_1 \Rightarrow \varphi_2)), (\forall x\, \varphi_1) \vdash_\Sigma (\forall x\, \varphi_2)$.

2. $((\forall x\, p(x)) \Rightarrow (\forall y\, p(y)))$.

5.2 Derivation as a fixed point

The objective of this section is to prove that, for a given set of hypotheses, a formula is derivable if and only if there is a derivation sequence for the formula. The proof is based on a fixed point construction, using the *one-step derivation operator*

$$D_\Sigma : \wp L_\Sigma \to \wp L_\Sigma$$

such that

$$D_\Sigma(\Psi) = \Psi \cup \{\beta : \alpha, (\alpha \Rightarrow \beta) \in \Psi\} \cup \{(\forall x\, \alpha) : x \in X \ \& \ \alpha \in \Psi\}.$$

Thus, $D_\Sigma(\Psi)$ is the one step closure of Ψ by modus ponens and generalization. Observe that D_Σ should be seen as an operator from the complete lattice[9] $(\wp L_\Sigma, \subseteq)$ to itself.

Proposition 4 Operator D_Σ is monotonic:

$$D_\Sigma(\Psi_1) \subseteq D_\Sigma(\Psi_2) \ \text{ whenever } \ \Psi_1 \subseteq \Psi_2.$$

Furthermore, D_Σ is extensive:

$$\Psi \subseteq D_\Sigma(\Psi).$$

But, it is not idempotent, since it is easy to find Ψ such that

$$D_\Sigma(D_\Sigma(\Psi)) \neq D_\Sigma(\Psi).$$

The idea now is to prove that the operator D_Σ has fixed points by showing first that it preserves unions of directed conglomerates in the sense detailed below.

A *conglomerate* is a family $\{\Psi_e\}_{e \in E}$ such that each $\Psi_e \subseteq L_\Sigma$. A conglomerate is said to be *directed* if, for every $e', e'' \in E$, there is $e \in E$ such that $\Psi_{e'} \cup \Psi_{e''} \subseteq \Psi_e$. In particular, if a conglomerate is closed for binary unions, then it is directed.

Proposition 5 (Continuity of D_Σ)
If the conglomerate $\{\Psi_e\}_{e \in E}$ is directed, then

$$D_\Sigma\left(\bigcup_{e \in E} \Psi_e\right) = \bigcup_{e \in E} D_\Sigma(\Psi_e).$$

---

[9] Partially ordered set in which all subsets have both a supremum (or join) and an infimum (or meet).

Proof:

(1) $\bigcup_{e \in E} D_\Sigma(\Psi_e) \subseteq D_\Sigma(\bigcup_{e \in E} \Psi_e)$:
For each $e \in E$, $\Psi_e \subseteq \bigcup_{e \in E} \Psi_e$, and, therefore, thanks to Proposition 4, $D_\Sigma(\Psi_e) \subseteq D_\Sigma(\bigcup_{e \in E} \Psi_e)$. Hence, $\bigcup_{e \in E} D_\Sigma(\Psi_e) \subseteq D_\Sigma(\bigcup_{e \in E} \Psi_e)$.

(2) $D_\Sigma(\bigcup_{e \in E} \Psi_e) \subseteq \bigcup_{e \in E} D_\Sigma(\Psi_e)$:
Take $\varphi \in D_\Sigma(\bigcup_{e \in E} \Psi_e)$. There are three cases to consider:
(i) $\varphi \in \bigcup_{e \in E} \Psi_e$. Then, there exists $e \in E$ such that $\varphi \in \Psi_e$. Therefore, there is $e \in E$ such that $\varphi \in D_\Sigma(\Psi_e)$ and, so, $\varphi \in \bigcup_{e \in E} D_\Sigma(\Psi_e)$.
(ii) $\varphi \in \{\beta : \alpha, (\alpha \Rightarrow \beta) \in \bigcup_{e \in E} \Psi_e\}$. Then, there is α such that $\alpha, (\alpha \Rightarrow \varphi) \in \bigcup_{e \in E} \Psi_e$. Therefore, there are $e_1, e_2 \in E$ for which $\alpha \in \Psi_{e_1}$ and $(\alpha \Rightarrow \varphi) \in \Psi_{e_2}$. Since $\{\Psi_e\}_{e \in E}$ is directed, there is $e \in E$ such that $\alpha, (\alpha \Rightarrow \varphi) \in \Psi_e$. Hence, there is $e \in E$ such that $\varphi \in D_\Sigma(\Psi_e)$, and, so, $\varphi \in \bigcup_{e \in E} D_\Sigma(\Psi_e)$.
(iii) $\varphi \in \{(\forall x\, \alpha) : x \in X \,\&\, \alpha \in \bigcup_{e \in E} \Psi_e\}$. Then, there is α such that $(\forall x\, \alpha)$ is φ for some variable x and there is $e \in E$ such that $\alpha \in \Psi_e$. Thus, there is $e \in E$ such that $(\forall x\, \alpha) \in D_\Sigma(\Psi_e)$, and, so, $\varphi \in \bigcup_{e \in E} D_\Sigma(\Psi_e)$.   QED

Note that the fact that the conglomerate is directed was only needed for dealing with modus ponens, since the set of premises is a singleton in the case of generalization.

It is also worthwhile to remark that the proof above would work for any other finitary Hilbert calculus, that is, as long as every inference rule has a finite set of premises.

A continuous operator in the sense above is also continuous in the topological sense by considering the so called Scott topology induced by the ordering in the lattice at hand. The interested reader on this issue should consult [19].

Observe that every continuous operator is monotonic. It is enough to consider the conglomerate $\{\Psi_1, \Psi_2\}$ with $\Psi_1 \subseteq \Psi_2$. Thus, $\Psi_2 = \Psi_1 \cup \Psi_2$ and the conglomerate is directed. Therefore,

$$D_\Sigma(\Psi_2) = D_\Sigma(\Psi_1 \cup \Psi_2) = D_\Sigma(\Psi_1) \cup D_\Sigma(\Psi_2)$$

and, so, $D_\Sigma(\Psi_1) \subseteq D_\Sigma(\Psi_2)$. On the other hand, justifying why D_Σ was proved first to be monotonic, it is very easy to prove that

$$\bigcup_{e \in E} D_\Sigma(\Psi_e) \subseteq D_\Sigma(\bigcup_{e \in E} \Psi_e)$$

knowing that D_Σ is monotonic.

Recall that a set $\Psi \subseteq L_\Sigma$ is said to be a *fixed point* of D_Σ if $D_\Sigma(\Psi) = \Psi$. The set of fixed points of D_Σ has a minimum, which is called the least fixed point of D_Σ. More generally, given a set $\Omega \subseteq L_\Sigma$, the set

$$\{\Psi \subseteq L_\Sigma : D_\Sigma(\Psi) = \Psi \text{ and } \Omega \subseteq \Psi\}$$

also has a minimum, usually denoted by

$$\text{lfp}(D_\Sigma, \Omega)$$

and known as the least fixed point of D_Σ containing Ω.

Proposition 6 Let $\Omega \subseteq L_\Sigma$. Then,

$$\text{lfp}(D_\Sigma, \Omega) = \bigcup_{k \in \mathbb{N}} D_\Sigma^k(\Omega).$$

Proof: The first step is to check that $\bigcup_{k \in \mathbb{N}} D_\Sigma^k(\Omega)$ is a fixed point of D_Σ containing Ω. Clearly, $\Omega \subseteq \bigcup_{k \in \mathbb{N}} D_\Sigma^k(\Omega)$. Moreover, $\{D_\Sigma^k(\Omega)\}_{k \in \mathbb{N}}$ is directed since, as the reader will be able to prove, $D_\Sigma^i(\Omega) \subseteq D_\Sigma^j(\Omega)$ for every $i \leq j$. Hence, by the continuity of D_Σ, we have:

$$D_\Sigma \left(\bigcup_{k \in \mathbb{N}} D_\Sigma^k(\Omega) \right) = \bigcup_{k \in \mathbb{N}} D_\Sigma(D_\Sigma^k(\Omega))$$

$$= \bigcup_{k \in \mathbb{N}} D_\Sigma^{k+1}(\Omega)$$

$$= \bigcup_{k \in \mathbb{N}^+} D_\Sigma^k(\Omega)$$

$$= \Omega \cup \bigcup_{k \in \mathbb{N}^+} D_\Sigma^k(\Omega)$$

$$= \bigcup_{k \in \mathbb{N}} D_\Sigma^k(\Omega).$$

It only remains to verify that $\bigcup_{k \in \mathbb{N}} D_\Sigma^k(\Omega)$ is the least fixed point of D_Σ containing Ω. Let Δ be a fixed point of D_Σ containing Ω. Then,

$$\bigcup_{k \in \mathbb{N}} D_\Sigma^k(\Omega) \subseteq \Delta$$

since $D_\Sigma^k(\Omega) \subseteq \Delta$ for every $k \in \mathbb{N}$, a fact proved by induction on k as follows:
(Basis) $D_\Sigma^0(\Omega) = \Omega \subseteq \Delta$.
(Step) Let $k = j + 1$. Then, $D_\Sigma^k(\Omega) = D_\Sigma(D_\Sigma^j(\Omega))$. By the induction hypothesis, $D_\Sigma^j(\Omega) \subseteq \Delta$. Therefore, by Proposition 4, $D_\Sigma(D_\Sigma^j(\Omega)) \subseteq D_\Sigma(\Delta)$, and, so, $D_\Sigma^k(\Omega) \subseteq \Delta$. QED

Observe that the inductive definition of $\Gamma^{\vdash_\Sigma}$ is equivalent to stating that $\Gamma^{\vdash_\Sigma}$ is the least fixed point of D_Σ containing $\Gamma \cup Ax_\Sigma$. Therefore, as a corollary of the previous proposition, the following result holds.

Proposition 7 Let $\Gamma \subseteq L_\Sigma$. Then,

$$\Gamma^{\vdash_\Sigma} = \bigcup_{k \in \mathbb{N}} D_\Sigma^k(\Gamma \cup \text{Ax}_\Sigma).$$

In other words, any derivable formula from Γ can be obtained by iterating D_Σ a finite number of times over $\Gamma \cup \text{Ax}_\Sigma$. Finally, it is possible to establish the equivalence of the two alternative notions of derivation.

Proposition 8 (Equivalence of derivation notions)
Let $\Gamma \cup \{\varphi\}$ be a subset of L_Σ. Then, $\Gamma \vdash_\Sigma \varphi$ if and only if there is a derivation sequence of φ from Γ.

Proof:

$(\rightarrow)$
It must first be shown that, for every $k \in \mathbb{N}$, if $\varphi \in D_\Sigma^k(\Gamma \cup \text{Ax}_\Sigma)$ then there is a derivation sequence for φ from Γ. The proof runs by induction on k:

(Basis) $\varphi \in D_\Sigma^0(\Gamma \cup \text{Ax}_\Sigma) = \Gamma \cup \text{Ax}_\Sigma$. There are two cases to consider.
(i) $\varphi \in \Gamma$. Simply take the derivation sequence (φ, Hyp).
(ii) $\varphi \in \text{Ax}j_\Sigma$. Simply take the derivation sequence $(\varphi, \text{Ax}j)$.

(Step) Let $k = j + 1$. Then, $\varphi \in D_\Sigma(D_\Sigma^j(\Gamma \cup \text{Ax}_\Sigma))$. There are three cases to consider:
(i) $\varphi \in D_\Sigma^j(\Gamma \cup \text{Ax}_\Sigma)$. Then, by the induction hypothesis, there is a derivation sequence of φ from Γ.
(ii) $\psi, (\psi \Rightarrow \varphi) \in D_\Sigma^j(\Gamma \cup \text{Ax}_\Sigma)$. It must be shown that there is a derivation sequence for φ from Γ. By the induction hypothesis, there are derivation sequences w and w' for ψ and $(\psi \Rightarrow \varphi)$ from Γ, respectively. Suppose that the length of w is r and the length of w' is s. Consider the sequence

$$w\, w_1'' \ldots w_s''\, (\varphi, \text{MP}\, r, r + s)$$

where each w_i'' is:

- w_i', if the justification for the latter is either Hyp or Ax;

- $(\delta_i, \text{MP}\, i_1 + r, i_2 + r)$ if w_i' is $(\delta_i, \text{MP}\, i_1, i_2)$;

- $(\delta_i, \text{Gen}\, i_0 + r)$ if w_i' is $(\delta_i, \text{Gen}\, i_0)$.

Clearly, $w\, w_1'' \ldots w_s''\, (\varphi, \text{MP}\, r, r + s)$ is a derivation sequence for φ from Γ.
(iii) $\psi \in D_\Sigma^j(\Gamma \cup \text{Ax}_\Sigma)$. It must be shown that there is a derivation sequence for $(\forall x\, \psi)$ from Γ for each variable x. This is left as an exercise.

Using the lemma just shown, the desired result is obtained as follows. Suppose that $\varphi \in \Gamma^{\vdash_\Sigma}$. Then, $\varphi \in \bigcup_{k \in \mathbb{N}} D_\Sigma^k(\Gamma \cup \text{Ax}_\Sigma)$ by Proposition 7. Hence, there exists $k \in \mathbb{N}$ such that $\varphi \in D_{\Sigma}^{k}(\Gamma \cup \mathrm{Ax}_{\Sigma})$. Therefore, invoking the lemma, there is a derivation sequence for φ from Γ.

$(\leftarrow)$

Suppose now that there exists a derivation sequence for φ from Γ. Let

$$(\psi_1, J_1) \ldots (\psi_n, J_n)$$

be such a derivation sequence. The thesis is an immediate consequence of

$$\varphi \in D_{\Sigma}^{n-1}(\Gamma \cup \mathrm{Ax}_{\Sigma}),$$

a fact proved by induction on n as follows:

(Basis) $n = 1$. Then, $\varphi \in (\Gamma \cup \mathrm{Ax}_{\Sigma}) = D_{\Sigma}^{0}(\Gamma \cup \mathrm{Ax}_{\Sigma})$.

(Step) Let $n = m + 1$. There are three cases to consider:

(i) J_n is Hyp or Ax. Then, $\varphi \in D_{\Sigma}^{0}(\Gamma \cup \mathrm{Ax}_{\Sigma}) \subseteq D_{\Sigma}^{m}(\Gamma \cup \mathrm{Ax}_{\Sigma})$.

(ii) J_n is MP i, j with $i, j \leq m$. In this case, $\psi_i \in D_{\Sigma}^{i-1}(\Gamma \cup \mathrm{Ax}_{\Sigma})$ and $\psi_j \in D_{\Sigma}^{j-1}(\Gamma \cup \mathrm{Ax}_{\Sigma})$ by the induction hypothesis. Therefore,

$$\varphi \in D_{\Sigma}^{\max\{i-1, j-1\}+1}(\Gamma \cup \mathrm{Ax}_{\Sigma}) = D_{\Sigma}^{\max\{i, j\}}(\Gamma \cup \mathrm{Ax}_{\Sigma}) \subseteq D_{\Sigma}^{m}(\Gamma \cup \mathrm{Ax}_{\Sigma}).$$

(iii) J_n is Gen i with $i \leq m$. Then, $\psi_i \in D_{\Sigma}^{i-1}(\Gamma \cup \mathrm{Ax}_{\Sigma})$ by the induction hypothesis. Therefore,

$$\varphi \in D_{\Sigma}^{i}(\Gamma \cup \mathrm{Ax}_{\Sigma}) \subseteq D_{\Sigma}^{m}(\Gamma \cup \mathrm{Ax}_{\Sigma}).$$

That is, in each case $\varphi \in D_{\Sigma}^{m}(\Gamma \cup \mathrm{Ax}_{\Sigma}) = D_{\Sigma}^{n-1}(\Gamma \cup \mathrm{Ax}_{\Sigma})$ as required.   QED

The following exercise shows that it is possible to make use of lemmas when making derivations.

Exercise 9 Show that the *Cut Law* (Cut) holds:

$$\frac{\Gamma \vdash_{\Sigma} \Delta \quad \Delta \vdash_{\Sigma} \varphi}{\Gamma \vdash_{\Sigma} \varphi}$$

where

$$\Gamma \vdash_{\Sigma} \Delta$$

should be taken to mean that $\Gamma \vdash_{\Sigma} \delta$ for every formula $\delta \in \Delta$.

5.3   Properties of derivation

In this section, several properties of the *derivation operator*

$$\lambda \Gamma . \Gamma^{\vdash_{\Sigma}} : (\wp L_{\Sigma}, \subseteq) \to (\wp L_{\Sigma}, \subseteq)$$

are established, allowing a better understanding of the calculus. The following result states the basic properties of the derivation operator that the interested reader should be able to prove using, when convenient, the equivalence of the two notions of derivation.

Proposition 10 For every $\Gamma, \Psi \subseteq L_\Sigma$:

1. *Extensivity*: $\Gamma \subseteq \Gamma^{\vdash_\Sigma}$.

2. *Monotonicity*: $\Psi^{\vdash_\Sigma} \subseteq \Gamma^{\vdash_\Sigma}$ whenever $\Psi \subseteq \Gamma$.

3. *Idempotence*: $(\Gamma^{\vdash_\Sigma})^{\vdash_\Sigma} \subseteq \Gamma^{\vdash_\Sigma}$.

4. *Compactness*: $\Gamma^{\vdash_\Sigma} = \displaystyle\bigcup_{\Phi \in \wp_{\text{fin}}\Gamma} \Phi^{\vdash_\Sigma}$.

Operator $\vdash_\Sigma$ is what is called a closure operator, since it is extensive, monotonic and idempotent. Observe that

$$(\Gamma^{\vdash_\Sigma})^{\vdash_\Sigma} = \Gamma^{\vdash_\Sigma}$$

holds, as it happens for any closure operator.

Compactness is very important. It amounts to saying that if a formula is derivable from a set of hypotheses then it can be derived using only a finite number of them.

Exercise 11 Justify or refute each of the following assertions:

- $\emptyset^{\vdash_\Sigma} \neq \emptyset$;

- $\Gamma_1^{\vdash_\Sigma} \cup \Gamma_2^{\vdash_\Sigma} \subseteq (\Gamma_1 \cup \Gamma_2)^{\vdash_\Sigma}$;

- There are Γ_1 and Γ_2 such that $\Gamma_1^{\vdash_\Sigma} \cup \Gamma_2^{\vdash_\Sigma} \neq (\Gamma_1 \cup \Gamma_2)^{\vdash_\Sigma}$.

The above exercise shows that $\vdash_\Sigma$ is not a Kuratowski closure operator (for more details see [31]).

It is worthwhile to mention here that the derivation operator is also closed for substitution (up to change of bound variables), as it will be proved at the end of Section 5.4.

Before proceeding with the study of the properties of the derivation operator, now from the point of view of Computability Theory, it is necessary to look again at the nature of the axioms. Several questions arise naturally. Are the axioms independent of each other? That is, is it possible to show that no axiom is derivable from the other axioms using the inference rules? The answer to this question is positive. The interested reader can see the proof, for instance, in [12]. Returning to the issue at hand, is the set of axioms computable? Furthermore, for each $j = 1, \ldots, 5$, is $\text{Ax}j_\Sigma$ computable? It is left as an exercise to prove the following answer to these computability questions.

Proposition 12 For every $j = 1, \ldots, 5$, the set $\mathrm{Ax}j_\Sigma$ is computable. Thus, the set Ax_Σ is also computable.

The goal now is to prove that

$$\Gamma^{\vdash_\Sigma} = \bigcup_{n \in \mathbb{N}} D_\Sigma^n(\Gamma \cup \mathrm{Ax}_\Sigma)$$

is computably enumerable when Γ is a computably enumerable set. To this end, it is convenient to use the following maps:

- $\mathrm{dmp}_\Sigma = \lambda\,\Psi\,.\,\Psi \cup \{\beta : \alpha, (\alpha \Rightarrow \beta) \in \Psi\} : \wp L_\Sigma \to \wp L_\Sigma;$

- $\mathrm{dgen}_\Sigma = \lambda\,\Psi\,.\,\Psi \cup \{(\forall x\,\alpha) : x \in X\ \&\ \alpha \in \Psi\} : \wp L_\Sigma \to \wp L_\Sigma.$

Sets $\mathrm{dmp}_\Sigma(\Psi)$ and $\mathrm{dgen}_\Sigma(\Psi)$ are the one-step closures of Ψ by modus ponens and generalization, respectively. Clearly

$$D_\Sigma(\Psi) = \mathrm{dmp}_\Sigma(\Psi) \cup \mathrm{dgen}_\Sigma(\Psi).$$

The following results show that maps dmp_Σ and dgen_Σ are "computable" in the sense discussed at the end of Section 3.7: it is possible to construct an enumeration of $\mathrm{dmp}_\Sigma(\Psi)$ and an enumeration of $\mathrm{dgen}_\Sigma(\Psi)$ from any given enumeration of Ψ.

Proposition 13 There is a computable map $s_{\mathrm{Gen}} : \mathbb{N} \to \mathbb{N}$ such that, if $i_{\mathcal{C}_1^1}$ is an enumeration of $\Psi \subseteq L_\Sigma$, then $s_{\mathrm{Gen}}(i)_{\mathcal{C}_1^1}$ is an enumeration of $\mathrm{dgen}(\Psi)$.

Proof: If $i_{\mathcal{C}_1^1}$ enumerates Ψ, then the map computed by the following algorithm p_{Gen}^i enumerates $\mathrm{dgen}_\Sigma(\Psi)$:

```
Function[k,
    If[EvenQ[k],
        a = (k/2)ℕ×ℕ[[1]];
        b = (k/2)ℕ×ℕ[[2]];
        "∀" <> "[" <> aX <> "," <> i_{C_1^1}[b] <> "]"
    ,
        i_{C_1^1}[(k − 1)/2]
    ]
]
```

Let f be the map computed by this algorithm. Clearly, for every $k \in \mathbb{N}$, $f(k) \in \mathrm{dgen}_\Sigma(\Psi)$. Indeed, if k is odd then the algorithm returns $i_{\mathcal{C}_1^1}[(k-1)/2]$ and, otherwise, it returns a generalization of a formula $(i_{\mathcal{C}_1^1}[b])$ in Ψ. It remains to check that, for every $w \in \mathrm{dgen}_\Sigma(\Psi)$, there is $k \in \mathbb{N}$ such that $f(k) = w$.

Assume that $w \in \mathrm{dgen}_\Sigma(\Psi)$. There are two cases to consider:

(i) $w \in \Psi$. Since $i_{\mathcal{C}_1^1}$ is an enumeration of Ψ, there is $m \in \mathbb{N}$ such that $i_{\mathcal{C}_1^1}(m) = w$. Take $k = 2m + 1$. Then, the execution of the algorithm on input k terminates with value $i_{\mathcal{C}_1^1}((k-1)/2) = i_{\mathcal{C}_1^1}(m) = w$, since k is a odd number.

(ii) $w \in \mathrm{dgen}_\Sigma(\Psi) \setminus \Psi$. Then, there are $w_1 \in X$ and $w_2 \in \Psi$ such that

$$w = \text{``}\forall\text{''} <> \text{``}[\text{''}} <> w_1 <> \text{``},\text{''} <> w_2 <> \text{``}]\text{''}.$$

Recalling that $\lambda\, k\,.\, k_X$ is an enumeration of X, there is $a \in \mathbb{N}$ such $a_X = w_1$. On the other hand, since $i_{\mathcal{C}_1^1}$ is an enumeration of Ψ, there is b such that $i_{\mathcal{C}_1^1}(b) = w_2$. Let m be the value assigned to the pair (a, b) by the bijection from $\mathbb{N} \times \mathbb{N}$ to $\mathbb{N}$. Take $k = 2m$. Then, the execution of the algorithm on input k returns w, since k is an even number.

Finally, let $r_{\mathrm{Gen}} : \mathbb{N} \to \mathcal{P}$ be the map taking each i to p_{Gen}^i as a sequence of characters. Clearly r_{Gen} is computable. Then, the desired map s_{Gen} is the map $\mathrm{mind} \circ r_{\mathrm{Gen}}$.   QED

Proposition 14 There is a computable map $s_{\mathrm{MP}} : \mathbb{N} \to \mathbb{N}$ such that, if $i_{\mathcal{C}_1^1}$ is an enumeration of $\Psi \subseteq L_\Sigma$, then $s_{\mathrm{MP}}(i)_{\mathcal{C}_1^1}$ is an enumeration of $\mathrm{dmp}(\Psi)$.

Proof: If $i_{\mathcal{C}_1^1}$ enumerates Ψ, then the map computed by the following algorithm p_{MP}^i enumerates $\mathrm{dmp}_\Sigma(\Psi)$:

```
Function[k,
   If[EvenQ[k],
       ea = ToExpression[i_{C_1^1}[(k/2)_{N×N}[[1]]]];
       ec = ToExpression[i_{C_1^1}[(k/2)_{N×N}[[2]]]];
       If[ec[[0]] === ⇒ && ec[[1]] === ea,
          ToString[ec[[2]]]

          ,
          ToString[ea]
       ]

       ,
       i_{C_1^1}[(k-1)/2]
   ]
]
```

Let f be the map computed by this algorithm. We have to check:

(a) For every $w \in \mathrm{dmp}_\Sigma(\Psi)$, there is $k \in \mathbb{N}$ such that $f(k) = w$.

Assume that $w \in \mathrm{dmp}_\Sigma(\Psi)$. There are two cases to consider:

(i) $w \in \Psi$. Since $i_{\mathcal{C}_1^1}$ is an enumeration of Ψ, there is $m \in \mathbb{N}$ such that $i_{\mathcal{C}_1^1}(m) = w$. Take $k = 2m + 1$. Then, the execution of the algorithm that computes f on input k terminates with value $i_{C_1^1}((k-1)/2) = i_{C_1^1}(m) = w$, since k is a odd number.

(ii) $w \in \mathrm{dmp}_\Sigma(\Psi) \setminus \Psi$. Then, there are $w_1, w_2 \in \Psi$ such that

$$w_2 = \text{``}{\Rightarrow}\text{''} <> \text{``[''} <> w_1 <> \text{``,''} <> w <> \text{``]''}.$$

Since $i_{C_1^1}$ is an enumeration of Ψ, there are m_1 and m_2 such that $i_{C_1^1}(m_1) = w_1$ and $i_{C_1^1}(m_2) = w_2$. Let m be the value assigned to the pair (m_1, m_2) by the bijection from $\mathbb{N} \times \mathbb{N}$ to $\mathbb{N}$. Take $k = 2m$. Then, the execution of the algorithm on input k terminates with output $\mathsf{ToString}[ec[[2]]] = w$, since k is even, $ec[[0]] = {\Rightarrow}$ and $ec[[1]] = ea$ (the formula corresponding to the string w_1).

(b) For every $k \in \mathbb{N}$, $f(k) \in \mathrm{dmp}_\Sigma(\Psi)$.
   If k is odd then the algorithm returns $i_{C_1^1}[(k-1)/2]$ in Ψ. If k is even and the condition

$$ec[[0]] === {\Rightarrow}\ \&\&\ ec[[1]] === ea$$

is false then the algorithm returns $\mathsf{ToString}[ea] = i_{C_1^1}[(k/2)_{\mathbb{N} \times \mathbb{N}}[[1]]]$ also in Ψ. If k is even and the condition

$$ec[[0]] === {\Rightarrow}\ \&\&\ ec[[1]] === ea$$

is true then the algorithm returns the consequent of an implication (ec) in Ψ with its antecedent (ea) also in Ψ.

Finally, let $r_{\mathrm{MP}} : \mathbb{N} \to \mathcal{P}$ be the map taking each i to program p_{MP}^i as a sequence of characters. The map r_{MP} is clearly computable. Then, the desired map $s_{\mathrm{MP}} : \mathbb{N} \to \mathbb{N}$ is $\mathrm{mind} \circ r_{\mathrm{MP}}$. QED

Exercise 15 Check that the algorithm p_{MP}^i given in the last proof can be simplified by removing the first If.

The following result is a simple corollary of the last two propositions. Note that it is a much weaker result.

Proposition 16 If $\Psi \subseteq L_\Sigma$ is computably enumerable, then $\mathrm{dmp}_\Sigma(\Psi)$ and $\mathrm{dgen}_\Sigma(\Psi)$ are also computably enumerable.

Proof: Recall that Ψ is computable enumerable if it is empty or if it allows a computable enumeration. In the latter case, one can immediately use Propositions 13 and 14 for concluding that $\mathrm{dgen}_\Sigma(\Psi)$ and $\mathrm{dmp}_\Sigma(\Psi)$ are computably enumerable. With respect to the former case, just observe that $\mathrm{dgen}_\Sigma(\emptyset) = \mathrm{dmp}_\Sigma(\emptyset) = \emptyset$. QED

From the properties of dgen_Σ and dmp_Σ stated in Propositions 13, 14 and 16, it is easy to establish similar results about D_Σ.

Proposition 17 There is a computable map $s_{D_\Sigma} : \mathbb{N} \to \mathbb{N}$ such that, if $i_{\mathcal{C}_1^1}$ is an enumeration of $\Psi \subseteq L_\Sigma$, then $s_{D_\Sigma}(i)_{\mathcal{C}_1^1}$ is an enumeration of $D_\Sigma(\Psi)$.

Proof: If $i_{\mathcal{C}_1^1}$ enumerates Ψ, then the map computed by the following algorithm $p_{D_\Sigma}^i$ enumerates $D_\Sigma(\Psi)$:

$$
\begin{aligned}
&\mathsf{Function}[k, \\
&\quad \mathsf{If}[\mathsf{EvenQ}[k], \\
&\qquad s_{\mathrm{MP}}[i]_{\mathcal{C}_1^1}[k/2] \\
&\quad , \\
&\qquad s_{\mathrm{Gen}}[i]_{\mathcal{C}_1^1}[(k-1)/2] \\
&\quad] \\
&]
\end{aligned}
$$

Let $r_{D_\Sigma} : \mathbb{N} \to \mathcal{P}$ be the map taking each i into $p_{D_\Sigma}^i$ as a sequence of characters. The map r_{D_Σ} is clearly computable. Then, the desired map $s_{D_\Sigma} : \mathbb{N} \to \mathbb{N}$ can be obtained as follows: $s_{D_\Sigma} = \text{mind} \circ r_{D_\Sigma}$.   QED

Proposition 18 If $\Psi \subseteq L_\Sigma$ is computably enumerable, then $D_\Sigma(\Psi)$ is also computably enumerable.

Proposition 17 is now extended to powers of D_Σ. In this case, the envisaged computable map has two arguments. The second argument is the chosen power of D_Σ.

Proposition 19 There is a computable map $\sigma : \mathbb{N} \times \mathbb{N} \to \mathbb{N}$ such that, if $i_{\mathcal{C}_1^1}$ is an enumeration of $\Psi \subseteq L_\Sigma$, then $\sigma(i, n)_{\mathcal{C}_1^1}$ is an enumeration of $D_\Sigma^n(\Psi)$.

Proof: It is enough to take $\sigma = \lambda i, n \,.\, s_{D_\Sigma}^n(i)$, computed, for example, as follows:[10]

$$
\begin{aligned}
&\mathsf{Function}[\{i, n\}, \\
&\quad \mathsf{Nest}[s_{D_\Sigma}, i, n] \\
&]
\end{aligned}
$$

QED

It is now possible to construct an enumeration of $\bigcup_{n \in \mathbb{N}} D_\Sigma^n(\Psi)$ from any given enumeration of Ψ.

---

[10]The *Mathematica* ternary primitive function Nest computes the power of the first argument to the third argument applied to the second argument. But the reader will have no difficulty in writing a program for computing σ without using Nest.

Proposition 20 There is a computable map $s : \mathbb{N} \to \mathbb{N}$ such that, if $i_{\mathcal{C}_1^1}$ is an enumeration of $\Psi \subseteq L_\Sigma$, then $s(i)_{\mathcal{C}_1^1}$ is an enumeration of $\bigcup_{n \in \mathbb{N}} D_\Sigma^n(\Psi)$.

Proof: If $i_{\mathcal{C}_1^1}$ enumerates Ψ, then the map computed by the following algorithm p^i enumerates $\bigcup_{n \in \mathbb{N}} D_\Sigma^n(\Psi)$:

$$
\begin{aligned}
&\text{Function}[k, \\
&\quad k_1 = k_{\mathbb{N} \times \mathbb{N}}[[1]]; \\
&\quad k_2 = k_{\mathbb{N} \times \mathbb{N}}[[2]]; \\
&\quad \sigma[i, k_1]_{\mathcal{C}_1^1}[k_2] \\
&]
\end{aligned}
$$

Let f be the map computed by this algorithm. Clearly, for every $k \in \mathbb{N}$, $f(k)$ is in $\bigcup_{n \in \mathbb{N}} D_\Sigma^n(\Psi)$. It remains to check that, for every $w \in \bigcup_{n \in \mathbb{N}} D_\Sigma^n(\Psi)$, there is $k \in \mathbb{N}$ such that $f(k) = w$. Assume that $w \in \bigcup_{n \in \mathbb{N}} D_\Sigma^n(\Psi)$. Then, there is n such that $w \in D_\Sigma^n(\Psi)$. Moreover, by Proposition 19, there is $m \in \mathbb{N}$ such that $\sigma(i, n)_{\mathcal{C}_1^1}(m) = w$. Take $k \in \mathbb{N}$ as the number assigned to the pair (n, m) by the bijection from $\mathbb{N} \times \mathbb{N}$ to $\mathbb{N}$. Then, the execution of p^i on input k terminates with the value w.

Finally, let $r : \mathbb{N} \to \mathcal{P}$ be the map taking each i to p^i as a sequence of characters. The map r is clearly computable. Then, the desired map $s : \mathbb{N} \to \mathbb{N}$ can be obtained as follows: $s = \text{mind} \circ r$. QED

Hence, if Ψ is computably enumerable, then $\bigcup_{n \in \mathbb{N}} D_\Sigma^n(\Psi)$ is also computably enumerable. Therefore, in particular:

Proposition 21 (Semi-decidability of Hilbert calculus)
If $\Gamma \subseteq L_\Sigma$ is computably enumerable, then its derivation closure $\Gamma^{\vdash_\Sigma}$ is computably enumerable.

Proof: Immediate consequence of Propositions 9 and 14 in Chapter 3 and Propositions 7, 12 and 20 in the present chapter. QED

Hence, *a fortiori*, $\Gamma^{\vdash_\Sigma}$ is computably enumerable whenever Γ is computable. The reader may wonder if this result can be strengthened towards establishing the decidability of $\Gamma^{\vdash_\Sigma}$. In fact, the decidability of Γ is not a sufficient condition for the decidability of $\Gamma^{\vdash_\Sigma}$. For instance, it is shown much later, in Section 13.3, that it is possible to find a signature Σ such that $\emptyset^{\vdash_\Sigma}$ is not a computable set. On the other hand, there are known conditions on the signature that ensure the decidability of the set of theorems. For example, if Σ only contains unary predicates and no function symbols.[11] then $\emptyset^{\vdash_\Sigma}$ is decidable

---

[11]Such a signature is said to be monadic.

5.4   Metatheorems and admissible rules

Several metatheorems and admissible rules are collected below that can help
when using the Hilbert calculus in practice. A metatheorem is a (meta)assertion
that guarantees the existence of a derivation when some other derivation or
derivations are known. An admissible rule is an additional inference rule that
can be added to the calculus without changing its power. In particular, the
rule inferring δ from Θ is admissible whenever $\delta \in \Theta^{\vdash_\Sigma}$ and Θ is finite. Such a
rule is said to be a *derivable admissible rule*.

Two important concepts related to dependence on hypotheses are first presented, since they are needed for several metatheorems and are frequently useful. Let

$$w = (\psi_1, J_1) \ldots (\psi_n, J_n)$$

be a derivation sequence for $\Gamma \vdash_\Sigma \varphi$. Then:

- Formula ψ_i *depends* on $\gamma \in \Gamma$ in w if:

  - either J_i is Hyp and ψ_i is γ;
  - or J_i is Gen k and ψ_k depends on γ;
  - or J_i is MP j, k and ψ_j or ψ_k depends on γ.

  Furthermore, given $\Delta \subseteq L_\Sigma$, let

  $$d_\Delta^w = \{\delta \in \Delta : \psi_n \text{ depends on } \delta \text{ in } w\}.$$

  This set is composed of the elements of Δ on which the conclusion of the
  derivation w depends. Clearly, d_Δ^w is finite for every choice of w and Δ.
  Moreover, $d_\Delta^w \subseteq \Gamma$ and, obviously, $d_\Delta^w \subseteq \Delta$.

- Formula ψ_i is an *essential generalization over a dependent* of $\gamma \in \Gamma$ in w
  if:

  - J_i is Gen k;
  - ψ_i is $(\forall x\, \psi_k)$;
  - ψ_k depends on γ;
  - and $x \in \mathrm{fv}_\Sigma(\gamma)$.

The following metatheorem rigorously states the intuition that one can
safely remove an hypothesis that is not used in a derivation to reach the same
conclusion.

Proposition 22 (Metatheorem of useless hypothesis)
If there is a derivation sequence w for

$$\Gamma \cup \{\eta\} \vdash_\Sigma \varphi$$

where φ does not depend on η, then there is a derivation sequence w' for

$$\Gamma \vdash_\Sigma \varphi$$

such that, for any $\gamma \in \Gamma$, if w' contains an essential generalization over a dependent of γ, then w already contains an essential generalization over a dependent of γ. Furthermore, $d_\Gamma^{w'} = d_\Gamma^w$.

Proof: Let $w = (\psi_1, J_1) \ldots (\psi_n, J_n)$ be a derivation sequence for $\Gamma \cup \{\eta\} \vdash_\Sigma \varphi$ where φ does not depend on η. First, it is shown by induction on n that it is possible to extract a subsequence[12] w' of w constituting a derivation sequence for $\Gamma \vdash_\Sigma \varphi$ and such that $d_\Gamma^{w'} = d_\Gamma^w$.

(Basis) $n = 1$: $\varphi \in \Gamma \cup \text{Ax}_\Sigma$ since φ does not depend on η. Therefore, $w' = w$ is a derivation sequence for $\Gamma \vdash_\Sigma \varphi$. Moreover, $d_\Gamma^{w'} = d_\Gamma^w = \emptyset$ if J_1 is Ax, and $d_\Gamma^{w'} = d_\Gamma^w = \{\varphi\}$ otherwise.

(Step) Let $n = m + 1$. There are three cases to consider.
(i) J_n is Hyp or Ax:
$\varphi \in \Gamma \cup \text{Ax}_\Sigma$ since φ does not depend on η. It suffices to take the one-step sequence (ψ_n, J_n) for w' in order to establish $\Gamma \vdash_\Sigma \varphi$. Clearly, $d_\Gamma^{w'} = d_\Gamma^w = \emptyset$ if J_n is Ax, and $d_\Gamma^{w'} = d_\Gamma^w = \{\varphi\}$ otherwise.
(ii) J_n is MP k, j:
$w_{1...k} = (\psi_1, J_1) \ldots (\psi_k, J_k)$ and $w_{1...j} = (\psi_1, J_1) \ldots (\psi_j, J_j)$ are derivation sequences for $\Gamma \cup \{\eta\} \vdash_\Sigma \psi_k$ and $\Gamma \cup \{\eta\} \vdash_\Sigma \psi_j$, respectively, where both ψ_k and ψ_j do not depend on η (if one of them did, then so would φ). Hence, by the induction hypothesis, there are subsequences w'' and w''' of $w_{1...k}$ and $w_{1...j}$ for $\Gamma \vdash_\Sigma \psi_k$ and $\Gamma \vdash_\Sigma \psi_j$, respectively, such that $d_\Gamma^{w''} = d_\Gamma^{w_{1..k}}$ and $d_\Gamma^{w'''} = d_\Gamma^{w_{1..j}}$. Let r and s be the lengths of w'' and w''', respectively. Then, appending the following step to the concatenation[13] of w'' and w'''

$$r + s + 1 \qquad \varphi \qquad \text{MP } r, r + s$$

yields a derivation sequence w' for $\Gamma \vdash_\Sigma \varphi$ that is still a subsequence of the original sequence w (up to the renumbering the steps). Furthermore,

$$d_\Gamma^{w'} = d_\Gamma^{w''} \cup d_\Gamma^{w'''} = d_\Gamma^{w_{1..k}} \cup d_\Gamma^{w_{1..j}} = d_\Gamma^w.$$

---

[12]Disregarding the changes in the justifications imposed by the renumbering of the steps.
[13]After the obvious changes in the justifications imposed by the renumbering of the steps.

(iii) J_n is Gen k:

$w_{1...k} = (\psi_1, J_1) \ldots (\psi_k, J_k)$ is a derivation sequence for $\Gamma \cup \{\eta\} \vdash_\Sigma \psi_k$ where ψ_k does not depend on η (if it did, then so would φ). Hence, by the induction hypothesis, there is a subsequence w'' of $w_{1...k}$ for $\Gamma \vdash_\Sigma \psi_k$ such that $d_\Gamma^{w''} = d_\Gamma^{w_{1..k}}$. Let r be the length of w''. Then, appending the following step to the sequence w''

$$r + 1 \qquad \varphi \qquad \text{Gen } r$$

yields a derivation sequence w' for $\Gamma \vdash_\Sigma \varphi$. Clearly, w' is still a a subsequence of the original sequence w (up to the renumbering the steps). Furthermore,

$$d_\Gamma^{w'} = d_\Gamma^{w''} = d_\Gamma^{w_{1..k}} = d_\Gamma^{w}.$$

Finally, observe that, since the derivation sequence w' thus constructed for $\Gamma \vdash_\Sigma \varphi$ is a subsequence of the original sequence w, no additional hypotheses were used and no generalizations were added. Therefore, *a fortiori*, no essential generalizations over dependents of elements of Γ were added.   QED

Note that the derivation sequence built in the proof of the metatheorem of useless hypothesis corresponds to removing from the original sequence all steps depending on useless hypotheses, including those depending on η.

The aim now is to state and prove the important metatheorem of deduction that states to what extent derivation can be internalized (expressed in the language L_Σ) by implication.

Proposition 23 (Metatheorem of deduction – MTD)
Given a derivation sequence w for

$$\Gamma \cup \{\eta\} \vdash_\Sigma \varphi$$

without essential generalizations over dependents of η, it is possible to build a derivation sequence w' for

$$\Gamma \vdash_\Sigma (\eta \Rightarrow \varphi)$$

such that, for every hypothesis $\gamma \in \Gamma$, if w' contains an essential generalization over a dependent of γ, then w already contains an essential generalization over a dependent of γ.

Proof: Let $w = (\psi_1, J_1) \ldots (\psi_n, J_n)$ be a derivation sequence for $\Gamma \cup \{\eta\} \vdash_\Sigma \varphi$ without essential generalizations over dependents of η. It is shown, by induction on n, that there is a derivation sequence w' for $\Gamma \vdash_\Sigma (\eta \Rightarrow \varphi)$ without additional essential generalizations over dependents of elements of Γ and such that $d_\Gamma^{w} = d_\Gamma^{w'}$.

(Basis) $n = 1$. There are three cases to consider:
(i) J_n is Hyp and $\varphi \in \Gamma$. Consider the following derivation sequence w':

$$
\begin{array}{lll}
1 & \varphi & \text{Hyp} \\
2 & (\varphi \Rightarrow (\eta \Rightarrow \varphi)) & \text{Ax1} \\
3 & (\eta \Rightarrow \varphi) & \text{MP } 1,2
\end{array}
$$

There are no additional essential generalizations over dependents of elements of Γ since there are no generalizations in w'. Furthermore, $d_\Gamma^w = d_\Gamma^{w'} = \{\varphi\}$.
(ii) J_n is Hyp and φ is η. Recall the derivation sequence for $\vdash_\Sigma (\eta \Rightarrow \eta)$ provided at the beginning of this chapter. Take this sequence as w'. There are no additional essential generalizations over dependents of elements of Γ since generalization is not used in w'. Furthermore, $d_\Gamma^w = d_\Gamma^{w'} = \emptyset$.
(iii) J_n is Ax. Consider the following derivation sequence w':

$$
\begin{array}{lll}
1 & \varphi & \text{Ax} \\
2 & (\varphi \Rightarrow (\eta \Rightarrow \varphi)) & \text{Ax1} \\
3 & (\eta \Rightarrow \varphi) & \text{MP } 1,2
\end{array}
$$

Again, generalization is not used and, so, there are no additional essential generalizations over dependents of elements of Γ in w'. Furthermore, $d_\Gamma^w = d_\Gamma^{w'} = \emptyset$.

(Step) $n = m + 1$. There are three cases to consider:
(i) J_n is Hyp or J_n is Ax. Then, the proof is similar to the corresponding case in the basis, and, so, it is left as an exercise.
(ii) J_n is MP i, k with $i, k \leq m$. Without loss of generality let ψ_k be the formula $(\psi_i \Rightarrow \varphi)$. Observe that $w_{1\ldots i} = (\psi_1, J_1) \ldots (\psi_i, J_i)$ and $w_{1\ldots k} = (\psi_1, J_1) \ldots (\psi_k, J_k)$ are derivation sequences for $\Gamma \cup \{\eta\} \vdash_\Sigma \psi_i$ and $\Gamma \cup \{\eta\} \vdash_\Sigma (\psi_i \Rightarrow \varphi)$, respectively, both without essential generalizations over dependents of η. Hence, by the induction hypothesis, there are derivation sequences w'' and w''' for $\Gamma \vdash_\Sigma (\eta \Rightarrow \psi_i)$ and $\Gamma \vdash_\Sigma (\eta \Rightarrow (\psi_i \Rightarrow \varphi))$, respectively, without additional essential generalizations over dependents of elements of Γ and such that $d_\Gamma^{w_{1\ldots i}} = d_\Gamma^{w''}$ and $d_\Gamma^{w_{1\ldots k}} = d_\Gamma^{w'''}$. Let r and s be their respective lengths. Consider the derivation sequence w' for $\Gamma \vdash_\Sigma (\eta \Rightarrow \varphi)$ obtained by appending the following steps to the concatenation[14] of w'' and w''':

$$
\begin{array}{lll}
r+s+1 & ((\eta \Rightarrow (\psi_i \Rightarrow \varphi)) \Rightarrow & \\
& \quad ((\eta \Rightarrow \psi_i) \Rightarrow (\eta \Rightarrow \varphi))) & \text{Ax2} \\
r+s+2 & ((\eta \Rightarrow \psi_i) \Rightarrow (\eta \Rightarrow \varphi)) & \text{MP } r+s, r+s+1 \\
r+s+3 & (\eta \Rightarrow \varphi) & \text{MP } r, r+s+2
\end{array}
$$

Clearly,

$$
d_\Gamma^w = d_\Gamma^{w_{1\ldots i}} \cup d_\Gamma^{w_{1\ldots k}} = d_\Gamma^{w''} \cup d_\Gamma^{w'''} = d_\Gamma^{w'}.
$$

Furthermore, each generalization in the steps 1 to r of w' is essential over a dependent of an element of Γ if and only if it is so in w'', and each generalization

---

[14]After the obvious changes in the justifications imposed by the renumbering of the steps.

in the steps $r + 1$ to $r + s$ of w' is essential over a dependent of an element of Γ if and only if it is so in w'''. Moreover, generalization is not used in the last three steps of w'. Hence, there are no additional essential generalizations over dependents of elements of Γ in the derivation sequence w'.

(iii) J_n is Gen k with $k \leq m$. Let φ be the formula $(\forall x\, \psi_k)$. Note that, by hypothesis, this generalization is not an essential one over a dependent of η. Hence, there are two cases to consider:

(a) ψ_k does not depend on η. Then, by the metatheorem of useless hypothesis, there is a subsequence w'' of $w_{1...k}$ constituting a derivation sequence for $\Gamma \vdash_\Sigma \psi_k$, without additional essential generalizations over dependents of elements of Γ and such that $d_\Gamma^{w''} = d_\Gamma^{w_{1...k}}$. Let r be the length of w''. Consider the following derivation sequence w' for $\Gamma \vdash_\Sigma (\eta \Rightarrow \varphi)$ obtained by appending the following steps to w'':

$$
\begin{array}{llr}
r + 1 & (\forall x\, \psi_k) & \text{Gen } r \\
r + 2 & ((\forall x\, \psi_k) \Rightarrow (\eta \Rightarrow (\forall x\, \psi_k))) & \text{Ax1} \\
r + 3 & (\eta \Rightarrow (\forall x\, \psi_k)) & \text{MP } r+1, r+2
\end{array}
$$

Clearly,

$$
d_\Gamma^w = d_\Gamma^{w_{1...k}} = d_\Gamma^{w''} = d_\Gamma^{w'}.
$$

Thus, the generalization at step $r + 1$ of w' is an essential one over a dependent of Γ if and only if so is the generalization at step n of w. Moreover, each generalization in the steps 1 to r of w' is essential over a dependent of an element of Γ if and only if it is so in w''. Hence, there are no additional essential generalizations over dependents of elements of Γ in the derivation sequence w'.

(b) $x \notin \mathrm{fv}_\Sigma(\eta)$. Observe that $(\psi_1, J_1) \ldots (\psi_k, J_k)$ is a derivation sequence for $\Gamma \cup \{\eta\} \vdash_\Sigma \psi_k$ without essential generalizations over dependents of η. Therefore, by the induction hypothesis, there is a derivation sequence w'' for $\Gamma \vdash_\Sigma (\eta \Rightarrow \psi_k)$ without additional essential generalizations over dependents of elements of Γ and such that $d_\Gamma^{w''} = d_\Gamma^{w_{1...k}}$. Let r be its length. Consider the derivation sequence w' for $\Gamma \vdash_\Sigma (\eta \Rightarrow \varphi)$ obtained by appending the following steps to w'':

$$
\begin{array}{llr}
r + 1 & (\forall x(\eta \Rightarrow \psi_k)) & \text{Gen } r \\
r + 2 & ((\forall x(\eta \Rightarrow \psi_k)) \Rightarrow (\eta \Rightarrow (\forall x\, \psi_k))) & \text{Ax5} \\
r + 3 & (\eta \Rightarrow (\forall x\, \psi_k)) & \text{MP } r+1, r+2
\end{array}
$$

Clearly,

$$
d_\Gamma^w = d_\Gamma^{w_{1...k}} = d_\Gamma^{w''} = d_\Gamma^{w'}
$$

and, therefore, the generalization at step $r + 1$ of w' is an essential one over a dependent of an element of Γ if and only if so is the generalization at step n of sequence w. Moreover, each generalization in the steps 1 to r of w' is essential over a dependent of an element of Γ if and only if it is so in w''. Thus, no additional essential generalizations over dependents of elements of Γ appear in the derivation sequence w'.   QED

Exercise 24 By resorting to the MTD establish:

1. $\vdash_\Sigma ((\forall x(\forall y\, \varphi)) \Rightarrow (\forall y(\forall x\, \varphi)));$

2. $\vdash_\Sigma ((\forall x(\varphi_1 \Rightarrow \varphi_2)) \Rightarrow ((\forall x\, \varphi_1) \Rightarrow (\forall x\, \varphi_2))).$

Exercise 25 Adapting the proof of the MTD, expand the derivation sequence given in Exercise 1 for $(\forall x(\forall y\, \varphi)) \vdash_\Sigma (\forall y(\forall x\, \varphi))$, in order to build a derivation sequence for $\vdash_\Sigma ((\forall x(\forall y\, \varphi)) \Rightarrow (\forall y(\forall x\, \varphi))).$

Results (a) to (j) below will be needed further on. The reader is invited to show the existence of derivations for them, using the MTD whenever convenient, before reading the proofs provided herein.

(a) $\vdash_\Sigma ((\neg(\neg\varphi)) \Rightarrow \varphi).$

Consider the following derivation sequence for $(\neg(\neg\varphi)) \vdash_\Sigma \varphi$:

1	$(\neg(\neg\varphi))$	Hyp
2	$((\neg(\neg\varphi)) \Rightarrow ((\neg(\neg(\neg(\neg\varphi)))) \Rightarrow (\neg(\neg\varphi))))$	Ax1
3	$((\neg(\neg(\neg(\neg\varphi)))) \Rightarrow (\neg(\neg\varphi)))$	MP 1,2
4	$(((\neg(\neg(\neg(\neg\varphi)))) \Rightarrow (\neg(\neg\varphi))) \Rightarrow ((\neg\varphi) \Rightarrow (\neg(\neg(\neg\varphi)))))$	Ax3
5	$((\neg\varphi) \Rightarrow (\neg(\neg(\neg\varphi))))$	MP 3,4
6	$(((\neg\varphi) \Rightarrow (\neg(\neg(\neg\varphi)))) \Rightarrow ((\neg(\neg\varphi)) \Rightarrow \varphi))$	Ax3
7	$((\neg(\neg\varphi)) \Rightarrow \varphi)$	MP 5,6
8	φ	MP 1,7

Then, applying the MTD, the desired result follows.

(b) $\vdash_\Sigma (((\neg\varphi) \Rightarrow \varphi) \Rightarrow \varphi).$

Consider the following derivation sequence for $((\neg\varphi) \Rightarrow \varphi) \vdash_\Sigma \varphi$:

1	$((\neg\varphi) \Rightarrow \varphi)$	Hyp
2	$((\neg\varphi) \Rightarrow ((\neg(\neg((\neg\varphi) \Rightarrow \varphi))) \Rightarrow (\neg\varphi)))$	Ax1
3	$(((\neg(\neg((\neg\varphi) \Rightarrow \varphi))) \Rightarrow (\neg\varphi)) \Rightarrow$ $(\varphi \Rightarrow (\neg((\neg\varphi) \Rightarrow \varphi))))$	Ax3
4	$((\neg\varphi) \Rightarrow (\varphi \Rightarrow (\neg((\neg\varphi) \Rightarrow \varphi))))$	HS 2,3
5	$(((\neg\varphi) \Rightarrow (\varphi \Rightarrow (\neg((\neg\varphi) \Rightarrow \varphi)))) \Rightarrow$ $(((\neg\varphi) \Rightarrow \varphi) \Rightarrow ((\neg\varphi) \Rightarrow (\neg((\neg\varphi) \Rightarrow \varphi)))))$	Ax2
6	$(((\neg\varphi) \Rightarrow \varphi) \Rightarrow ((\neg\varphi) \Rightarrow (\neg((\neg\varphi) \Rightarrow \varphi))))$	MP 4,5
7	$((\neg\varphi) \Rightarrow (\neg((\neg\varphi) \Rightarrow \varphi)))$	MP 1,6
8	$(((\neg\varphi) \Rightarrow (\neg((\neg\varphi) \Rightarrow \varphi))) \Rightarrow (((\neg\varphi) \Rightarrow \varphi) \Rightarrow \varphi))$	Ax3
9	$(((\neg\varphi) \Rightarrow \varphi) \Rightarrow \varphi)$	MP 7,8
10	φ	MP 9,1

Then, applying the MTD, the desired result follows.

(c) $\vdash_\Sigma (((\neg\varphi) \Rightarrow (\neg\psi)) \Rightarrow (((\neg\varphi) \Rightarrow \psi) \Rightarrow \varphi))$.

Consider the derivation sequence for $((\neg\varphi) \Rightarrow (\neg\psi)), ((\neg\varphi) \Rightarrow \psi) \vdash_\Sigma \varphi$ where justification Thm stands for previously derived theorem:

1	$((\neg\varphi) \Rightarrow (\neg\psi))$	Hyp
2	$((\neg\varphi) \Rightarrow \psi)$	Hyp
3	$(((\neg\varphi) \Rightarrow (\neg\psi)) \Rightarrow (\psi \Rightarrow \varphi))$	Ax3
4	$(\psi \Rightarrow \varphi)$	MP 1,3
5	$((\neg\varphi) \Rightarrow \varphi)$	HS 2,4
6	$(((\neg\varphi) \Rightarrow \varphi) \Rightarrow \varphi)$	Thm (b)
7	φ	MP 5,6

Then, applying the MTD twice, the desired result follows.

(d) $\vdash_\Sigma (\varphi \Rightarrow (\neg(\neg\varphi)))$.

Consider the following derivation sequence:

1	$(((\neg(\neg(\neg\varphi))) \Rightarrow (\neg\varphi)) \Rightarrow$ $(((\neg(\neg(\neg\varphi))) \Rightarrow \varphi) \Rightarrow (\neg(\neg\varphi))))$	Thm (c)
2	$((\neg(\neg(\neg\varphi))) \Rightarrow (\neg\varphi))$	Thm (a)
3	$(((\neg(\neg(\neg\varphi))) \Rightarrow \varphi) \Rightarrow (\neg(\neg\varphi)))$	MP 2,1
4	$(\varphi \Rightarrow ((\neg(\neg(\neg\varphi))) \Rightarrow \varphi))$	Ax1
5	$(\varphi \Rightarrow (\neg(\neg\varphi)))$	HS 4,3

(e) $\vdash_\Sigma ((\varphi \Rightarrow (\neg\psi)) \Rightarrow (\psi \Rightarrow (\neg\varphi)))$.

Consider the derivation sequence for $(\varphi \Rightarrow (\neg\psi)) \vdash_\Sigma (\psi \Rightarrow (\neg\varphi))$:

1	$(\varphi \Rightarrow (\neg\psi))$	Hyp
2	$((\neg(\neg\varphi)) \Rightarrow \varphi)$	Thm (a)
3	$((\neg(\neg\varphi)) \Rightarrow (\neg\psi))$	HS 2,1
4	$(((\neg(\neg\varphi)) \Rightarrow (\neg\psi)) \Rightarrow (\psi \Rightarrow (\neg\varphi)))$	Ax3
5	$(\psi \Rightarrow (\neg\varphi))$	MP 3,4

Then, applying the MTD, the desired result follows.
(f) $\vdash_\Sigma ((\varphi \Rightarrow \psi) \Rightarrow ((\neg\psi) \Rightarrow (\neg\varphi)))$.

Consider the following derivation sequence for $(\varphi \Rightarrow \psi) \vdash_\Sigma ((\neg\psi) \Rightarrow (\neg\varphi))$:

1	$(\varphi \Rightarrow \psi)$	Hyp
2	$((\neg(\neg\varphi)) \Rightarrow \varphi)$	Thm (a)

$$
\begin{array}{lll}
3 & ((\neg(\neg\varphi)) \Rightarrow \psi) & \text{HS } 2,1 \\
4 & (\psi \Rightarrow (\neg(\neg\psi))) & \text{Thm (d)} \\
5 & ((\neg(\neg\varphi)) \Rightarrow (\neg(\neg\psi))) & \text{HS } 3,4 \\
6 & (((\neg(\neg\varphi)) \Rightarrow (\neg(\neg\psi))) \Rightarrow ((\neg\psi) \Rightarrow (\neg\varphi))) & \text{Ax3} \\
7 & ((\neg\psi) \Rightarrow (\neg\varphi)) & \text{MP } 5,6
\end{array}
$$

Then, applying the MTD, the desired result follows.
(g) $\vdash_\Sigma ((\neg\varphi) \Rightarrow (\varphi \Rightarrow \psi))$.

Consider the following derivation sequence for $(\neg\varphi) \vdash_\Sigma (\varphi \Rightarrow \psi)$:

$$
\begin{array}{lll}
1 & (\neg\varphi) & \text{Hyp} \\
2 & ((\neg\varphi) \Rightarrow ((\neg\psi) \Rightarrow (\neg\varphi))) & \text{Ax1} \\
3 & ((\neg\psi) \Rightarrow (\neg\varphi)) & \text{MP } 1,2 \\
4 & (((\neg\psi) \Rightarrow (\neg\varphi)) \Rightarrow (\varphi \Rightarrow \psi)) & \text{Ax3} \\
5 & (\varphi \Rightarrow \psi) & \text{MP } 3,4
\end{array}
$$

Then, applying the MTD, the desired result follows.

(h1) $\vdash_\Sigma ((\varphi \wedge \psi) \Rightarrow \varphi)$.

Consider the following derivation sequence for $(\neg(\varphi \Rightarrow (\neg\psi))) \vdash_\Sigma \varphi$:

$$
\begin{array}{lll}
1 & (\neg(\varphi \Rightarrow (\neg\psi))) & \text{Hyp} \\
2 & ((\neg\varphi) \Rightarrow (\varphi \Rightarrow (\neg\psi))) & \text{Thm (g)} \\
3 & (((\neg\varphi) \Rightarrow (\varphi \Rightarrow (\neg\psi))) \Rightarrow ((\neg(\varphi \Rightarrow (\neg\psi))) \Rightarrow (\neg(\neg\varphi)))) & \text{Thm (f)} \\
4 & ((\neg(\varphi \Rightarrow (\neg\psi))) \Rightarrow (\neg(\neg\varphi))) & \text{MP } 2,3 \\
5 & (\neg(\neg\varphi)) & \text{MP } 1,4 \\
6 & ((\neg(\neg\varphi)) \Rightarrow \varphi) & \text{Thm (a)} \\
7 & \varphi & \text{MP } 5,6
\end{array}
$$

Then, applying the MTD, the desired result follows.

(h2) $\vdash_\Sigma ((\varphi \wedge \psi) \Rightarrow \psi)$.

Consider the following derivation sequence for $(\neg(\varphi \Rightarrow (\neg\psi))) \vdash_\Sigma \psi$:

$$
\begin{array}{lll}
1 & (\neg(\varphi \Rightarrow (\neg\psi))) & \text{Hyp} \\
2 & ((\neg\psi) \Rightarrow (\varphi \Rightarrow (\neg\psi))) & \text{Ax1} \\
3 & ((\varphi \Rightarrow (\neg\psi)) \Rightarrow (\neg(\neg(\varphi \Rightarrow (\neg\psi))))) & \text{Thm (d)} \\
4 & ((\neg\psi) \Rightarrow (\neg(\neg(\varphi \Rightarrow (\neg\psi))))) & \text{HS } 2,3 \\
5 & (((\neg\psi) \Rightarrow (\neg(\neg(\varphi \Rightarrow (\neg\psi))))) \Rightarrow \\
  & \qquad ((\neg(\varphi \Rightarrow (\neg\psi))) \Rightarrow (\neg(\neg\psi)))) & \text{Thm (e)} \\
6 & ((\neg(\varphi \Rightarrow (\neg\psi))) \Rightarrow (\neg(\neg\psi))) & \text{MP } 4,5 \\
7 & (\neg(\neg\psi)) & \text{MP } 1,6 \\
8 & ((\neg(\neg\psi)) \Rightarrow \psi) & \text{Thm (a)} \\
9 & \psi & \text{MP } 7,8
\end{array}
$$

Then, applying the MTD, the desired result follows.

(i) $\vdash_\Sigma (\varphi \Rightarrow (\psi \Rightarrow (\varphi \wedge \psi)))$.

Note that $\varphi, (\varphi \Rightarrow (\neg\psi)) \vdash_\Sigma (\neg\psi)$. So, by the MTD, $\varphi \vdash_\Sigma ((\varphi \Rightarrow (\neg\psi)) \Rightarrow (\neg\psi))$. Let w be a derivation sequence for this formula, and let r be its length. Consider the derivation sequence for $\varphi \vdash_\Sigma (\psi \Rightarrow (\varphi \wedge \psi))$ obtained from the concatenation of w with the following sequence:

$$
\begin{array}{lll}
r+1 & (((\varphi \Rightarrow (\neg\psi)) \Rightarrow (\neg\psi)) \Rightarrow (\psi \Rightarrow (\neg(\varphi \Rightarrow (\neg\psi))))) & \text{Thm(e)} \\
r+2 & (\psi \Rightarrow (\neg(\varphi \Rightarrow (\neg\psi)))) & \text{MP } r,r+1
\end{array}
$$

Then, applying the MTD, the desired result follows.

(j) $(\psi \Rightarrow \psi') \vdash_\Sigma ((\forall x\, \psi) \Rightarrow (\forall x\, \psi'))$.

Consider the following derivation sequence:

$$
\begin{array}{lll}
1 & (\psi \Rightarrow \psi') & \text{Hyp} \\
2 & (\forall x(\psi \Rightarrow \psi')) & \text{Gen 1} \\
3 & ((\forall x(\psi \Rightarrow \psi')) \Rightarrow ((\forall x\, \psi) \Rightarrow (\forall x\, \psi'))) & \text{Exercise 24(2)} \\
4 & ((\forall x\, \psi) \Rightarrow (\forall x\, \psi')) & \text{MP 2,3}
\end{array}
$$

The metatheorem of deduction has some interesting and useful corollaries, namely the following two metatheorems.

Proposition 26 (Metatheorem of contradiction – MTC)
If there exists a derivation sequence for

$$\Gamma \cup \{\varphi\} \vdash_\Sigma (\eta \wedge (\neg\eta))$$

without essential generalizations over dependents of φ, then

$$\Gamma \vdash_\Sigma (\neg\varphi).$$

Proof: Suppose that there is a derivation sequence for $\Gamma \cup \{\varphi\} \vdash_\Sigma (\eta \wedge (\neg\eta))$ without essential generalizations over dependents of φ. Then, applying the MTD,

$$\Gamma \vdash_\Sigma (\varphi \Rightarrow (\eta \wedge (\neg\eta)))$$

and, so, by expanding the conjunction abbreviation,

$$(\dagger) \quad \Gamma \vdash_\Sigma (\varphi \Rightarrow (\neg(\eta \Rightarrow (\neg(\neg\eta))))).$$

Consider now the derivation sequence

$$
\begin{array}{lll}
1 & (\varphi \Rightarrow (\neg(\eta \Rightarrow (\neg(\neg\eta))))) & \text{Hyp} \\
2 & ((\neg(\neg\varphi)) \Rightarrow \varphi) & \text{Thm (a)} \\
3 & ((\neg(\neg\varphi)) \Rightarrow (\neg(\eta \Rightarrow (\neg(\neg\eta))))) & \text{HS 2,1} \\
4 & (((\neg(\neg\varphi)) \Rightarrow (\neg(\eta \Rightarrow (\neg(\neg\eta))))) \Rightarrow \\
  & \qquad\qquad\qquad ((\eta \Rightarrow (\neg(\neg\eta))) \Rightarrow (\neg\varphi))) & \text{Ax3} \\
5 & ((\eta \Rightarrow (\neg(\neg\eta))) \Rightarrow (\neg\varphi)) & \text{MP 3,4} \\
6 & (\eta \Rightarrow (\neg(\neg\eta))) & \text{Thm (d)} \\
7 & (\neg\varphi) & \text{MP 5,6}
\end{array}
$$

for

$$(\dagger\dagger) \qquad (\varphi \Rightarrow (\neg(\eta \Rightarrow (\neg(\neg\eta))))) \vdash_{\Sigma} (\neg\varphi).$$

Finally, one concludes

$$\Gamma \vdash_{\Sigma} (\neg\varphi)$$

by applying the cut law to ($\dagger$) and ($\dagger\dagger$). QED

Observe that the derivation provided by the MTC does not introduce additional essential generalizations over dependents of elements of Γ. In other words, for every $\gamma \in \Gamma$, if the derivation sequence built for $\Gamma \vdash_{\Sigma} (\neg\varphi)$ contains an essential generalization over a dependent of γ, then the original sequence for $\Gamma \cup \{\varphi\} \vdash_{\Sigma} (\eta \wedge (\neg\eta))$ already contains an essential generalization over a dependent of that γ.

Proposition 27 (Metatheorem of contraposition – MTCP)
If there exists a derivation sequence for

$$\Gamma \cup \{\varphi\} \vdash_{\Sigma} (\neg\eta)$$

without essential generalizations over dependents of φ, then

$$\Gamma \cup \{\eta\} \vdash_{\Sigma} (\neg\varphi).$$

Proof: Suppose that there exists a derivation sequence for $\Gamma \cup \{\varphi\} \vdash_{\Sigma} (\neg\eta)$ without essential generalizations over dependents of φ. Then, applying the MTD, one obtains $\Gamma \vdash_{\Sigma} (\varphi \Rightarrow (\neg\eta))$. Therefore, using (e) and the cut law, one concludes $\Gamma \vdash_{\Sigma} (\eta \Rightarrow (\neg\varphi))$, and the thesis follows by MP. QED

Note that the derivation provided by the MTCP also does not introduce additional essential generalizations over dependents of elements of Γ.

The following proposition is a very simple first example of a derivable admissible rule. The proof is left as an exercise.

Proposition 28 (Instantiation rule – IR)
Let t be a term free for variable x in formula φ. Then,

$$(\forall x\, \varphi) \vdash_{\Sigma} [\varphi]^{x}_{t}.$$

The next two results present two very useful derivable admissible rules concerning existential quantifications.

Proposition 29 (Existence rule 1 – ExR1)
Let t be a term free for x in φ. Then,

$$[\varphi]^x_t \vdash_\Sigma (\exists x\,\varphi).$$

Proof: The reader will have no trouble in providing a proof directly from the axiomatization or using the metatheorem of contradiction.   QED

Proposition 30 (Existence rule 2 – ExR2)
Let x be a variable that is not free in ψ. Then,

$$(\forall x\,(\varphi \Rightarrow \psi)) \vdash_\Sigma ((\exists x\,\varphi) \Rightarrow \psi).$$

Proof: Consider the following derivation sequence:

1	$(\forall x(\varphi \Rightarrow \psi))$	Hyp
2	$((\forall x(\varphi \Rightarrow \psi)) \Rightarrow (\varphi \Rightarrow \psi))$	Ax4
3	$(\varphi \Rightarrow \psi)$	MP 1,2
4	$((\varphi \Rightarrow \psi) \Rightarrow ((\neg\,\psi) \Rightarrow (\neg\,\varphi)))$	Thm (f)
5	$((\neg\,\psi) \Rightarrow (\neg\,\varphi))$	MP 3,4
6	$(\forall x((\neg\,\psi) \Rightarrow (\neg\,\varphi)))$	Gen 5
7	$((\forall x((\neg\,\psi) \Rightarrow (\neg\,\varphi))) \Rightarrow ((\neg\,\psi) \Rightarrow (\forall x(\neg\,\varphi))))$	Ax 5
8	$((\neg\,\psi) \Rightarrow (\forall x(\neg\,\varphi)))$	MP 6,7
9	$(((\neg\,\psi) \Rightarrow (\forall x(\neg\,\varphi))) \Rightarrow ((\neg(\forall x(\neg\,\varphi))) \Rightarrow (\neg(\neg\,\psi))))$	Thm (f)
10	$((\neg(\forall x(\neg\,\varphi))) \Rightarrow (\neg(\neg\,\psi)))$	MP 8,9
11	$((\neg(\neg\,\psi)) \Rightarrow \psi)$	Thm (a)
12	$((\neg(\forall x(\neg\,\varphi))) \Rightarrow \psi)$	HS 10,11

Observe that at step 7 axiom 5 can be used since, by hypothesis, $x \notin \mathrm{fv}_\Sigma(\psi)$ and, so, $x \notin \mathrm{fv}_\Sigma((\neg\,\psi))$.   QED

The next metatheorem states a very common form of reasoning: if in a formula φ a subformula is replaced by an equivalent one, then the resulting formula is equivalent to φ. But, first, the reader is invited to define rigorously the notion of subformula.

Exercise 31 Define inductively the map $\mathrm{sbf}_\Sigma : L_\Sigma \to \wp L_\Sigma$ that assigns to each formula the set of its subformulas (it is convenient to consider that a formula is a subformula of itself).

It is worthwhile to observe that

$$\mathrm{sbf}_\Sigma(\eta) \subseteq \mathrm{sbf}_\Sigma(\varphi) \quad \text{whenever } \eta \in \mathrm{sbf}_\Sigma(\varphi).$$

Proposition 32
(Metatheorem of substitution by equivalent – MTSE)
Let $\varphi, \varphi', \eta, \eta' \in L_\Sigma$ be such that:

- $\eta \in \mathrm{sbf}_\Sigma(\varphi)$;

- $\vdash_\Sigma (\eta \Leftrightarrow \eta')$;

- φ' can be obtained from φ by replacing zero, one or more occurrences of η by η'.

Then, $\vdash_\Sigma (\varphi \Leftrightarrow \varphi')$.

Proof: The proof proceeds by induction on the structure of φ.
(Basis) φ is $p(t_1, \ldots, t_n)$. Then:
Formula η is $p(t_1, \ldots, t_n)$ and φ' is:

- either some η' equivalent to $p(t_1, \ldots, t_n)$;

- or $p(t_1, \ldots, t_n)$.

In the first case, $\vdash_\Sigma (p(t_1, \ldots, t_n) \Leftrightarrow \eta')$ holds by hypothesis. In the second case, since $(p(t_1, \ldots, t_n) \Leftrightarrow p(t_1, \ldots, t_n))$ is an abbreviation of

$$((p(t_1, \ldots, t_n) \Rightarrow p(t_1, \ldots, t_n)) \wedge (p(t_1, \ldots, t_n) \Rightarrow p(t_1, \ldots, t_n))),$$

the thesis $\vdash_\Sigma (p(t_1, \ldots, t_n) \Leftrightarrow p(t_1, \ldots, t_n))$ follows from result (i) and $\vdash_\Sigma (p(t_1, \ldots, t_n) \Rightarrow p(t_1, \ldots, t_n))$, by applying MP twice.
(Step) There are three cases to consider:
(1) φ is $(\neg\psi)$. Observe that, by the induction hypothesis, $\vdash_\Sigma (\psi \Leftrightarrow \psi')$, where ψ' is such that φ' is $(\neg\psi')$. Therefore, expanding the equivalence abbreviation and using (h1) and (h2), $\vdash_\Sigma (\psi \Rightarrow \psi')$ and $\vdash_\Sigma (\psi' \Rightarrow \psi)$. Hence, by result (f) above, $\vdash_\Sigma ((\neg\psi') \Rightarrow (\neg\psi))$ and $\vdash_\Sigma ((\neg\psi) \Rightarrow (\neg\psi'))$, whence, using (i), one establishes $\vdash_\Sigma ((\neg\psi) \Leftrightarrow (\neg\psi'))$.
(2) φ is $(\psi \Rightarrow \delta)$. Observe that, by the induction hypothesis, $\vdash_\Sigma (\psi \Leftrightarrow \psi')$ and $\vdash_\Sigma (\delta \Leftrightarrow \delta')$, where ψ' and δ' are such that φ' is $(\psi' \Rightarrow \delta')$. Consider the following derivation sequence for $(\psi \Rightarrow \psi'), (\delta' \Rightarrow \delta), (\psi' \Rightarrow \delta') \vdash_\Sigma (\psi \Rightarrow \delta)$:

1	$(\psi \Rightarrow \psi')$	Hyp
2	$(\delta' \Rightarrow \delta)$	Hyp
3	$(\psi' \Rightarrow \delta')$	Hyp
4	$(\psi \Rightarrow \delta')$	HS 1,3
4	$(\psi \Rightarrow \delta)$	HS 2,4

Therefore, by the MTD, $(\psi \Rightarrow \psi'), (\delta' \Rightarrow \delta) \vdash_\Sigma ((\psi' \Rightarrow \delta') \Rightarrow (\psi \Rightarrow \delta))$. In a similar way one can conclude $(\psi' \Rightarrow \psi), (\delta \Rightarrow \delta') \vdash_\Sigma ((\psi \Rightarrow \delta) \Rightarrow (\psi' \Rightarrow \delta'))$. Finally, using (h1), (h2) and (i), one obtains $\vdash_\Sigma ((\psi \Rightarrow \delta) \Leftrightarrow (\psi' \Rightarrow \delta'))$.

(3) φ is $(\forall x\, \psi)$. This case is left as an exercise: it is similar to that of negation, but using (j) instead of (f).   QED

The next result[15] provides a sufficient condition for replacing the quantification variable in a universally quantified formula.

Proposition 33 (Rule of substitution – RS)
Let $x, y \in X$ and $\varphi \in L_\Sigma$ be such that:

- $y \notin \mathrm{fv}_\Sigma(\varphi)$;

- $y \triangleright_\Sigma x : \varphi$.

Then, $\vdash_\Sigma ((\forall x\, \varphi) \Leftrightarrow (\forall y\, [\varphi]_y^x))$.

Proof:
(1) $\vdash_\Sigma ((\forall x\, \varphi) \Rightarrow (\forall y\, [\varphi]_y^x))$. Consider the following derivation sequence:

1	$((\forall x\, \varphi) \Rightarrow [\varphi]_y^x)$	Ax4, $y \triangleright_\Sigma x : \varphi$
2	$(\forall y\, ((\forall x\, \varphi) \Rightarrow [\varphi]_y^x))$	Gen 1
3	$((\forall y\, ((\forall x\, \varphi) \Rightarrow [\varphi]_y^x)) \Rightarrow ((\forall x\, \varphi) \Rightarrow (\forall y\, [\varphi]_y^x)))$	Ax5, $y \notin \mathrm{fv}_\Sigma((\forall x\, \varphi))$
4	$((\forall x\, \varphi) \Rightarrow (\forall y\, [\varphi]_y^x))$	MP 2,3

(2) $\vdash_\Sigma ((\forall y\, [\varphi]_y^x) \Rightarrow (\forall x\, \varphi))$. Consider the following derivation sequence:

1	$((\forall y\, [\varphi]_y^x) \Rightarrow \varphi)$	Ax4 (†)
2	$(\forall x\, ((\forall y\, [\varphi]_y^x) \Rightarrow \varphi))$	Gen 1
3	$((\forall x\, ((\forall y\, [\varphi]_y^x) \Rightarrow \varphi)) \Rightarrow ((\forall y\, [\varphi]_y^x) \Rightarrow (\forall x\, \varphi)))$	Ax5, $x \notin \mathrm{fv}_\Sigma((\forall y\, [\varphi]_y^x))$
4	$((\forall y\, [\varphi]_y^x) \Rightarrow (\forall x\, \varphi))$	MP 2,3

(†) Note that, at step 1, Ax4 is used in a valid form since, as $y \triangleright_\Sigma x : \varphi$, it is also the case that $x \triangleright_\Sigma y : [\varphi]_y^x$, thanks to Exercise 20 in Chapter 4. Furthermore, $[[\varphi]_y^x]_x^y$ is φ because $y \notin \mathrm{fv}_\Sigma(\varphi)$.
Finally, using (i), the desired result follows.   QED

The following example illustrates the application of the RS and is representative of its frequent usage in other chapters, namely in Chapter 10.

---

[15]Usually known as the rule of susbtitution (of a bound variable), although it has no premises.

Example 34 Let x and y be distinct variables. Even when $y \in \mathrm{var}_\Sigma(t)$,

$$(\forall x\,(\forall y\,p(y,x))) \Rightarrow p(y,t)$$

is a theorem. Indeed, choosing $z \notin \mathrm{var}_\Sigma(t) \cup \{x,y\}$, the sequence

1	$(\forall x\,(\forall y\,p(y,x)))$	Hyp
2	$((\forall y\,p(y,x)) \Rightarrow (\forall z\,p(z,x)))$	RS
3	$(\forall x\,((\forall y\,p(y,x)) \Rightarrow (\forall z\,p(z,x))))$	Gen 2
4	$((\forall x\,((\forall y\,p(y,x)) \Rightarrow (\forall z\,p(z,x)))) \Rightarrow$	
	$\quad\quad ((\forall x\,(\forall y\,p(y,x))) \Rightarrow (\forall x\,(\forall z\,p(z,x)))))$	Exercise 24
5	$((\forall x\,(\forall y\,p(y,x))) \Rightarrow (\forall x\,(\forall z\,p(z,x))))$	MP 3,4
6	$(\forall x\,(\forall z\,p(z,x)))$	MP 1,5
7	$((\forall x\,(\forall z\,p(z,x))) \Rightarrow (\forall z\,p(z,t)))$	Ax4
8	$(\forall z\,p(z,t))$	MP 6,7
9	$((\forall z\,p(z,t)) \Rightarrow p(y,t))$	Ax4
10	$p(y,t)$	MP 8,9

is a derivation for

$$(\forall x\,(\forall y\,p(y,x))) \vdash_\Sigma p(y,t).$$

Using the MTD (the only generalization is over variable x that does not occur free in $(\forall x\,(\forall y\,p(y,x)))$), one concludes that $(\forall x\,(\forall y\,p(y,x))) \Rightarrow p(y,t)$ is a theorem. Observe that the proof could not have been done directly (without invoking RS for replacing y by z) because t is not free for x in $(\forall y\,p(y,x))$ when $y \in \mathrm{var}_\Sigma(t)$.

The rule of substitution of bound variable is also helpful when showing that the operator $\vdash_\Sigma$ is closed for substitution up to change of bound variables, a fact the reader is asked to prove in the next exercise.

Exercise 35 Let $[\Gamma]^x_t = \{[\gamma]^x_t : \gamma \in \Gamma\}$. Show that $[\varphi']^x_t \in ([\Gamma]^x_t)^{\vdash_\Sigma}$ whenever $\varphi \in \Gamma^{\vdash_\Sigma}$, for some φ' obtained from φ by an appropriate substitution of bound variables.

The proof that closure for substitution does not hold in general is delayed until Exercise 34 of Section 7.4 since it relies on the soundness of the calculus.

Exercise 36 Give derivation sequences for:

1. $\vdash_\Sigma (((\varphi_1 \Rightarrow \varphi_2) \wedge (\varphi_3 \Rightarrow \varphi_4)) \Rightarrow ((\varphi_1 \vee \varphi_3) \Rightarrow (\varphi_2 \vee \varphi_4)))$.

2. $\vdash_\Sigma ((\forall x(\varphi_1 \wedge \varphi_2)) \Leftrightarrow ((\forall x\,\varphi_1) \wedge (\forall x\,\varphi_2)))$.

3. $\vdash_\Sigma ((\exists x\,(\varphi_1 \vee \varphi_2)) \Leftrightarrow ((\exists x\,\varphi_1) \vee (\exists x\,\varphi_2)))$.

4. $\vdash_\Sigma ((\forall x\, \varphi) \Rightarrow (\exists x\, \varphi))$.

5. $\vdash_\Sigma ((\forall x(\varphi_1 \vee \varphi_2)) \Rightarrow ((\forall x\, \varphi_1) \vee (\exists x\, \varphi_2)))$.

6. $\vdash_\Sigma (\exists x(\varphi \Rightarrow (\forall x\, \varphi)))$.

7. $\vdash_\Sigma ((\exists x(\forall y\, \varphi)) \Rightarrow (\forall y(\exists x\, \varphi)))$.

Theorems 2 and 3 above state that universal quantification distributes over conjunction and that existential quantification distributes over disjunction, respectively.

Chapter 6

Theories and presentations

The concept of first-order theory is motivated in Section 6.1 by the idea in Hilbert's programme of axiomatizing fragments of Mathematics. In Section 6.2, several results are established concerning important properties of theories, like consistency, exhaustiveness, decidability and semi-decidability, including a necessary and sufficient condition for a theory to be axiomatizable. The chapter ends with a brief introduction in Section 6.3 to the constructive variant of the technique of quantifier elimination, a quite useful technique for proving the decidability of a theory.

6.1 Formalizing mathematical concepts

As an illustration of the idea of formalizing mathematical concepts, consider the case of groups, a simple but important notion in Algebra. What can one do to use first-order logic to reason about groups? After choosing the signature needed for setting up the envisaged first-order language (recall Exercise 1 of Chapter 4), one would like to be able to decide which formulas express true facts about groups. According to Hilbert's programme, one should be able to identify a computable set of axioms that would allow the derivation of all true facts about groups. Hence, the set of all such true formulas would be closed for derivation, motivating the following definition.

A *theory* over a signature Σ is a set $\Theta \subseteq L_\Sigma$ such that $\Theta^{\vdash_\Sigma} = \Theta$. Given a theory Θ, one may write $\vdash_\Theta \varphi$ in place of $\varphi \in \Theta$. In this case, φ is said to be a *theorem* of Θ.

As a first example, observe that, for every signature Σ, the set L_Σ is a theory (the improper one) over Σ. On the other hand, the set $\emptyset$ is not a theory.

Returning to the problem of formalizing in first-order logic the concept of group, let Σ_G be the signature of the language of groups chosen in Exercise 1 of Chapter 4. What should be the theory Θ_G of groups? It should contain every formula in L_{Σ_G} expressing a true fact about groups. For instance, the uniqueness of the inverse should be a theorem of the theory of groups.

Exercise 1 Given a signature Σ, prove or refute by providing a counter-example the following assertions:

- The intersection of theories over Σ is a theory over Σ.

- The union of theories over Σ is a theory over Σ.

A theory Θ is said to be *consistent* if $cL_\Sigma \not\subseteq \Theta$, and it is said to be *exhaustive* or *complete* if $\Theta \cap \{\varphi, (\neg \varphi)\} \neq \emptyset$ for every $\varphi \in cL_\Sigma$.

Exercise 2 Show that if a theory is not consistent, then it is exhaustive.

Observe that it is possible for a theory to be consistent but not exhaustive. For instance, for each signature Σ, theory $\emptyset^{\vdash_\Sigma}$ is consistent but not exhaustive, as it will be proved later on. On the other hand, theory L_Σ is exhaustive and not consistent. In fact, L_Σ is the unique inconsistent theory. As the reader might expect, those theories that are both consistent and exhaustive are of particular interest.

Axiomatization is another important issue. Returning to the formalization of the notion of group, it would be nice to be able to identify a computable set of formulas about groups that when used as (extra) axioms would allow the derivation of all formulas expressing true facts about groups. This idea motivates the following definitions.

A *presentation* over a signature Σ is a computable set Ξ of closed formulas. The elements of Ξ are called *specific axioms* or *proper axioms* of the presentation. The elements of the set $\Xi^{\vdash_\Sigma}$ are called *theorems* of the presentation. The set $\Xi^{\vdash_\Sigma}$ is the *theory axiomatized* by the presentation. Furthermore, Ξ is said to be an *axiomatization* or a *presentation* of theory $\Xi^{\vdash_\Sigma}$.

A theory Θ over Σ is said to be *axiomatizable* if it is axiomatized by some presentation Ξ over Σ, that is, if there is a computable set Ξ of closed formulas such that $\Xi^{\vdash_\Sigma} = \Theta$.

Unfortunately, not every theory is axiomatizable. That is the case of the theory of natural numbers as stated in Proposition 15 of Section 13.2, a corollary of Gödel's first incompleteness theorem.

Example 3 Let $\Sigma_\leq$ be the signature with no function symbols and only one binary predicate symbol $\leq$. Consider the following two binary predicates introduced as abbreviations:

- $(x < y)$ for $(\neg(y \leq x))$;

- $(x \cong y)$ for $((x \leq y) \wedge (y \leq x))$.

In order to formalize the concept of dense order without left and right end-points, consider the theory Θ_D over $\Sigma_{\leq}$ with the following specific axioms:

- Equality:

  - $(\forall x_1 (x_1 \cong x_1))$;
  - $(\forall x_1 (\forall x_2 (\forall x_3 (\forall x_4 ((x_1 \cong x_3) \Rightarrow ((x_2 \cong x_4) \Rightarrow$
    $$((x_1 \leq x_2) \Rightarrow (x_3 \leq x_4)))))))).$

- Strict order:

  - $(\forall x_1 (\forall x_2 (\neg((x_1 < x_2) \wedge (x_2 < x_1)))))$;
  - $(\forall x_1 (\forall x_2 (\forall x_3(((x_1 < x_2) \wedge (x_2 < x_3)) \Rightarrow (x_1 < x_3))))).$

- Total order:

  - $(\forall x_1 (\forall x_2 ((x_1 < x_2) \vee (x_1 \cong x_2) \vee (x_2 < x_1)))).$

- Denseness:

  - $(\forall x_1 (\forall x_2 ((x_1 < x_2) \Rightarrow ((\exists x_3 ((x_1 < x_3) \wedge (x_3 < x_2))))))).$

- No left endpoint:

  - $(\forall x_1 (\exists x_2 (x_2 < x_1))).$

- No right endpoint:

  - $(\forall x_1 (\exists x_2 (x_1 < x_2))).$

As an illustration, the reader can prove that the predicate $<$ is irreflexive and asymmetric in dense orders without left and right endpoints. That is:

- $\vdash_{\Theta_D} (\forall x_1 (\neg(x_1 < x_1)))$;

- $\vdash_{\Theta_D} (\forall x_1 (\forall x_2 ((x_1 < x_2) \Rightarrow (\neg(x_2 < x_1))))).$

Exercise 4 Axiomatize the theory of partial orders over signature $\Sigma_{\leq}$.

Exercise 5 Axiomatize the theory of groups over signature Σ_G. Show that uniqueness of the inverse is a theorem.

The following exercise provides a useful characterization of exhaustiveness.

Exercise 6 Show that a theory Θ over Σ is exhaustive if and only if, for all closed formulas φ and ψ, the following holds:

$$\text{if } \Theta \vdash_{\Sigma} (\varphi \vee \psi), \text{ then } \Theta \vdash_{\Sigma} \varphi \text{ or } \Theta \vdash_{\Sigma} \psi.$$

Given the great importance of exhaustiveness, it is worthwhile mentioning Vaught's theorem that provides a sufficient condition for showing that a theory is exhaustive when it is known that the theory has no finite models. The interested reader should consult for instance [12], but only after mastering the material in Chapters 7 and 8.

6.2 Decidability and semi-decidability

Since theories are sets of formulas, it makes sense to ask whether theories are computable (decidable) or computably enumerable (semi-decidable) sets. The following result provides a sufficient condition for a computably enumerable theory to be computable.

Proposition 7 Every computably enumerable and exhaustive theory is computable.

Proof:
Two cases are considered.
(i) If theory Θ over Σ is not consistent, then, from result (g) in Section 5.4, $\Theta^{\vdash_{\Sigma}} = L_{\Sigma}$, and hence $\Theta = \Theta^{\vdash_{\Sigma}}$ is computable.

(ii) Otherwise, it is still possible to give an algorithm to compute χ_{Θ} using the fact that Θ is computably enumerable and exhaustive. Consider the following algorithm:

```
Function[w,
    If[χ_{L_Σ}[w] == 0,
        0
        ,
        φ₁ = a[w];
        φ₂ = b[w];
        k = 0;
        φ = f[0];
        While[(φ =!= φ₁) && (φ =!= φ₂),
            k = k + 1;
            φ = f[k]
        ];
        If[φ === φ₁, 1, 0]
    ]
]
```

where

- f is a computable enumeration of Θ that exists because Θ is computably enumerable and non empty;

- $a = \lambda\psi.(\forall\psi) : L_\Sigma \to cL_\Sigma$ and $b = \lambda\psi.(\neg(\forall\psi)) : L_\Sigma \to cL_\Sigma$ that are easily shown to be computable.

The justification of the algorithm is as follows. Recall that $w \in \Theta$ if and only if $(\forall w) \in \Theta$. Given w assume that $w \in L_\Sigma$ (if not, the algorithm returns 0 as envisaged).

Then, either $(\forall w) \in \Theta$ or $(\neg(\forall w)) \in \Theta$ but not both. Indeed, since Θ is exhaustive, either $(\forall w) \in \Theta$ or $(\neg(\forall w)) \in \Theta$. Moreover, since theory Θ is also consistent, exactly one of the two formulas $(\forall w)$ and $(\neg(\forall w))$ is in Θ.

Finally since f is an enumeration of Θ there must exist k such that $f(k) = (\forall w)$ if $(\forall w) \in \Theta$ or $f(k) = (\neg(\forall w))$ otherwise. In the first case the algorithm returns 1 and in the second case it returns 0, precisely as required. QED

Proposition 8 Every axiomatizable theory is computably enumerable.

Proof: If Θ is axiomatizable, then there exists a computable set Ξ such that $\Xi^{\vdash_\Sigma} = \Theta$. Note that Ξ is computably enumerable due to Proposition 9 in Chapter 3. Then, by the semi-decidability of the Hilbert calculus (Proposition 21 in Chapter 5), it follows that $\Xi^{\vdash_\Sigma}$ is a computably enumerable set. QED

The following result is a direct corollary of Proposition 7 and Proposition 8.

Proposition 9 Every axiomatizable and exhaustive theory is computable.

The following theorem was first stated by William Craig (see [13]). It is the converse of Proposition 8. Together they provide a necessary and sufficient condition for a theory to be semi-decidable.

Proposition 10 (Craig's theorem)
Every computably enumerable theory is axiomatizable.

Proof: Let Θ be a computably enumerable theory over Σ. Observe that $\Theta \neq \emptyset$, since $\Theta = \Theta^{\vdash_\Sigma}$. Therefore, there exists a computable enumeration f of Θ.
Consider the map $u : \Theta \to cL_\Sigma$ such that $u(\theta)$ is the closed formula

$$(\forall x_0(\ldots(\forall x_{m_\theta}\,\theta)\ldots))$$

where m_θ is the least upper bound of the indexes of the variables occurring free in θ. Note that $u(\theta)$ is $(\forall x_0\,\theta)$ when θ is closed. Map u is clearly computable (the interested reader should have no difficulty in presenting an algorithm for computing it).

Observe that $\Theta = u(\Theta)^{\vdash_\Sigma}$, although $u(\Theta)$ is no longer closed under derivation. Note also that the set $u(\Theta)$ is computably enumerated by $g = u \circ f$.

Finally, consider the set

$$\Xi = \{(\wedge_{i=1}^{n+1} g(n)) : n \in \mathbb{N}\},$$

where formula $(\wedge_{i=1}^{n}\psi)$, known as the *n-conjunctive replication* of ψ, is defined for $n \in \mathbb{N}^+$ as follows:

- $(\wedge_{i=1}^{1}\psi)$ is ψ;

- $(\wedge_{i=1}^{k+1}\psi)$ is $((\wedge_{i=1}^{k}\psi) \wedge \psi)$, that is, $(\neg((\wedge_{i=1}^{k}\psi) \Rightarrow (\neg\,\psi)))$.

When $n > 1$, formula $(\wedge_{i=1}^{n}\psi)$ is said to be a *proper conjunctive replication* of ψ.

Since $\Theta = \Xi^{\vdash_\Sigma}$ and $\Xi \subseteq cL_\Sigma$, it only remains to show that Ξ is computable. To this end, it is worthwhile to look at the nature of the elements of Ξ as illustrated in the following table:

Θ	$u(\Theta)$	Ξ
$f(0) = \theta_0$	$\overline{\theta}_0 = u(\theta_0)$	$\overline{\theta}_0$
$f(1) = \theta_1$	$\overline{\theta}_1 = u(\theta_1)$	$(\overline{\theta}_1 \wedge \overline{\theta}_1)$
$f(2) = \theta_2$	$\overline{\theta}_2 = u(\theta_2)$	$((\overline{\theta}_2 \wedge \overline{\theta}_2) \wedge \overline{\theta}_2)$
$\cdots$	$\cdots$	$\cdots$

Observe that every element of $u(\Theta)$ is of the form $(\forall x_0\,\gamma)$. Hence, for deciding if a formula is in Ξ or not it is sufficient to check if it is the $n + 1$-conjunctive replication of $g(n) = u(f(n))$ for some $n \in \mathbb{N}$. Accordingly, the following algorithm computes χ_Ξ:

```
Function[w,
    If[χ_{L_Σ}[w] == 0,
        0
        ,
        φ = ToExpression[w];
        q[φ, c[φ], ToExpression[g[c[φ] − 1]]]]
    ]
]
```

where maps q and c are defined as follows, writing EL_Σ to denote the set of *Mathematica* expressions that correspond to formulas. The map

$$q : EL_\Sigma \times \mathbb{N}^+ \times EL_\Sigma \to \{0, 1\},$$

when applied to (φ, n, ψ), returns:

- 1 if φ has the form $(\wedge_{i=1}^{n} \psi)$;

- 0 otherwise.

The map

$$c : EL_\Sigma \to \mathbb{N}^+,$$

when applied to φ, returns the unique $n \in \mathbb{N}^+$ such that $q(\varphi, n, \psi) = 1$ for some $\psi \in EL_\Sigma$.

Map q can be computed as follows:

```
Function[{φ, n, ψ},
    If[n == 1,
        If[φ === ψ, 1, 0]
    ,
        If[(φ[[0]] === ¬) && (φ[[1, 0]] === ⇒) &&
                (φ[[1, 2, 0]] === ¬) && (φ[[1, 2, 1]] === ψ),
            q[φ[[1, 1]], n − 1, ψ]
        ,
            0
        ]
    ]
]
```

Map c is computed by the following algorithm:

```
Function[φ,
    If[(φ[[0]] === ¬) && (φ[[1, 0]] === ⇒) &&
            (φ[[1, 2, 0]] === ¬) && e[φ[[1, 1]], φ[[1, 2, 1]]]],
        c[φ[[1, 1]]] + 1
    ,
        1
    ]
]
```

where map

$$e : EL_\Sigma \times EL_\Sigma \to \{0, 1\},$$

when applied to (φ, ψ), returns:

- 1 if there is $n \in \mathbb{N}^+$ such that $q(\varphi, n, \psi) = 1$;

- 0 otherwise.

Map e can be computed as follows:

```
Function[{φ, ψ},
   Which[
      (φ === ψ),
      1
   ,
      (φ[[0]] === ¬) && (φ[[1, 0]] === ⇒) &&
         (φ[[1, 2, 0]] === ¬) && (φ[[1, 2, 1]] === ψ),
      e[φ[[1, 1]], ψ]
   ,
      True,
      0
   ]
]
```

Note that, in the computation of $\chi_\Xi(w)$, for the correct counting of the number of replications, it is essential that no formula in the image of u is a proper conjunctive replication (because it is of the form $(\forall x_0\, \gamma)$).

For example, if $g(0)$ were the closed formula $(p(d) \wedge p(d))$ and $g(1) \neq g(0)$, then $(p(d) \wedge p(d))$ would be incorrectly classified as not being in Ξ. Indeed, $c((p(d) \wedge p(d))) = 2$ and $g(2 - 1) \neq (p(d) \wedge p(d))$.   QED

6.3   Quantifier elimination

Notwithstanding Gödel's first incompleteness theorem, there are many useful mathematical theories that are known to be decidable. Namely, the theories of dense orders, Abelian groups, real closed fields, algebraic closed fields and Presburger arithmetic.

Several techniques have been used for proving that a theory is computable. Among them quantifier elimination deserves special attention (for a survey see [42]). The technique of quantifier elimination has two main variants. The constructive variant (also known as symbolic or syntactic) is illustrated below for proving the decidability of Θ_D (the theory of dense orders without left and right endpoints). The semantic variant is used in Chapter 15 for proving the decidability of Presburger arithmetic. The reader interested in the semantic variant should also consult [53]. Semantic quantifier elimination was used by Alfred Tarski [47, 50] to establish the decidability of the theory of real closed fields. Another important technique for proving decidability of theories relies on Vaught's theorem (see for instance [12]).

A set Ψ of formulas is said to be *closed for Boolean combinations* if Ψ is closed for negation and implication. That is, Ψ is such that:

- $(\neg \psi) \in \Psi$ whenever $\psi \in \Psi$;

- $(\psi_1 \Rightarrow \psi_2) \in \Psi$ whenever $\psi_1, \psi_2 \in \Psi$.

Let Γ^{B} denote the closure for Boolean combinations of set Γ. That is, Γ^{B} is the smallest set closed for Boolean combinations that contains Γ. A formula φ is said to be a *Boolean combination* of the formulas in Γ if $\varphi \in \Gamma^{\mathsf{B}}$.

An *elimination set* (or *elimination kernel*) for a theory Θ is a set of formulas Δ such that, for every formula $\varphi \in L_\Sigma$, there is $\varphi^* \in \Delta^{\mathsf{B}}$ such that:

- $\mathrm{fv}_\Sigma(\varphi) = \mathrm{fv}_\Sigma(\varphi^*)$;

- $\Theta \vdash_\Sigma (\varphi \Leftrightarrow \varphi^*)$.

The following result is a sufficient condition for concluding that a set of formulas is an elimination kernel for a theory.[1]

Proposition 11 Let Θ be a theory over Σ and $\Delta \subseteq L_\Sigma$. Assume that

- $B_\Sigma \subseteq \Delta$;

- $\Delta^{\mathsf{B}} = \Delta$;

- for each formula $\delta \in \Delta$ with $\mathrm{fv}_\Sigma(\delta) = \{x_1, \ldots, x_n, x\}$, there is $\delta' \in \Delta$ with $\mathrm{fv}_\Sigma(\delta') = \{x_1, \ldots, x_n\}$ such that

$$\Theta \vdash_\Sigma ((\exists x\, \delta) \Leftrightarrow \delta').$$

Then, Δ is an elimination set for Θ.

Proof: Let $\varphi \in L_\Sigma$. The proof follows by induction on the structure of φ.

Base: $\varphi \in B_\Sigma$. Then, $\varphi \in \Delta$ and so it is enough to take φ^* as φ.

Step: In this case, it is easier in one of the induction steps to consider existential quantification instead of universal quantification.

(a) φ is $(\neg \psi)$.

By the induction hypothesis, there is $\psi^* \in \Delta$ such that $\mathrm{fv}_\Sigma(\psi) = \mathrm{fv}_\Sigma(\psi^*)$ and

$$\Theta \vdash_\Sigma (\psi \Leftrightarrow \psi^*).$$

---

[1] Recall that B_Σ is the set of atomic formulas over signature Σ.

Then, $\mathrm{fv}_\Sigma((\neg\,\psi)) = \mathrm{fv}_\Sigma((\neg\,\psi^*))$ and

$$\Theta \vdash_\Sigma ((\neg\,\psi) \Leftrightarrow (\neg\,\psi^*)).$$

Moreover, $(\neg\,\psi^*) \in \Delta$, since Δ is closed for Boolean combinations.
(b) φ is $(\exists x\,\psi)$.
By the induction hypothesis, there is $\psi^* \in \Delta$ such that $\mathrm{fv}_\Sigma(\psi) = \mathrm{fv}_\Sigma(\psi^*)$ and

$$\Theta \vdash_\Sigma (\psi \Leftrightarrow \psi^*).$$

Hence

$$\Theta \vdash_\Sigma ((\exists x\,\psi) \Leftrightarrow (\exists x\,\psi^*)).$$

On the other hand, using the third hypothesis on ψ^*, there is $\psi^{*\prime} \in \Delta$ such that $\mathrm{fv}_\Sigma(\psi^{*\prime}) = \mathrm{fv}_\Sigma(\psi^*) \setminus \{x\}$ and

$$\Theta \vdash_\Sigma ((\exists x\,\psi^*) \Leftrightarrow \psi^{*\prime}).$$

Therefore, $\mathrm{fv}_\Sigma(\psi^{*\prime}) = \mathrm{fv}_\Sigma(\varphi)$ and

$$\Theta \vdash_\Sigma ((\exists x\,\psi) \Leftrightarrow \psi^{*\prime}).$$

The proof for implication is left as an exercise.   QED

In the sequel, $\overline{\Psi}$ is used to denote the set $\{(\neg\,\psi) : \psi \in \Psi\}$.

Proposition 12 Let Θ be a theory over Σ and $\Delta \subseteq L_\Sigma$. Assume that

- $B_\Sigma \subseteq \Delta$;

- for each formula $(\exists x\,\varphi)$ such that

    - φ is a conjunction of formulas in $\Delta \cup \overline{\Delta}$;
    - $\mathrm{fv}_\Sigma(\varphi) = \{x_1, \ldots, x_n, x\}$;

  there is a formula $\psi \in L_\Sigma$ such that

    - ψ is a Boolean combination of the formulas in Δ;
    - $\mathrm{fv}_\Sigma(\psi) = \{x_1, \ldots, x_n\}$;
    - $\Theta \vdash_\Sigma ((\exists x\varphi) \Leftrightarrow \psi)$.

Then, Δ is an elimination set for Θ.

Proof: Use Proposition 11.   QED

A theory Θ allows *quantifier elimination* if the set of quantifier free formulas is an elimination kernel for Θ. Proposition 12 also provides a sufficient condition for concluding that a theory has quantifier elimination. Indeed, the set of atomic formulas is contained in the set of quantifier free formulas. Hence, only the second condition has to be verified.

Before using this technique for proving the decidability of the theory of dense orders without left and right endpoints introduced in the previous section, it is necessary to introduce the following concept.

A formula $\psi \in L_\Sigma$ is said to be in *disjunctive normal form* if ψ is a finite disjunction of formulas $\delta_1, \ldots, \delta_n$ and each δ_i is a finite conjunction of formulas in $B_\Sigma \cup \overline{B}_\Sigma$.

Exercise 13 Show that for every formula φ without quantifiers there is a formula ψ in disjunctive normal form such that $\vdash_\Sigma (\varphi \Leftrightarrow \psi)$.

Proposition 14 Theory Θ_D of dense orders without left and right endpoints has quantifier elimination.

Proof: The proof relies on Proposition 12. By Exercise 13, every formula without quantifiers is equivalent to a formula in disjunctive normal form. Moreover

- $\Theta_D \vdash_\Sigma ((\neg(x \cong y)) \Leftrightarrow ((x < y) \vee (y < x)))$;

- $\Theta_D \vdash_\Sigma ((\neg(x < y)) \Leftrightarrow ((x \cong y) \vee (y < x)))$.

Hence, every formula without quantifiers is, in this case, equivalent to a formula in disjunctive normal form where the elements of the conjunctions are atomic formulas.

Let $\top$ be an abbreviation of $(\forall x(x \cong x))$ and $\bot$ be $(\neg \top)$.

Observe also that formulas of the form $(x \cong x)$ and $(x_i \cong x_i)$ are equivalent to $\top$ and that formulas of the form $(x < x)$ and $(x_i < x_i)$ are equivalent to $\bot$. Moreover, since

$$\vdash_\Sigma ((\exists x (\varphi_1 \vee \varphi_2)) \Leftrightarrow ((\exists x \varphi_1) \vee (\exists x \varphi_2))),$$

it is enough to consider formulas that are conjunctions of atomic formulas of the form

$$(*) \quad (x \cong x_i), \quad (x_i \cong x_j), \quad (x < x_i), \quad (x_i < x), \quad (x_i < x_j).$$

Assume that γ is a conjunction of formulas of the form in $(*)$. Let $B_\Sigma(\gamma)$ be the set of atomic formulas that occur in γ. There are two cases to consider.
(a) $(x \cong x_i) \in B_\Sigma(\gamma)$ for some $i = 1, \ldots, n$. Then, take ψ as

$$[\gamma]^x_{x_i}.$$

(b) Otherwise, let γ_1 and γ_2 be formulas such that γ_1 is the conjunction of atomic formulas of the form $(x_i \cong x_j), (x_i < x_j)$ that occur in γ and γ_2 is the conjunction of atomic formulas of the form $(x < x_j), (x_i < x)$ that occur in γ. Of course γ is $(\gamma_1 \wedge \gamma_2)$. Then, $(\exists x \gamma)$ is equivalent to the formula

$$(\gamma_1 \wedge (\exists x \, \gamma_2))$$

and the formula $(\exists x \gamma_2)$ can be written as

$$(\dagger) \quad \left(\exists x \left(\left(\bigwedge_{i \in I}(x_i < x) \right) \wedge \left(\bigwedge_{j \in J}(x < x_j) \right) \right) \right)$$

where $I = \{i : (x_i < x) \in B_\Sigma(\gamma_2)\}$ and $J = \{j : (x < x_j) \in B_\Sigma(\gamma_2)\}$. There are four cases to consider:

(i) $I \cap J \neq \emptyset$. Then, $(\dagger)$ is equivalent to $\bot$ and so taking ψ as $\bot$

$$\Theta_D \vdash_\Sigma ((\exists x \, \gamma) \Leftrightarrow \psi).$$

(ii) $I \cap J = \emptyset$ and $I, J \neq \emptyset$. Take ψ as

$$\left(\bigwedge_{i \in I, j \in J} (x_i < x_j) \right).$$

It remains to show that

$$\Theta_D \vdash_\Sigma ((\exists x \, \gamma_2) \Leftrightarrow \psi).$$

(a) $\Theta_D \vdash_\Sigma ((\exists x \, \gamma_2) \Rightarrow \psi)$
By transitivity of $<$, for every $i \in I$ and $j \in J$,

$$\Theta_D \vdash_\Sigma ((\neg(x_i < x_j)) \Rightarrow (\neg(\exists x \, ((x_i < x) \wedge (x < x_j)))))$$

and so, by Exercise 36, from Θ_D one can derive

$$\left(\left(\bigvee_{i \in I, j \in J} (\neg(x_i < x_j)) \right) \Rightarrow \left(\bigvee_{i \in I, j \in J} (\neg(\exists x \, ((x_i < x) \wedge (x < x_j)))) \right) \right).$$

Therefore, from Θ_D one can get

$$\left(\left(\bigvee_{i \in I, j \in J} (\neg(x_i < x_j)) \right) \right) \Rightarrow \left(\neg \left(\bigwedge_{i \in I, j \in J} (\exists x \, ((x_i < x) \wedge (x < x_j))) \right) \right)$$

and so, from Θ_D one can get

$$\left(\left(\bigvee_{i\in I, j\in J}(\neg(x_i < x_j))\right)\right) \Rightarrow \left(\neg\left(\exists x\left(\bigwedge_{i\in I, j\in J}((x_i < x) \wedge (x < x_j)))\right)\right)\right).$$

(b) $\Theta_D \vdash_\Sigma (\psi \Rightarrow (\exists x\, \gamma_2))$

By denseness of $<$, for every $i \in I$ and $j \in J$,

$$\Theta_D \vdash_\Sigma ((x_i < x_j) \Rightarrow (\exists x\,((x_i < x) \wedge (x < x_j))))$$

and so

$$\Theta_D \vdash_\Sigma \left(\left(\bigwedge_{i\in I, j\in J}(x_i < x_j)\right) \Rightarrow \left(\bigwedge_{i\in I, j\in J}(\exists x((x_i < x) \wedge (x < x_j)))\right)\right)$$

Observe that, in this particular case,

$$\left(\exists x\left(\left(\bigwedge_{i\in I}(x_i < x)\right) \wedge \left(\bigwedge_{j\in J}(x < x_j)\right)\right)\right) \Leftrightarrow \left(\bigwedge_{\substack{i\in I \\ j\in J}}(\exists x((x_i < x) \wedge (x < x_j)))\right)$$

since $x_i < x_j$ for all possible combinations of $i \in I$ and $j \in J$.

(iii) $I = \emptyset$. Taking ψ as $\top$ the equivalence is a theorem using the fact that the order does not have left endpoint.

(iv) $J = \emptyset$. Taking ψ as $\top$ the equivalence is a theorem using the fact that the order does not have right endpoint. QED

For other examples of theories that allow quantifier elimination, the interested reader should consult [29, 14].

Proposition 15 Let Θ be a theory over Σ. Assume that

- Θ has quantifier elimination;

- for each $\alpha \in (B_\Sigma \cap cL_\Sigma)$, either $\Theta \vdash_\Sigma \alpha$ or $\Theta \vdash_\Sigma (\neg\alpha)$.

Then, Θ is exhaustive.

Proof: Let φ be a closed formula. Then, since Θ has quantifier elimination, there is φ^* such that φ^* is a Boolean combination of formulas in Δ, $fv_\Sigma(\varphi) = fv_\Sigma(\varphi^*) = \emptyset$ and $\Theta \vdash_\Sigma (\varphi \Leftrightarrow \varphi^*)$. It can be proved that if ψ is a closed formula and a Boolean combination of formulas in Δ, then either $\Theta \vdash_\Sigma \psi$ or $\Theta \vdash_\Sigma (\neg\psi)$. The proof is by induction on the structure of ψ. Hence either $\Theta \vdash_\Sigma \varphi$ or $\Theta \vdash_\Sigma (\neg\varphi)$. QED

Proposition 16 Let Θ be a theory satisfying the conditions of Proposition 15. If Θ is axiomatizable, then Θ is computable.

Proof: From Proposition 15, Θ is exhaustive. Hence, by Proposition 9 is computable.   QED

Proposition 17 Theory Θ_D of dense orders without left and right endpoints is computable.

Chapter 7

Semantics

Intuitively speaking, first-order formulas make assertions about the intended universe of discourse. For instance, recall signature Σ_G (the signature of the language appropriate for the universe of a group). Intuitively, the formula

$$(\forall x_1 \, ((x_1 \, \mathbf{o} \, \mathbf{e}) \cong x_1))$$

means that, for every element x_1 of the group at hand, by applying operation $\mathbf{o}$ to x_1 and to the element $\mathbf{e}$ of the group one recovers x_1. One should expect that every theorem in the theory of groups should be interpreted as true in any group (soundness of the theory). Furthermore one might hope that the converse holds (completeness of the theory).

This intuitive approach to providing meaning to terms and formulas was made precise by Alfred Tarski (in the so called semantic theory of truth), see [45]. The main objective of this chapter is to give a modern account of the Tarskian semantics of first-order logic. Terms and formulas are given meanings in interpretation structures (corresponding to universes of discourse). An interpretation structure provides a domain of values (also known as individuals), an n-ary operation for each n-ary function symbol, and an n-ary relation for each n-ary predicate symbol. Given an assignment to variables, each term denotes an element of the domain and each formula denotes a Boolean value.

Although outside the scope of this book, it should be mentioned that the Tarskian semantics is not the only way to provide meaning to terms and formulas. The so called algebraic semantics is very interesting from the theoretical point of view. In the case of first-order logic, the algebraic approach was also started by Alfred Tarski who introduced cylindric algebras for the purpose (see [24]). The role of cylindric algebras is comparable to the role that Boolean algebras play for propositional logic (see for instance [11]), and to the role that Heyting algebras play for intuitionistic logic (see for instance [39]).

Section 7.1 presents interpretation structures and assignments. In Section 7.2 the denotation of terms, local satisfaction and (global) satisfaction of formulas are introduced. Afterward, the notions of valid formula and semantic entailment are defined and illustrated. It is also shown that semantic entailment induces a closure operator. Section 7.3 contains several technical lemmas concerning absent variables, closed formulas, substitution and tautologies. Soundness of the Hilbert calculus is proved in Section 7.4. Section 7.5 introduces the key notion of theory induced by an interpretation structure. Finally, Section 7.6 addresses important relationships between interpretation structures, using a modicum of Category Theory.

7.1 Interpretation structures

Let $\Sigma = (F, P, \tau)$ be a first-order signature. An *interpretation structure* over Σ is a triple

$$I = (D, \_^{\mathsf{F}}, \_^{\mathsf{P}})$$

where:

- D is a non empty set (*domain*);

- $\_^{\mathsf{F}} = \lambda\, n \in \mathbb{N} \,.\, \_^{\mathsf{F}}_n$ where each $\_^{\mathsf{F}}_n : F_n \to (D^n \to D)$;

- $\_^{\mathsf{P}} = \lambda\, n \in \mathbb{N}^+ \,.\, \_^{\mathsf{P}}_n$ where each $\_^{\mathsf{P}}_n : P_n \to (D^n \to \{0, 1\})$.

For setting up an interpretation structure, a non empty set of values should be chosen (the domain), a n-ary map f_n^{F} should be chosen for each function symbol f of arity n (the denotation of the function symbol), and a n-ary relation p_n^{P} should be chosen for each predicate symbol p of arity n (taking its characteristic map as the denotation of the predicate symbol). Hence, for the same signature, there are many possible interpretation structures with different domains and different denotations of function and predicate symbols.

The restriction to non empty domains leads to a more well behaved logic. Readers interested in the case where empty domains are allowed should consult [35].

Example 1 Let Σ_G be the signature of groups in Exercise 1 of Chapter 4. Consider the following interpretation structures over Σ_G:

- $\langle \mathbb{Z}, \_^{\mathsf{F}}, \_^{\mathsf{P}} \rangle$ where $\mathbf{o}_2^{\mathsf{F}}$ is $+$, $\mathbf{e}_0^{\mathsf{F}}$ is 0, $\mathbf{i}_1^{\mathsf{F}}$ is $\lambda\, d \,.\, -d$, and $\cong_2^{\mathsf{P}}$ is $=$.

- $\langle \mathbb{Q} \setminus \{0\}, \_^{\mathsf{F}}, \_^{\mathsf{P}} \rangle$ where $\mathbf{o}_2^{\mathsf{F}}$ is $\times$, $\mathbf{e}_0^{\mathsf{F}}$ is 1, $\mathbf{i}_1^{\mathsf{F}}$ is $\lambda\, d \,.\, d^{-1}$, and $\cong_2^{\mathsf{P}}$ is $=$.

- In general, each group $\langle G, \circ, e, i \rangle$ induces the interpretation structure $\langle G, \_^{\mathsf{F}}, \_^{\mathsf{P}} \rangle$ where $\mathbf{o}_2^{\mathsf{F}}$ is $\circ$, $\mathbf{e}_0^{\mathsf{F}}$ is e, $\mathbf{i}_1^{\mathsf{F}}$ is i, and $\cong_2^{\mathsf{P}}$ is $=$.

Exercise 2 Find an interpretation structure over Σ_G that is not a group.

Exercise 3 Define an interpretation structure over Σ_G based on a vector space.

Exercise 4 Define interpretation structures for $\Sigma_\leq$.

The following is an example of an interpretation structure with a symbolic domain. The underlying idea plays an important role in the proof of the completeness of the Hilbert calculus (see Chapter 8).

Exercise 5 Let $\Sigma = (F, P, \tau)$ be a signature with $F_0 \neq \emptyset$. Define an interpretation structure over Σ with domain cT_Σ and

$$f_n^{\mathsf{F}} = \lambda\, d_1, \ldots, d_n \,.\, f(d_1, \ldots, d_n) : cT_\Sigma{}^n \to cT_\Sigma$$

for each $n \in \mathbb{N}$ and $f \in F_n$.

An interpretation structure I provides the denotation of the function and predicate symbols. In order to determine the denotation of terms and formulas one needs an additional ingredient. An *assignment* into I is a map

$$\rho : X \to D$$

providing a value to each variable.

An interpretation structure and an assignment are enough to determine the meaning of all terms and formulas, as explained in Section 7.2. To this end, the following notion of equivalence of assignments (barring on a set of variables where they may differ) is needed for interpreting universally quantified formulas.

Let Y be a subset of X. Two assignments ρ, ρ' into I are said to be *Y-equivalent*, which can be written as $\rho \equiv_Y \rho'$, if $\rho(x) = \rho'(x)$ for each $x \notin Y$, that is, if ρ and ρ' differ at most in Y. It is convenient to use Y-equiv(ρ) for denoting the set of assignments that are Y-equivalent to ρ. One may write y-equivalent instead of $\{y\}$-equivalent, and y-equiv(ρ) for $\{y\}$-equiv(ρ).

When D is a finite set, there are only a finite number of assignments that are y-equivalent to a given assignment (if the number of elements of D is n then there are exactly n assignments y-equivalent to the given one).

Exercise 6 Discuss the cardinality of the set of assignments over an interpretation structure I.

Exercise 7 For each $Y \subseteq X$, define Y-equivalence as a binary relation over the set of assignments. Identify the properties of this family of relations.

7.2   Satisfaction, entailment and validity

The first objective of this section is to define satisfaction of a formula by an interpretation structure. To this end, it is necessary first to show how terms are interpreted.

Given an interpretation structure I over signature Σ and an assignment ρ into I, the *term denotation* map

$$[\![\_]\!]_{\Sigma}^{I\rho} : T_{\Sigma} \to D$$

is inductively defined as follows:

- $[\![x]\!]_{\Sigma}^{I\rho} = \rho(x);$

- $[\![c]\!]_{\Sigma}^{I\rho} = c_0^{\mathsf{F}};$

- $[\![f(t_1,\ldots,t_n)]\!]_{\Sigma}^{I\rho} = f_n^{\mathsf{F}}([\![t_1]\!]_{\Sigma}^{I\rho},\ldots,[\![t_n]\!]_{\Sigma}^{I\rho}).$

This denotation map is structural in the sense that the denotation of a term is functionally dependent on the denotation of its subterms. Indeed, the denotation of a term $f(t_1,\ldots,t_n)$ is the value that is obtained by applying the denotation f_n^{F} of the function symbol f to the denotations of terms $t_1,\ldots,t_n$. The denotation of a variable seen as an atomic term is the value that the assignment gives to the variable.

Satisfaction can be defined at two levels: local satisfaction when an interpretation structure and an assignment are given and global satisfaction when only an interpretation structure is given. Global satisfaction holds when local satisfaction holds for all possible assignments.

Local satisfaction of a formula by an interpretation structure and an assignment is inductively defined as follows:

- $I\rho \Vdash_{\Sigma} p(t_1,\ldots,t_n)$ if $p_n^{\mathsf{P}}([\![t_1]\!]_{\Sigma}^{I\rho},\ldots,[\![t_n]\!]_{\Sigma}^{I\rho}) = 1;$

- $I\rho \Vdash_{\Sigma} (\neg\, \varphi)$ if $I\rho \not\Vdash_{\Sigma} \varphi;$

- $I\rho \Vdash_{\Sigma} (\varphi \Rightarrow \psi)$ if $I\rho \not\Vdash_{\Sigma} \varphi$ or $I\rho \Vdash_{\Sigma} \psi;$

- $I\rho \Vdash_{\Sigma} (\forall x\, \varphi)$ if $I\rho' \Vdash_{\Sigma} \varphi$ for every assignment $\rho' \in x\text{-equiv}(\rho).$

When $I\rho \Vdash_{\Sigma} \varphi$ one says that I together with ρ *satisfies* φ. In this situation one may also say that φ is *true* at I for ρ.

For a given interpretation structure I over Σ, the *formula denotation* map

$$[\![\_]\!]_{\Sigma}^{I} : L_{\Sigma} \to D^X$$

is easily defined using local satisfaction:

$$\llbracket \varphi \rrbracket_\Sigma^I = \{\rho \in D^X : I\rho \Vdash_\Sigma \varphi\}.$$

This denotation map is structural. Indeed:

- $\llbracket (\neg\varphi) \rrbracket_\Sigma^I = D^X \setminus \llbracket \varphi \rrbracket_\Sigma^I$;

- $\llbracket (\varphi \Rightarrow \psi) \rrbracket_\Sigma^I = (D^X \setminus \llbracket \varphi \rrbracket_\Sigma^I) \cup \llbracket \psi \rrbracket_\Sigma^I$;

- $\llbracket (\forall x\, \varphi) \rrbracket_\Sigma^I = \{\rho \in D^X : x\text{-equiv}(\rho) \subseteq \llbracket \varphi \rrbracket_\Sigma^I\}$.

Exercise 8 Using the abbreviations define local satisfaction and formula denotation for conjunction, disjunction and existential quantification.

Global satisfaction (in short, *satisfaction*) of a formula by an interpretation structure is defined as follows: I is said to satisfy φ, written

$$I \Vdash_\Sigma \varphi,$$

if $I\rho \Vdash_\Sigma \varphi$ for every assignment ρ into I. In this situation, I is said to be a *model* of φ, or, more colloquially, one may say that φ is *true* at I. These notions extend to sets of formulas in the expected way (satisfaction of every formula in the set).

A formula φ is said to be *valid*, written

$$\vDash_\Sigma \varphi,$$

if $I \Vdash_\Sigma \varphi$ for every interpretation structure I over signature Σ. In this situation one may say that φ is universally true, since it is true for all possible interpretations of the function and predicate symbols and for all possible values of the variables.

Exercise 9 Show that the following formulas are valid:

1. $((\forall x(\varphi_1 \wedge \varphi_2)) \Leftrightarrow ((\forall x\, \varphi_1) \wedge (\forall x\, \varphi_2)))$.

2. $((\exists x(\varphi_1 \vee \varphi_2)) \Leftrightarrow ((\exists x\, \varphi_1) \vee (\exists x\, \varphi_2)))$.

3. $((\forall x\, \varphi) \Rightarrow (\exists x\, \varphi))$.

4. $((\forall x(\varphi_1 \vee \varphi_2)) \Rightarrow ((\forall x\, \varphi_1) \vee (\exists x\, \varphi_2)))$.

5. $(\exists x(\varphi \Rightarrow (\forall x\, \varphi)))$.

6. $((\exists x(\forall y\, \varphi)) \Rightarrow (\forall y(\exists x\, \varphi)))$.

7. $((\forall x(\varphi_1 \Rightarrow \varphi_2)) \Rightarrow ((\forall x\, \varphi_1) \Rightarrow (\forall x\, \varphi_2)))$

Exercise 10 Given $\Gamma \subseteq L_\Sigma$, let $\mathcal{M}(\Gamma)$ be the class of interpretation structures over Σ that satisfy Γ.

- Show that $\mathcal{M}(\Gamma_2) \subseteq \mathcal{M}(\Gamma_1)$ whenever $\Gamma_1 \subseteq \Gamma_2$.

- Evaluate $\mathcal{M}(\Gamma)$ in terms of $\mathcal{M}(\{\gamma\})$ for each $\gamma \in \Gamma$.

The following exercises concern formulas that are not valid. Showing that a formula is not valid requires that a particular interpretation structure must be provided that does not satisfy the formula for a particular value of the variables.

Exercise 11 Show that the following formulas are not valid:

1. $((\exists x(\varphi_1 \wedge \varphi_2)) \Leftrightarrow ((\exists x\, \varphi_1) \wedge (\exists x\, \varphi_2)))$.

2. $((\forall x(\varphi_1 \vee \varphi_2)) \Leftrightarrow ((\forall x\, \varphi_1) \vee (\forall x\, \varphi_2)))$.

3. $((\forall y(\exists x\, \varphi)) \Rightarrow (\exists x(\forall y\, \varphi)))$.

Exercise 12 Show that if the empty domain were allowed, then the formula

$$((\forall x\, \varphi) \Rightarrow (\exists x\, \varphi))$$

would no longer be valid.

A formula φ over Σ is (globally) *entailed* by a set Γ of formulas over the same signature, written

$$\Gamma \vDash_\Sigma \varphi,$$

if, for every interpretation structure I over Σ, $I \Vdash_\Sigma \varphi$ whenever $I \Vdash_\Sigma \gamma$ for each $\gamma \in \Gamma$.

This notion of entailment is based on global satisfaction. It is possible to introduce a local notion of entailment based on local satisfaction (see Exercise 36 in Section 7.4). As will be seen in the remainder of this chapter and in Chapter 8, (global) entailment is the semantic counterpart of the symbolic notion of derivability.

Clearly, $\vDash_\Sigma \varphi$ if and only if $\emptyset \vDash_\Sigma \varphi$. One may write

$$\gamma_1, \ldots, \gamma_n \vDash_\Sigma \varphi$$

instead of $\{\gamma_1, \ldots, \gamma_n\} \vDash_\Sigma \varphi$.

The *semantic closure* of a set $\Gamma \subseteq L_\Sigma$ is the set

$$\Gamma^{\vDash_\Sigma} = \{\varphi \in L_\Sigma : \Gamma \vDash_\Sigma \varphi\}$$

of its entailed formulas. The following proposition shows that $\lambda\, \Gamma\,.\, \Gamma^{\vDash_\Sigma}$ is indeed a closure operator.

Proposition 13 For every $\Gamma, \Psi \subseteq L_\Sigma$:

1. *Extensivity*: $\Gamma \subseteq \Gamma^{\vDash_\Sigma}$.

2. *Idempotence*: $(\Gamma^{\vDash_\Sigma})^{\vDash_\Sigma} \subseteq \Gamma^{\vDash_\Sigma}$.

3. *Monotonicity*: $\Psi^{\vDash_\Sigma} \subseteq \Gamma^{\vDash_\Sigma}$ whenever $\Psi \subseteq \Gamma$.

The semantic entailment is also a compact operator as will be seen in Chapter 8, using the soundness and completeness of the Hilbert calculus. It is possible to establish the compactness of entailment by more direct means (using ultraproducts), but that falls outside the scope of this book. The interested reader should consult for instance [2, 35].

Exercise 14 Show that the operator $\vDash_\Sigma$ is not a Kuratowski operator.

7.3 Basic lemmas

Some semantic lemmas, mostly of technical nature, are stated and proved in this section. They pave the way for proving the soundness of the Hilbert calculus with respect to the Tarskian semantics. The first lemma states a very intuitive fact: the value given by an assignment to a variable that does not occur in a term should not affect the denotation of the term.

Proposition 15 (Lemma of absent variables in term)
For every $(X \setminus \mathrm{var}_\Sigma(t))$-equivalent assignments ρ, ρ',

$$\llbracket t \rrbracket_\Sigma^{I\rho'} = \llbracket t \rrbracket_\Sigma^{I\rho}.$$

Proof: The proof is easily carried out by induction on the structure of the term and is left as an exercise to the reader.   QED

It is worthwhile to remark here that two assignments are $(X \setminus Y)$-equivalent iff they coincide at least on Y.

The second lemma is also intuitive: the local satisfaction of a formula should not be affected by the value that the assignment gives to the variables that do not occur free in the formula.

Proposition 16 (Lemma of absent free variables in formula)
For every $(X \setminus \mathrm{fv}_\Sigma(\varphi))$-equivalent assignments ρ, ρ',

$$I\rho' \Vdash_\Sigma \varphi \quad \text{if and only if} \quad I\rho \Vdash_\Sigma \varphi.$$

Proof: The proof is by induction on the structure of φ.

(Basis) φ is $p(t_1, \ldots, t_n)$. Then:

$$\begin{aligned}
&I\rho' \Vdash_\Sigma p(t_1, \ldots, t_n) &&\text{if and only if} \\
&p_n^{\mathsf{P}}(\llbracket t_1 \rrbracket_\Sigma^{I\rho'}, \ldots, \llbracket t_n \rrbracket_\Sigma^{I\rho'}) = 1 &&\text{if and only if } (*) \\
&p_n^{\mathsf{P}}(\llbracket t_1 \rrbracket_\Sigma^{I\rho}, \ldots, \llbracket t_n \rrbracket_\Sigma^{I\rho}) = 1 &&\text{if and only if} \\
&I\rho \Vdash_\Sigma p(t_1, \ldots, t_n)\,.
\end{aligned}$$

(*) The lemma of absent variables in terms can be applied since the assignments ρ and ρ' are $(X \setminus \mathrm{var}_\Sigma(t_i))$-equivalent for each $i = 1, \ldots, n$. Indeed, $\mathrm{fv}_\Sigma(p(t_1, \ldots, t_n)) = \bigcup_{i=1}^n \mathrm{var}_\Sigma(t_i)$.

(Step) There are three cases to consider:
(1) φ is $(\neg\,\psi)$. This is left as an exercise.
(2) φ is $(\psi_1 \Rightarrow \psi_2)$. This is left as an exercise.
(3) φ is $(\forall x\, \psi)$. It is only shown that if $I\rho' \Vdash_\Sigma (\forall x\, \psi)$ then $I\rho \Vdash_\Sigma (\forall x\, \psi)$. The proof of the converse is left as an exercise. Suppose that $I\rho' \Vdash_\Sigma (\forall x\, \psi)$. In order to conclude that $I\rho \Vdash_\Sigma (\forall x\, \psi)$, one needs to show that $I\sigma \Vdash_\Sigma \psi$ for every assignment σ in $x\text{-equiv}(\rho)$.

Let σ be an arbitrary assignment in $x\text{-equiv}(\rho)$. Consider the assignment

$$\sigma' = \lambda\, y\,.\; \begin{cases} \rho'(y) & \text{if } y \text{ is not } x \\ \sigma(x) & \text{otherwise.} \end{cases}$$

Clearly, σ' is x-equivalent to ρ' and, so,

$$(\dagger) \qquad I\sigma' \Vdash_\Sigma \psi$$

since $I\rho' \Vdash_\Sigma (\forall x\, \psi)$. Furthermore:

- σ' is $(X \setminus \mathrm{fv}_\Sigma(\varphi))$-equivalent to ρ', since $\{x\} \subset (X \setminus \mathrm{fv}_\Sigma(\varphi))$ and σ' is x-equivalent to ρ';

- ρ' is $(X \setminus \mathrm{fv}_\Sigma(\varphi))$-equivalent to ρ, by hypothesis;

- ρ is $(X \setminus \mathrm{fv}_\Sigma(\varphi))$-equivalent to σ, since $\{x\} \subset (X \setminus \mathrm{fv}_\Sigma(\varphi))$ and ρ is x-equivalent to σ.

Hence, by transitivity, σ' is $(X \setminus \mathrm{fv}_\Sigma(\varphi))$-equivalent to σ and, so,

$$(\ddagger) \qquad \sigma' \text{ is } (X \setminus \mathrm{fv}_\Sigma(\psi))\text{-equivalent to } \sigma$$

since

- $\sigma'(x) = \sigma(x)$, by construction of σ';

- $\mathrm{fv}_\Sigma(\psi) \subseteq \mathrm{fv}_\Sigma(\varphi) \cup \{x\}$.

Therefore, from (†) and (‡), using the induction hypothesis, $I\sigma \Vdash_\Sigma \psi$, as envisaged. QED

The lemma of the absent free variables in formula will be used in the next section to prove the soundness of the axioms in Ax5$_\Sigma$.

Given an interpretation structure I and a formula φ it is possible that $I \nVdash \varphi$ and $I \nVdash (\neg\varphi)$. However, when φ is a closed formula, either $I \Vdash \varphi$ or $I \Vdash (\neg\varphi)$, but not both. The following lemma establishes several such semantic properties of closed formulas that are not shared by formulas with free variables.

Proposition 17 (Lemma of closed formula – LCF)
Let φ and ψ be closed formulas. Then:

1. $I\rho \Vdash_\Sigma \varphi$ if and only if $I\rho' \Vdash_\Sigma \varphi$ for every pair of assignments ρ and ρ' into I;

2. $I \Vdash_\Sigma \varphi$ if and only if $I\rho \Vdash_\Sigma \varphi$ for some assignment ρ into I;

3. $I \Vdash_\Sigma \varphi$ or and only or $I \Vdash_\Sigma (\neg\varphi)$;

4. $I \Vdash_\Sigma (\neg\varphi)$ if and only if $I \nVdash_\Sigma \varphi$;

5. $I \Vdash_\Sigma (\varphi \Rightarrow \psi)$ if and only if $I \nVdash_\Sigma \varphi$ or $I \Vdash_\Sigma \psi$.

Proof: Observe that 1 is a direct corollary of the the lemma of absent free variables in formula. The rest of the proof is left as an exercise. QED

In other words, the law of excluded middle holds for (global) satisfaction of closed formulas in first-order logic. However, it does not hold in general. The reader is invited to provide counter-examples. On the other hand, it is immediate from the definition that the law of excluded middle holds for local satisfaction of any formula.

The following lemmas state technical results on the interplay between substitutions and assignments for terms and formulas. The result on formulas will be essential to prove the soundness of the axioms in Ax4$_\Sigma$.

Proposition 18 (Lemma of substitution in term)
Let $y_1, \ldots, y_m$ be distinct variables. Then:

$$[\![t]\!]_\Sigma^{I\rho'} = [\![t]_{u_1,\ldots,u_m}^{y_1,\ldots,y_m}]\!]_\Sigma^{I\rho}$$

for all $\{y_1, \ldots, y_m\}$-equivalent assignments ρ, ρ' such that

$$\rho'(y_i) = [\![u_i]\!]_\Sigma^{I\rho}$$

for each $1 \le i \le m$.

Proof: The result is proved by straightforward induction on the structure of term t and is left as an exercise to the interested reader.   QED

Proposition 19 (Lemma of substitution in formula)
Let $y_1, \ldots, y_m$ be distinct variables and $u_i \rhd_\Sigma y_i : \varphi$ for each $i = 1, \ldots, m$. Then:

$$I\rho' \Vdash_\Sigma \varphi \text{ if and only if } I\rho \Vdash_\Sigma [\varphi]^{y_1,\ldots,y_m}_{u_1,\ldots,u_m}$$

for every $\{y_1, \ldots, y_m\}$-equivalent assignments ρ, ρ' such that

$$\rho'(y_i) = [\![u_i]\!]^{I\rho}_\Sigma$$

for each $1 \leq i \leq m$.

Proof: The result is shown by induction on the structure of φ. Only the case $m = 1$ is addressed for simplicity of notation.
(Basis) φ is $p(t_1, \ldots, t_n)$:

$$
\begin{array}{ll}
I\rho' \Vdash_\Sigma p(t_1, \ldots, t_n) & \text{if and only if} \\
p_n^\mathsf{P}([\![t_1]\!]^{I\rho'}_\Sigma, \ldots, [\![t_n]\!]^{I\rho'}_\Sigma) = 1 & \text{if and only if } (*) \\
p_n^\mathsf{P}([\![[t_1]^y_u]\!]^{I\rho}_\Sigma, \ldots, [\![[t_n]^y_u]\!]^{I\rho}_\Sigma) = 1 & \text{if and only if} \\
I\rho \Vdash_\Sigma [p(t_1, \ldots, t_n)]^y_u &
\end{array}
$$

(*) Using the lemma of substitution in term.
(Step) There are three steps to consider.
(1) φ is $(\neg\,\psi)$:

$$
\begin{array}{ll}
I\rho' \Vdash_\Sigma (\neg\,\psi) & \text{if and only if} \\
I\rho' \nVdash_\Sigma \psi & \text{if and only if } (\text{IH}) \\
I\rho \nVdash_\Sigma [\psi]^y_u & \text{if and only if} \\
I\rho \Vdash_\Sigma (\neg[\psi]^y_u) & \text{if and only if} \\
I\rho \Vdash_\Sigma [(\neg\,\psi)]^y_u &
\end{array}
$$

(2) φ is $(\psi_1 \Rightarrow \psi_2)$:

$$
\begin{array}{ll}
I\rho' \Vdash_\Sigma (\psi_1 \Rightarrow \psi_2) & \text{if and only if} \\
I\rho' \nVdash_\Sigma \psi_1 \text{ or } I\rho' \Vdash_\Sigma \psi_2 & \text{if and only if } (\text{IH}) \\
I\rho \nVdash_\Sigma [\psi_1]^y_u \text{ or } I\rho \Vdash_\Sigma [\psi_2]^y_u & \text{if and only if} \\
I\rho \Vdash_\Sigma ([\psi_1]^y_u \Rightarrow [\psi_2]^y_u) & \text{if and only if} \\
I\rho \Vdash_\Sigma [(\psi_1 \Rightarrow \psi_2)]^y_u &
\end{array}
$$

(3) φ is $(\forall x\,\psi)$:
Then, since $u \rhd_\Sigma y : (\forall x\,\psi)$, there are three cases to consider:
(i) y is x:

$$
\begin{array}{ll}
I\rho \Vdash_\Sigma [(\forall x\,\psi)]^y_u & \text{if and only if} \\
I\rho \Vdash_\Sigma (\forall x\,\psi) & \text{if and only if } (*) \\
I\rho' \Vdash_\Sigma (\forall x\,\psi) &
\end{array}
$$

(*) By the lemma of absent free variables in formula, since $y \notin \mathrm{fv}_\Sigma((\forall x\,\psi))$ in this case.

(ii) y is not x, $u \vartriangleright_\Sigma y : \psi$ and $y \notin \mathrm{fv}_\Sigma(\psi)$:

$$\begin{array}{lll} I\rho \Vdash_\Sigma [(\forall x\,\psi)]^y_u & \text{if and only if} & (\text{since } y \text{ is not } x) \\ I\rho \Vdash_\Sigma (\forall x\,[\psi]^y_u) & \text{if and only if} & (\text{since } y \notin \mathrm{fv}_\Sigma(\psi)) \\ I\rho \Vdash_\Sigma (\forall x\,\psi) & \text{if and only if} & (*) \\ I\rho' \Vdash_\Sigma (\forall x\,\psi) & & \end{array}$$

(*) By the lemma of absent free variables in formula, since $y \notin \mathrm{fv}_\Sigma((\forall x\,\psi))$ in this case.

(iii) y is not x, $u \vartriangleright_\Sigma y : \psi$ and $x \notin \mathrm{var}_\Sigma(u)$:

($\rightarrow$) Towards showing by contraposition that if $I\rho' \Vdash_\Sigma (\forall x\,\psi)$ then $I\rho \Vdash_\Sigma [(\forall x\,\psi)]^y_u$, suppose that $I\rho \nVdash_\Sigma [(\forall x\,\psi)]^y_u$. Then, $I\rho \nVdash_\Sigma (\forall x[\psi]^y_u)$, since in this case y is not x, and, so, one can find an assignment σ in x-equiv(ρ) such that $I\sigma \nVdash_\Sigma [\psi]^y_u$.

Let σ' be the assignment y-equivalent to σ such that

$$\sigma'(y) = \llbracket u \rrbracket^{I\sigma}_\Sigma.$$

Therefore, by the induction hypothesis, $I\sigma' \nVdash_\Sigma \psi$, since in this case $u \vartriangleright_\Sigma y : \psi$.

Finally, it must be shown that σ' is x-equivalent to ρ', which allows one to conclude that $I\rho' \nVdash_\Sigma (\forall x\,\psi)$. Indeed, the following hold:

- σ' is $\{x, y\}$-equivalent to σ, since σ' is y-equivalent to σ;

- σ is $\{x, y\}$-equivalent to ρ, since σ is x-equivalent to ρ;

- ρ is $\{x, y\}$-equivalent to ρ', since by hypothesis ρ' is y-equivalent to ρ.

Hence, by transitivity, σ' is $\{x, y\}$-equivalent to ρ'. Moreover,

$$\rho'(y) = \llbracket u \rrbracket^{I\rho}_\Sigma = \llbracket u \rrbracket^{I\sigma}_\Sigma = \sigma'(y)$$

due to $\llbracket u \rrbracket^{I\rho}_\Sigma = \llbracket u \rrbracket^{I\sigma}_\Sigma$, which holds by the lemma of absent variables in term, since in this case $x \notin \mathrm{var}_\Sigma(u)$. Thus, as required, σ' is x-equivalent to ρ'.

($\leftarrow$) The proof of the converse assertion is done in a similar way, again by contraposition. Suppose that $I\rho' \nVdash_\Sigma (\forall x\,\psi)$. Then, one can find an assignment σ' in x-equiv(ρ') such that $I\sigma' \nVdash'_\Sigma \psi$. Note that

$$\sigma'(y) = \rho'(y) = \llbracket u \rrbracket^{I\rho}_\Sigma$$

because in this case y is not x.

Let σ be the assignment y-equivalent to σ' such that $\sigma(y) = \rho(y)$. Then, σ is x-equivalent to ρ. Furthermore, $\sigma'(y) = [\![u]\!]_\Sigma^{I\sigma}$, since $[\![u]\!]_\Sigma^{I\sigma} = [\![u]\!]_\Sigma^{I\rho}$, applying the lemma of absent variables in term, which is allowed since in this case $x \notin \mathrm{var}_\Sigma(u)$. Hence, by the induction hypothesis, it follows that $I\sigma \not\Vdash_\Sigma [\psi]_u^y$.

Finally, since σ is x-equivalent to ρ, $I\rho \not\Vdash_\Sigma (\forall x[\psi]_u^y)$, whence $I\rho \not\Vdash_\Sigma [(\forall x\psi)]_u^y$, taking again into account that in this case y is not x.   QED

The following exercise concerns a sufficient condition for the legitimate substitution of bound variables. It is the semantic counterpart of Proposition 33 of Chapter 5.

Exercise 20
Let $x, y \in X$ and $\varphi \in L_\Sigma$ be such that:

- $y \notin \mathrm{fv}_\Sigma(\varphi)$;

- $y \rhd_\Sigma x : \varphi$.

Then, the formula $((\forall x\,\varphi) \Leftrightarrow (\forall y\,[\varphi]_y^x))$ is valid.

The objective now is to show that certain formulas of first-order logic are instances of propositional formulas. This fact has two main advantages. One is to see that first-order logic extends propositional logic. But, more importantly, some knowledge about true propositional formulas can be imported.

A signature Π of propositional logic is a countable nonempty set (of *propositional variables*). The *alphabet* induced by Π contains Π, the parentheses and the connectives $\neg$ and $\Rightarrow$. The set L_Π of (propositional) *formulas* over Π is inductively defined as expected:[1]

- $\pi \in L_\Pi$ for each $\pi \in \Pi$;

- $(\neg\beta) \in L_\Pi$ whenever $\beta \in L_\Pi$;

- $(\beta_1 \Rightarrow \beta_2) \in L_\Pi$ whenever $\beta_1, \beta_2 \in L_\Pi$.

As in first-order logic, the connectives $\wedge$, $\vee$ and $\Leftrightarrow$ can be introduced as abbreviations.

Propositional interpretations, known as valuations, are very simple: they assign truth values to the propositional variables. More precisely, a *valuation* over a propositional signature Π is a map $v : \Pi \to \{0, 1\}$. *Satisfaction* of a formula $\beta \in L_\Pi$ by a valuation v, written $v \Vdash_\Pi \beta$, is defined, inductively, as follows:

- $v \Vdash_\Pi \pi$ if $v(\pi) = 1$;

---

[1] Adopting the traditional notation used by logicians.

- $v \Vdash_\Pi (\neg \beta)$ if $v \nVdash_\Pi \beta$;

- $v \Vdash_\Pi (\beta_1 \Rightarrow \beta_2)$ if $v \nVdash_\Pi \beta_1$ or $v \Vdash_\Pi \beta_2$.

A formula $\beta \in L_\Pi$ is said to be a *tautology*, written $\vDash_\Pi \beta$, when $v \Vdash_\Pi \beta$ for every valuation v over Π.

Let $\mathrm{var}_\Pi(\beta)$ be the set of propositional variables that occur in β. As the reader might expect, a result on absent propositional variables holds and is easily proved by induction on the structure of the formula:

Proposition 21 Let v_1 and v_2 be two valuations over a propositional signature Π and $\beta \in L_\Pi$. Assume that $v_1(\pi) = v_2(\pi)$ for every $\pi \in \mathrm{var}_\Pi(\beta)$. Then,

$$v_1 \Vdash_\Pi \beta \quad \text{if and only if} \quad v_2 \Vdash_\Pi \beta.$$

As a consequence of this result, for testing if β is a tautology it is necessary to consider only the restriction of the valuations to $\mathrm{var}_\Pi(\beta)$. Clearly, the number of restrictions that have to be considered is $2^{\#\mathrm{var}_\Pi(\beta)}$.

For instance, $(\pi_1 \Rightarrow (\pi_2 \Rightarrow \pi_1))$ is a tautology because it is satisfied by the four possible restrictions to $\{\pi_1, \pi_2\}$ of the valuations.

Since it is so simple to check if a propositional formula is a tautology, it is worthwhile to be able to recognize when a first-order formula is an instance of a propositional formula, that is, when it is obtained by uniformly replacing in a propositional formula each propositional variable by a first-order formula. Indeed, as it will be established in due course, a first-order formula is valid whenever it is an instance of a tautology.

Let Π be a propositional signature. An *instantiation* of Π for Σ is a map $\mu : \Pi \to L_\Sigma$. Such an instantiation is canonically extended to $\hat{\mu} : L_\Pi \to L_\Sigma$ as follows:

- $\hat{\mu}(\pi) = \mu(\pi)$;

- $\hat{\mu}((\neg \alpha)) = (\neg \hat{\mu}(\alpha))$;

- $\hat{\mu}((\alpha \Rightarrow \beta)) = (\hat{\mu}(\alpha) \Rightarrow \hat{\mu}(\beta))$.

In first-order logic, a formula φ over Σ is said to be a *tautological formula* if there are a propositional signature Π, an instantiation μ of Π for Σ and a tautology α over Π such that $\hat{\mu}(\alpha) = \varphi$.

For example, consider some propositional signature Π containing π_1 and π_2, the tautology $(\pi_1 \Rightarrow (\pi_2 \Rightarrow \pi_1)) \in L_\Pi$ and an instantiation $\mu : \Pi \to L_\Sigma$ such that $\mu(\pi_1) = (\forall x_1\, p(x_1))$ and $\mu(\pi_2) = q(x_1, x_2, x_3)$. Then,

$$\hat{\mu}(\pi_1 \Rightarrow (\pi_2 \Rightarrow \pi_1)) = ((\forall x_1\, p(x_1)) \Rightarrow (q(x_1, x_2, x_3) \Rightarrow (\forall x_1\, p(x_1)))),$$

showing that the latter is a tautological formula of L_Σ.

Proposition 22 (Lemma of tautological formula)
Every tautological formula is valid.

Proof: Suppose that φ is a tautological formula. Then, there exist a propositional signature Π, an instantiation $\mu : \Pi \to L_\Sigma$ and a tautology α over Π such that $\hat{\mu}(\alpha) = \varphi$. Let I be an arbitrary interpretation structure over Σ and ρ be an arbitrary assignment into I. Consider the valuation $v_{I\rho} : \Pi \to \{0, 1\}$ such that:

$$v_{I\rho}(\pi) = \begin{cases} 1 & \text{if } I\rho \Vdash_\Sigma \mu(\pi) \\ 0 & \text{otherwise.} \end{cases}$$

Then, $I\rho \Vdash_\Sigma \hat{\mu}(\beta)$ if and only if $v_{I\rho} \Vdash_\Pi \beta$ for every $\beta \in L_\Pi$. The proof, by induction on the structure of β, is left as an exercise. Since α is a tautology, $v_{I\rho} \Vdash_\Pi \alpha$ and, so, $I\rho \Vdash_\Sigma \hat{\mu}(\alpha)$. Therefore, $I\rho \Vdash_\Sigma \varphi$ for arbitrary I and ρ. Thus, φ is valid.   QED

Exercise 23 Show that every axiom in $\text{Ax}1_\Sigma \cup \text{Ax}2_\Sigma \cup \text{Ax}3_\Sigma$ is valid.

Exercise 24 (Lemma of modus ponens)
Show that:

$$\varphi, (\varphi \Rightarrow \psi) \vDash_\Sigma \psi \,.$$

Exercise 25 (Lemmas of instantiation and generalization)
Show that:

- $I \Vdash_\Sigma \varphi$ if and only if $I \Vdash_\Sigma (\forall x \, \varphi)$ if and only if $I \Vdash_\Sigma (\forall \, \varphi)$;

- $(\forall x \, \varphi) \vDash_\Sigma \varphi$;

- $\varphi \vDash_\Sigma (\forall x \, \varphi)$.

Exercise 26 Show, by providing counter-examples, that the following results do not hold in general:

- $\vDash_\Sigma ((\forall x \, \varphi) \Rightarrow [\varphi]_t^x)$;

- $\vDash_\Sigma ((\forall x \, (\varphi \Rightarrow \psi)) \Rightarrow (\varphi \Rightarrow (\forall x \, \psi)))$;

- $\varphi \vDash_\Sigma \psi$ if and only if $\vDash_\Sigma (\varphi \Rightarrow \psi)$.

7.4   Soundness of the Hilbert calculus

By soundness of a calculus it is meant that what is derived should be entailed. Soundness of the Hilbert calculus presented in Chapter 5 is established below as follows:

- first, the axioms are proved to be valid – soundness of the axioms;

- second, for each inference rule, the conclusion is proved to be entailed by the set of premises – soundness of the inference rules;

- finally, $\Gamma^{\vdash_\Sigma} \subseteq \Gamma^{\vDash_\Sigma}$ is proved for arbitrary $\Gamma \subseteq L_\Sigma$ – soundness of the calculus.

Proposition 27 (Lemma of soundness of the axioms)
The axioms of the Hilbert calculus are valid:

1. $\vDash_\Sigma (\varphi \Rightarrow (\psi \Rightarrow \varphi))$;

2. $\vDash_\Sigma ((\varphi \Rightarrow (\psi \Rightarrow \delta)) \Rightarrow ((\varphi \Rightarrow \psi) \Rightarrow (\varphi \Rightarrow \delta)))$;

3. $\vDash_\Sigma (((\neg \varphi) \Rightarrow (\neg \psi)) \Rightarrow (\psi \Rightarrow \varphi))$;

4. $\vDash_\Sigma ((\forall x\, \varphi) \Rightarrow [\varphi]_t^x)$ provided that $t \rhd_\Sigma x : \varphi$;

5. $\vDash_\Sigma ((\forall x\, (\varphi \Rightarrow \psi)) \Rightarrow (\varphi \Rightarrow (\forall x\, \psi)))$ provided that $x \notin \mathrm{fv}_\Sigma(\varphi)$.

Proof: Thanks to Exercise 23, it remains only to prove the validity of the axioms in $\mathrm{Ax4}_\Sigma \cup \mathrm{Ax5}_\Sigma$.

(Ax4) Let I be an arbitrary interpretation structure over Σ and ρ be an arbitrary assignment into I. One has to show that $I\rho \Vdash_\Sigma ((\forall x\, \varphi) \Rightarrow [\varphi]_t^x)$ if t is free for x in φ. There are two cases to consider:

(i) $I\rho \nVdash_\Sigma (\forall x\, \varphi)$. Then, $I\rho \Vdash_\Sigma ((\forall x\, \varphi) \Rightarrow [\varphi]_t^x)$.

(ii) $I\rho \Vdash_\Sigma (\forall x\, \varphi)$. In this case it is necessary to establish that $I\rho \Vdash_\Sigma [\varphi]_t^x$ in order to conclude that $I\rho \Vdash_\Sigma ((\forall x\, \varphi) \Rightarrow [\varphi]_t^x)$. To this end, consider the assignment ρ' that is x-equivalent to ρ and such that $\rho'(x) = [\![t]\!]_\Sigma^{I\rho}$. Then, $I\rho' \Vdash_\Sigma \varphi$, whence, using the lemma of substitution in formula, which applies since t is free for x in φ, it follows that $I\rho \Vdash_\Sigma [\varphi]_t^x$.

(Ax5) The proof is by contradiction. Assume that there exist an interpretation structure I over Σ and an assignment ρ into I such that

$$I\rho \nVdash_\Sigma ((\forall x\, (\varphi \Rightarrow \psi)) \Rightarrow (\varphi \Rightarrow (\forall x\, \psi)))$$

where $x \notin \mathrm{fv}_\Sigma(\varphi)$. Then:

$$\begin{cases} (1) & I\rho \Vdash_\Sigma (\forall x\, (\varphi \Rightarrow \psi)) \\ (2) & I\rho \nVdash_\Sigma (\varphi \Rightarrow (\forall x\, \psi)). \end{cases}$$

From (2) one concludes that

$$\begin{cases} (2.1) & I\rho \Vdash_\Sigma \varphi \\ (2.2) & I\rho \nVdash_\Sigma (\forall x\, \psi). \end{cases}$$

From (2.2) it follows that there is an assignment σ x-equivalent to ρ such that

$$(2.2')\ I\sigma \not\Vdash_\Sigma \psi\,.$$

Since $x \notin \mathrm{fv}_\Sigma(\varphi)$, applying the lemma of absent free variables in formula to (2.1) yields

$$(2.1')\ I\sigma \Vdash_\Sigma \varphi\,.$$

Therefore, from (2.1') and (2.2') one obtains

$$I\sigma \not\Vdash_\Sigma (\varphi \Rightarrow \psi)$$

and hence

$$I\rho \not\Vdash_\Sigma (\forall x\,(\varphi \Rightarrow \psi)),$$

contradicting (1).   QED

Observe that the validity of the axioms in $Ax4_\Sigma$ would be at stake without the proviso $t \rhd_\Sigma x : \varphi$, as the following exercise illustrates.

Exercise 28 Show that $((\forall x(\exists y\, p(x, y))) \Rightarrow (\exists y\, p(y, y)))$ is not valid. Hint: Use an interpretation structure I such that:

- D is $\mathbb{N}$;

- $p_2^{\mathrm{P}} : \mathbb{N}^2 \to \{0, 1\}$ is such that $p_2^{\mathrm{P}}(d_0, d_1) = 1$ if and only if $d_0 < d_1$.

A similar observation can be made concerning $Ax5_\Sigma$. The reader should be able to find a counter-example for illustrating that the validity of the axioms in $Ax5_\Sigma$ would be in question without the proviso $x \notin \mathrm{fv}_\Sigma(\varphi)$.

Taking into account the relevant results of the previous sections, the proofs of the next two lemmas are easy and left as exercises to the interested reader.

Proposition 29 (Lemma of soundness of the rules)
The rules of the Hilbert calculus preserve entailment:

1. if $\Gamma \vDash_\Sigma \varphi$ and $\Gamma \vDash_\Sigma (\varphi \Rightarrow \psi)$, then $\Gamma \vDash_\Sigma \psi$;

2. if $\Gamma \vDash_\Sigma \varphi$, then $\Gamma \vDash_\Sigma (\forall x\, \varphi)$.

Proposition 30 (Lemma of soundness of using hypotheses)
If $\varphi \in \Gamma$, then $\Gamma \vDash_\Sigma \varphi$.

Capitalizing on the previous results and on Proposition 8 of Chapter 5, it is now possible to establish the soundness of the calculus.

Proposition 31 (Soundness of Hilbert calculus)

$$\text{If } \Gamma \vdash_\Sigma \varphi, \text{ then } \Gamma \vDash_\Sigma \varphi.$$

Proof: This result is established by induction on the length of a derivation sequence for $\Gamma \vdash_\Sigma \varphi$, using the previous propositions. QED

The following implication is a direct corollary of the soundness of the Hilbert calculus:

$$\text{If } \varphi \text{ is a theorem, then } \varphi \text{ is a valid formula.}$$

This weaker result is known as weak soundness of the Hilbert calculus. For this reason, some authors refer to Proposition 31 as the strong soundness of the Hilbert calculus.

The goal of the next chapter is to prove the converses of the two previous results: every entailment can be derived and validity implies theoremhood (strong and weak completeness of the Hilbert calculus with respect to the Tarskian semantics).

Soundness can be used for proving that a formula is not derivable from a given set of hypotheses, as illustrated in the following exercises.

Exercise 32 Show that

$$((\forall x \, \varphi) \Rightarrow [\varphi]^x_t) \in \emptyset^{\vdash_\Sigma}$$

does not hold in general.

Exercise 33 Recall the metatheorem of deduction. Show that

$$(\eta \Rightarrow \varphi) \in \Gamma^{\vdash_\Sigma} \text{ whenever } \varphi \in (\Gamma \cup \{\eta\})^{\vdash_\Sigma}$$

does not hold in general.

Exercise 34 Recall Exercise 35 of Section 5.4. Show that

$$[\varphi]^x_t \in ([\Gamma]^x_t)^{\vdash_\Sigma} \text{ whenever } \varphi \in \Gamma^{\vdash_\Sigma}$$

does not hold in general.

Soundness provides the means for a semantic proof of the consistency of the Hilbert calculus, as shown in the next exercise.

Exercise 35 (Consistency of Hilbert calculus)

Show that the Hilbert calculus is consistent by proving that there is no formula φ such that

$$\begin{cases} \emptyset \vdash_\Sigma \varphi \\ \emptyset \vdash_\Sigma (\neg\varphi). \end{cases}$$

Discuss this proof of consistency of the Hilbert calculus from the perspective of Hilbert's programme.

A purely symbolic proof of the consistency of the Hilbert calculus is the main objective of Chapter 9.

At this point, the reader may wonder if it is feasible to define a semantic entailment based on local satisfaction. The next exercise addresses this issue.

Exercise 36 Formula φ is said to be locally entailed by Γ, written

$$\Gamma \vDash_\Sigma^\ell \varphi,$$

if $I\rho \Vdash_\Sigma \Gamma$ implies that $I\rho \Vdash_\Sigma \varphi$ for every interpretation structure I over Σ and assignment ρ into I. Check that the (global) Hilbert calculus introduced in Chapter 5 is not sound with respect to local entailment. Propose a sound local Hilbert calculus. Establish the metatheorem of deduction for the local calculus. Hint: Define $\Gamma \vdash_\Sigma^\ell$ like $\Gamma \vdash_\Sigma$ but without the generalization rule and including as axioms all the generalizations of the axioms in Ax_Σ.

7.5   Theory of interpretation structure

Given an interpretation structure I over Σ, the set

$$\mathrm{Th}_\Sigma(I) = \{\gamma \in L_\Sigma : I \Vdash_\Sigma \gamma\}$$

is known as the *theory of* I, since $\mathrm{Th}_\Sigma(I)^{\vdash_\Sigma} = \mathrm{Th}_\Sigma(I)$.

Exercise 37 Show that $\mathrm{Th}_\Sigma(I)$ is a consistent and exhaustive theory.

One of the main goals of this book is to prove that the theory of the interpretation structure corresponding to (a sufficiently rich) arithmetic over the natural numbers is not axiomatizable, since it is not computably enumerable. This fact (a corollary of the first incompleteness theorem by Gödel) is proved in Chapter 13.

Exercise 38 Let $\mathrm{Th}_\Sigma(\mathcal{I})$, where $\mathcal{I}$ is a class of interpretation structures over Σ, be the set of formulas that are satisfied by every interpretation structure in $\mathcal{I}$. Recall the definition of $\mathcal{M}(\Gamma)$ in Exercise 10. Show that

1. $\mathrm{Th}_\Sigma(\mathcal{I})$ is a theory;

2. $\mathrm{Th}_\Sigma(\mathcal{I}') \subseteq \mathrm{Th}_\Sigma(\mathcal{I})$ whenever $\mathcal{I} \subseteq \mathcal{I}'$;

3. $\Gamma \subseteq \mathrm{Th}_\Sigma(\mathcal{I})$ if and only if $\mathcal{I} \subseteq \mathcal{M}(\Gamma)$;

4. $\Gamma \subseteq \mathrm{Th}_\Sigma(\mathcal{M}(\Gamma))$ and $\mathcal{I} \subseteq \mathcal{M}(\mathrm{Th}_\Sigma(\mathcal{I}))$.

Apply the statements above to the class of all groups and the class of all Abelian groups.

Fact 3 in the exercise above is known as the Galois connection between theories and interpretation structures.

7.6 Relating interpretation structures

The objective of this section is to relate different interpretation structures, over the same signature or even over different signatures. Several interpretation structures introduced in this book are seen to be related in significant ways, namely in Chapters 8 and 15.

Given two interpretation structures I and I' over the same signature Σ, one says that:

- I is a *substructure* of I', written

$$I \subseteq I',$$

  if $D \subseteq D'$, $f_n^{\mathsf{F}} = f_n^{\mathsf{F}'}|_D$ and $p_m^{\mathsf{P}} = p_m^{\mathsf{P}'}|_D$ for every $n \in \mathbb{N}$, $m \in \mathbb{N}^+$, $f \in F_n$ and $p \in P_m$.

- I is an *elementary substructure* of I', written

$$I \leq I',$$

  if $I \subseteq I'$ and, for every φ and ρ into I, $I\rho \Vdash_\Sigma \varphi$ if and only if $I'\rho \Vdash_\Sigma \varphi$.

Exercise 39 Give examples of interpretation structures I and I' over the signature of groups discussed in Chapter 4, Exercise 1, such that:

1. $I \subseteq I'$;

2. $I \not\leq I'$.

Exercise 40 Show that the relation of elementary substructure is reflexive and transitive.

Exercise 41 Show that if $I' \leq I$, $I'' \leq I$ and $I' \subseteq I''$, then $I' \leq I''$.

Two interpretation structures I and I' over the same signature Σ are said to be *equivalent*, written

$$I \equiv I',$$

if $\mathrm{Th}_\Sigma(I) = \mathrm{Th}_\Sigma(I')$.

Exercise 42 Show that if $I \leq I'$, then $I \equiv I'$.

The notions above on inclusions are generalized to maps as follows. Given two interpretation structures I and I' over the same signature Σ, a map

$$h : D \to D'$$

is said to be:

- an *homomorphism* from I to I' (written $h : I \to I'$) if:

  - $h(f_n^{\mathsf{F}}(d_1, \ldots, d_n)) = f_n^{\mathsf{F}'}(h(d_1), \ldots, h(d_n))$;
  - $p_n^{\mathsf{P}}(d_1, \ldots, d_n) = p_n^{\mathsf{P}'}(h(d_1), \ldots, h(d_n))$;

- *elementary* if the following condition holds:

$$I\rho \Vdash_\Sigma \varphi \text{ if and only if } I'(h \circ \rho) \Vdash_\Sigma \varphi.$$

The homomorphism condition can be seen as the commutativity of the diagrams in Figures 7.1 and 7.2.

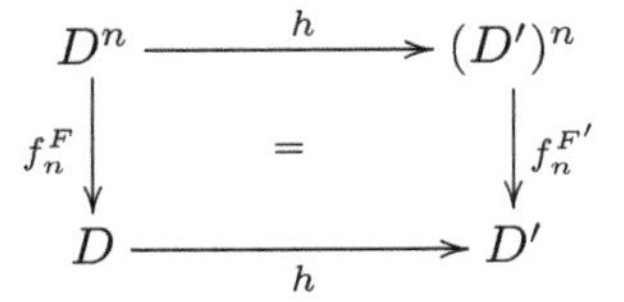

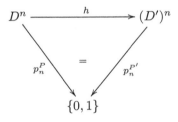

Figure 7.1: Homomorphism condition for function symbol f.

$$
\begin{array}{ccc}
D^n & \xrightarrow{\quad h \quad} & (D')^n \\
 & & \\
p_n^P \searrow & = & \swarrow p_n^{P'} \\
 & \{0,1\} &
\end{array}
$$

Figure 7.2: Homomorphism condition for predicate symbol p.

Exercise 43 Show that not every map that is an homomorphism is elementary.

The next exercise builds a category of interpretation structures over the same signature and establish some of its basic properties. The interested reader should consult, for instance, [34] for an introduction to categories.

Exercise 44 Prove the following results:

1. Interpretation structures over a fixed signature and homomorphisms between them form a category,[2] in the sequel denoted by Int_Σ and known as the *category of interpretation structures* over Σ.

---

[2]Composition of homomorphisms is associative. There is an identity homomorphism for each interpretation structure that is left and right zero for composition.

2. In Int$_\Sigma$, monomorphisms[3], epimorphisms[4] and isomorphisms[5] are the homomorphic maps that are, respectively, injective, surjective and bijective.

The next exercise provides *inter alia* a sufficient condition for an homomorphism $h : I \to I'$ to be elementary, and, assuming that I and I' are over the same signature, a sufficient condition for the image $h(I)$ to be an elementary substructure of I'.

Exercise 45 Prove the following results:

- If $I \subseteq I'$, then the inclusion $D \hookrightarrow D'$ is a homomorphism from I to I'.

- If $h : I \to I'$ is an epimorphism, then the map $h : D \to D'$ is elementary.

- If $h : D \to D'$ is elementary and injective, then the interpretation structure
$$h(I) = (h(D), \_^{h(\mathsf{F})}, \_^{h(\mathsf{P})})$$

where:

  - $f_n^{h(\mathsf{F})}(h(d_1), \ldots, h(d_n)) = h(f_n^{\mathsf{F}}(d_1, \ldots, d_n))$;
  - $p_n^{h(\mathsf{P})}(h(d_1), \ldots, h(d_n)) = p_n^{\mathsf{P}}(d_1, \ldots, d_n)$.

  is an elementary substructure of I'. Furthermore, $h : D \to h(D)$ is an isomorphism.

- If there exists an elementary $h : D \to D'$, then $I \equiv I'$.

The following result is later used for proving the cardinality theorem (see Proposition 17 of Chapter 8). It shows how to construct an interpretation structure I' that extends the domain of a given interpretation structure I while ensuring that $I \leq I'$.

Exercise 46 Let $I = (D, \_^{\mathsf{F}}, \_^{\mathsf{P}})$ be an interpretation structure over Σ, $D' \supseteq D$ and $a \in D$. Consider the map

$$(\_)_a : D' \to D$$

such that

$$(d')_a = \begin{cases} d' & \text{whenever } d' \in D \\ a & \text{otherwise} \end{cases}$$

---

[3] A homomorphism h is a monomorphism if $h \circ f_1 = h \circ f_2$ implies $f_1 = f_2$.
[4] A homomorphism h is an epimorphism if $f_1 \circ h = f_2 \circ h$ implies $f_1 = f_2$.
[5] A homomorphism $h : I \to I'$ is an isomorphism if there exists $h' : I' \to I$ such that $h' \circ h = \mathrm{id}_I$ and $h \circ h' = \mathrm{id}_{I'}$.

1. Consider the triple $I' = (D', \_^{\mathsf{F}'}, \_^{\mathsf{P}'})$ where

   - $f_n^{\mathsf{F}'}(d_1', \dots, d_n') = f_n^{\mathsf{F}}((d_1')_a, \dots, (d_n')_a)$;
   - $p_n^{\mathsf{P}'}(d_1', \dots, d_n') = p_n^{\mathsf{P}}((d_1')_a, \dots, (d_n')_a)$.

   Show that I' is an interpretation structure over Σ.

2. Show that:

   - $f_n^{\mathsf{F}'}|_D = f_n^{\mathsf{F}}$;
   - $p_n^{\mathsf{P}'}|_D = p_n^{\mathsf{P}}$.

3. Given $\rho' : X \to D'$, let $\rho_a' : X \to D$ be such that $\rho_a'(x) = (\rho'(x))_a$. Show that:

   (a) $[\![t]\!]_\Sigma^{I\rho_a'} = ([\![t]\!]_\Sigma^{I'\rho'})_a$;

   (b) $I\rho_a' \Vdash_\Sigma \varphi$ if and only if $I'\rho' \Vdash_\Sigma \varphi$.

4. Classify the relationship between I and I'.

It is also possible to define interesting relationships between interpretation structures over different signatures, namely the concept of cryptomorphism or heteromorphism that will be used in Chapter 8. The notion of cryptomorphism was first introduced by Garrett Birkhoff in [3] for algebras. Herein, it is generalized to interpretation structures.

To this end, it is necessary first to define what one means by a signature morphism. Let $\Sigma = (F, P, \tau)$ and $\Sigma' = (F', P', \tau')$ be signatures. A pair $(\sigma_{\mathsf{F}} : F \to F', \sigma_{\mathsf{P}} : P \to P')$ is said to be a *signature morphism* $\sigma : \Sigma \to \Sigma'$ if $\tau' \circ \sigma_{\mathsf{F}}(f) = \tau(f)$ and $\tau' \circ \sigma_{\mathsf{P}}(p) = \tau(p)$ for every $f \in F$ and $p \in P$.

The *reduct* along $\sigma : \Sigma \to \Sigma'$ of an interpretation structure $I' = (D', \_^{\mathsf{F}'}, \_^{\mathsf{P}'})$ over Σ' is the following interpretation structure over Σ:

$$I'|_\sigma = (D', \_^{\mathsf{F}}, \_^{\mathsf{P}})$$

where

- $f_n^{\mathsf{F}} = (\sigma(f))_n^{\mathsf{F}'}$, for every $f \in F_n$;

- $p_n^{\mathsf{P}} = (\sigma(p))_n^{\mathsf{P}'}$, for every $p \in P_n$.

If σ is an inclusion, one may write $I'|_\Sigma$ for $I'|_\sigma$.

Given interpretation structures I and I' over Σ and Σ', respectively, a *cryptomorphism* from I to I' is a pair (σ, h) where $\sigma : \Sigma \to \Sigma'$ is a signature morphism and $h : I \to I'|_\sigma$ is an homomorphism.

Exercise 47 Given a signature morphism $\sigma : \Sigma \to \Sigma'$, define $\sigma_T : T_\Sigma \to T_{\Sigma'}$ and $\sigma_L : L_\Sigma \to L_{\Sigma'}$ as the canonical extensions of σ to terms and formulas, respectively.

Map σ_T is known as the *term translation map* induced by σ, and σ_L as the *formula translation map* induced by σ. One may write $\sigma(t)$ and $\sigma(\varphi)$ for $\sigma_T(t)$ and $\sigma_L(\varphi)$, respectively.

Exercise 48 Let $\sigma : \Sigma \to \Sigma'$ be a signature morphism and I' an interpretation structure over Σ'. Show that:

1. For every assignment ρ' into I',

   - $[\![\sigma(t)]\!]_{\Sigma'}^{I'\rho'} = [\![t]\!]_{\Sigma}^{I'|_\sigma\rho'}$;
   - $I'\rho' \Vdash_{\Sigma'} \sigma(\varphi)$ if and only if $I'|_\sigma\rho' \Vdash_\Sigma \varphi$.

2. $I' \Vdash_{\Sigma'} \sigma(\varphi)$ if and only if $I'|_\sigma \Vdash_\Sigma \varphi$.

Assertion 2 is called the *satisfaction condition* or the *Barwise condition*. In fact, it was John Barwise that first stated the condition (see [1]).

Exercise 49 Let (σ, h) be a cryptomorphism from I to I'. State and prove a sufficient condition for:

$$I' \Vdash_{\Sigma'} \sigma(\varphi) \text{ if and only if } I \Vdash_\Sigma \varphi.$$

The interested reader should consult [21, 40, 9] for further developments in the context of the theory of institutions, in particular, and universal logic, in general.

Chapter 8

Completeness

The aim of this chapter is to prove the completeness of the Hilbert calculus introduced in Chapter 5 with respect to the Tarskian semantics presented in Chapter 7:

$$\text{if } \Gamma \vDash_\Sigma \varphi, \text{ then } \Gamma \vdash_\Sigma \varphi.$$

In other words, assuming that φ is semantically entailed by Γ one wants to provide a derivation sequence of φ from Γ, a daunting task at first sight. It seems easier to look instead for a proof of the contraposition:

$$\text{if } \Gamma \nvdash_\Sigma \varphi, \text{ then } \Gamma \nvDash_\Sigma \varphi.$$

That is, assuming that there is no derivation of φ from Γ, the objective is to find an interpretation structure I such that I is a model of Γ but not of φ.

The proof of this result relies on a construction by Leon Henkin [23] wherefore, given a consistent set Γ of formulas, one of its models is built as follows:

1. An enriched signature Σ^+ is obtained from the original signature Σ by adding a denumerable set of new constants, and a consistent enrichment $\Gamma^\exists$ of Γ is defined by imposing that for each assertion $(\exists x\, \varphi) \in \Gamma^{\vdash_\Sigma}$ with $\mathrm{fv}_\Sigma(\varphi) = \{x\}$ there is such a constant c such that $[\varphi]_c^x \in (\Gamma^\exists)^{\vdash_{\Sigma^+}}$.

2. An exhaustive and still consistent extension $\widehat{\Gamma}$ of $\Gamma^\exists$ is obtained (using a technique by Adolf Lindenbaum).

3. An interpretation structure I_Γ over Σ^+ is built from $\widehat{\Gamma}$, taking as domain the set of closed terms.

4. The structure I_Γ is shown to satisfy every formula of Γ.

125

5. Finally, the intended model of Γ over the original signature Σ is obtained from I_Γ by forgetting the denotation of the extra constants.

The desired completeness result then follows easily. Indeed, if $\Gamma \not\vDash_\Sigma \varphi$, then

$$\Gamma \cup \{(\neg(\forall\varphi))\}$$

is consistent, and, hence, the Henkin construction is applicable, yielding a model of Γ that does not satisfy φ.

It should be stressed that Henkin's proof of completeness is essentially non constructive, since no derivation sequence is provided for obtaining φ from Γ when $\Gamma \vDash_\Sigma \varphi$. Instead, an interpretation structure I_Γ is built for showing that $\Gamma \not\vDash_\Sigma \varphi$ when $\Gamma \not\vdash_\Sigma \varphi$.

Note also that I_Γ is obtained by Henkin's method in a non effective way. That is, no algorithm is presented for producing it from Γ. The analysis of the computability of Henkin's construction falls outside the scope of this book since it requires a wider knowledge of Computatibility Theory, beyond what was provided in Chapter 3.

Section 8.1 concentrates on the proof of the Lindenbaum lemma starting with some other useful lemmas. Section 8.2 is dedicated to the Henkin construction itself. The major step is the definition of the envisaged interpretation structure I_Γ over Σ^+, from the exhaustive extension $\widehat{\Gamma} \subseteq L_{\Sigma^+}$ of the given consistent set $\Gamma \subseteq L_\Sigma$, so that satisfaction by I_Γ mimics derivability from $\widehat{\Gamma}$ for closed formulas. Section 8.3 presents the proof of the completeness of the Hilbert calculus using the model existence lemma established at the end of Section 8.2. Some corollaries of the completeness theorem are also established, namely the Skolem–Löwenheim theorem, the cardinality theorem and the compactness theorem.

8.1   Lindenbaum's lemma

Before proceeding, it is convenient to extend to arbitrary sets of formulas the notions of consistency and exhaustiveness introduced for theories in Chapter 6.

A set Γ of formulas over Σ is said to be *consistent* with respect to Σ (in short, consistent w.r.t. Σ) if the theory $\Gamma^{\vdash_\Sigma}$ over Σ is consistent, that is, if there exists a closed formula φ over Σ such that $\Gamma \not\vdash_\Sigma \varphi$.

A set Γ of formulas over Σ is said to be *exhaustive* with respect to Σ (in short, exhaustive w.r.t. Σ) if the theory $\Gamma^{\vdash_\Sigma}$ over Σ is exhaustive, that is, if for each closed formula φ over Σ it is the case that either $\Gamma \vdash_\Sigma \varphi$ or $\Gamma \vdash_\Sigma (\neg\varphi)$.

When the signature at hand is clear from the context, one may say that a set of formulas is consistent or exhaustive without stating with respect to which signature.

For any closed formula φ, consistency imposes that φ and $(\neg \varphi)$ cannot both be in $\Gamma^{\vdash_\Sigma}$ while exhaustiveness imposes that either $\varphi \in \Gamma^{\vdash_\Sigma}$ or $(\neg \varphi) \in \Gamma^{\vdash_\Sigma}$. For example, $\emptyset$ is consistent but not exhaustive and the language L_Σ is exhaustive but not consistent.

Proposition 1 Set Γ is not consistent if and only if $\Gamma^{\vdash_\Sigma} = L_\Sigma$.

Proof:

($\leftarrow$) Immediate by definition of consistent set.

($\rightarrow$) By contraposition, suppose that $\Gamma^{\vdash_\Sigma} \neq L_\Sigma$. Then, there exists $\varphi \in L_\Sigma$ such that $\Gamma \nvdash_\Sigma \varphi$. Hence, by Exercise 2 of Chapter 5, $\Gamma \nvdash_\Sigma (\forall \varphi)$, whence it follows, by definition, that Γ is consistent. QED

Therefore, L_Σ is the unique inconsistent theory over Σ. However, L_Σ is not the unique inconsistent set w.r.t. Σ. Indeed, a set its inconsistent whenever its derivation closure is L_Σ. The reader should expect inconsistency to arise from the derivation of a formula and its negation. This intuition is captured by the next result.

Proposition 2 Set Γ is not consistent if and only if there exists ψ such that $\Gamma \vdash_\Sigma \psi$ and $\Gamma \vdash_\Sigma (\neg \psi)$.

Proof:

($\rightarrow$) If Γ is not consistent, then Proposition 1 entails that $\Gamma^{\vdash_\Sigma} = L_\Sigma$. Therefore, since $L_\Sigma \neq \emptyset$, there is ψ such that $\psi, (\neg \psi) \in \Gamma^{\vdash_\Sigma}$.

($\leftarrow$) Let ψ be such that $\psi, (\neg \psi) \in \Gamma^{\vdash_\Sigma}$. By result (g) in Section 5.4 and monotonicity of derivation, it follows that, for every η, $((\neg \psi) \Rightarrow (\psi \Rightarrow \eta)) \in \Gamma^{\vdash_\Sigma}$. Then, by MP, one obtains $\eta \in \Gamma^{\vdash_\Sigma}$ for every η. Hence, $\Gamma^{\vdash_\Sigma} = L_\Sigma$, whence it follows, again by Proposition 1, that Γ is not consistent. QED

The next result shows how to get a still consistent superset of a consistent set by adding with some caution a closed formula to the original set.

Proposition 3 If $\Gamma \subseteq L_\Sigma$, $\varphi \in cL_\Sigma$ and $\Gamma \nvdash_\Sigma (\neg \varphi)$, then $\Gamma \cup \{\varphi\}$ is consistent.

Proof: The proof is done by contradiction. Assume that $\Gamma \subseteq L_\Sigma$, $\varphi \in cL_\Sigma$, $\Gamma \nvdash_\Sigma (\neg \varphi)$ and $\Gamma \cup \{\varphi\}$ is not consistent w.r.t. Σ.

Since $\Gamma \cup \{\varphi\}$ is not consistent, $\Gamma \cup \{\varphi\} \vdash_\Sigma (\neg \varphi)$. From the equivalence between the two notions of derivation, there exists a derivation sequence w for $\Gamma \cup \{\varphi\} \vdash_\Sigma (\neg \varphi)$. Since φ is closed, w does not contain any essential generalizations over dependents of φ. Therefore, using the MTD, it follows that

$$(1) \quad \Gamma \vdash_\Sigma (\varphi \Rightarrow (\neg \varphi)).$$

On the other hand, $\vdash_\Sigma ((\varphi \Rightarrow (\neg \varphi)) \Rightarrow (\neg \varphi))$, the proof of which is left as an exercise, whence, by monotonicity of derivation,

$$(2) \quad \Gamma \vdash_\Sigma ((\varphi \Rightarrow (\neg \varphi)) \Rightarrow (\neg \varphi)).$$

Then, applying MP to (1) and (2), it follows that $\Gamma \vdash_\Sigma (\neg \varphi)$, contradicting the assumption that $\Gamma \not\vdash_\Sigma (\neg \varphi)$.   QED

Although not needed in the sequel, the following exercise introduces an interesting notion that provides a characterization of consistent and exhaustive sets.

Exercise 4 A set Γ of formulas is said to be *maximal consistent* w.r.t. Σ if it is consistent w.r.t. Σ and, for every formula ψ over Σ, if $\Gamma \not\vdash_\Sigma \psi$, then $\Gamma \cup \{\psi\}$ is not consistent w.r.t. Σ. Show that a set is maximal consistent if and only if it is consistent and exhaustive.

Proposition 3 provides the means for extending a consistent set to an exhaustive but still consistent set: iterate the application of the proposition through all closed formulas.

Let $e : \mathbb{N} \to cL_\Sigma$ be an injective enumeration of the closed formulas over Σ. Given a set Γ of formulas over Σ, let

$$\Gamma^e = \bigcup_{k \in \mathbb{N}} \Gamma^e_k$$

where the sequence $\lambda k \,.\, \Gamma^e_k$ is defined as follows:

- $\Gamma^e_0 = \Gamma$;

- $\Gamma^e_{k+1} = \begin{cases} \Gamma^e_k \cup \{e(k)\} & \text{if } \Gamma^e_k \not\vdash_\Sigma (\neg e(k)) \\ \Gamma^e_k & \text{otherwise.} \end{cases}$

Observe that by construction each Γ^e_k contains Γ^e_j for every $j = 0, \dots, k$.

Proposition 5 (Lindenbaum's lemma)
Given a consistent set Γ of formulas, regardless of the injective enumeration e of cL_Σ, the set Γ^e is a consistent and exhaustive extension of Γ.

Proof: The first step is to show by induction that Γ^e_k is consistent for each k in $\mathbb{N}$:
(Basis) $\Gamma^e_0 = \Gamma$, hence consistent by hypothesis.
(Step) $k = k' + 1$. There are two cases to consider:
(i) $\Gamma^e_{k'} \vdash_\Sigma (\neg e(k'))$, in which case $\Gamma^e_k = \Gamma^e_{k'}$, hence consistent by the induction hypothesis.

(ii) $\Gamma_{k'}^e \not\vdash_\Sigma (\neg e(k'))$, in which case $\Gamma_k^e = \Gamma_{k'}^e \cup \{e(k')\}$. Since $\Gamma_{k'}^e \not\vdash_\Sigma (\neg e(k'))$, calling upon Proposition 3 above, again one concludes that Γ_k^e is consistent.

Next, one proves by contradiction that Γ^e is consistent. Indeed, suppose that Γ^e is not consistent. Then, by Proposition 2, there exists a formula ψ such that $\Gamma^e \vdash_\Sigma \psi$ and $\Gamma^e \vdash_\Sigma (\neg \psi)$. Hence, by compactness of derivation, there are finite sets $\Delta_1, \Delta_2 \subset \Gamma^e$ such that $\Delta_1 \vdash_\Sigma \psi$ and $\Delta_2 \vdash_\Sigma (\neg \psi)$. Since these sets are finite, there is $k \in \mathbb{N}$ for which $\Delta_1 \cup \Delta_2 \subseteq \Gamma_k^e$. Thus, there is $k \in \mathbb{N}$ such that $\Gamma_k^e \vdash_\Sigma \psi$ and $\Gamma_k^e \vdash_\Sigma (\neg \psi)$. Hence, by Proposition 2, there is $k \in \mathbb{N}$ such that Γ_k^e is not consistent, contradicting the result established at the beginning of this proof.

Finally, one checks that Γ^e is exhaustive. Indeed, for every $\delta \in cL_\Sigma$, one knows that δ is $e(k)$ for some (unique) k. Observe that

- either $\Gamma_k^e \not\vdash_\Sigma (\neg \delta)$ in which case $\delta \in \Gamma_{k+1}^e$ and, so, $\delta \in \Gamma^e$,

- or $\Gamma_k^e \vdash_\Sigma (\neg \delta)$ in which case $\Gamma^e \vdash_\Sigma (\neg \delta)$.

So, for each closed formula δ, either $\Gamma^e \vdash_\Sigma \delta$ or $\Gamma^e \vdash_\Sigma (\neg \delta)$. Therefore, Γ^e is exhaustive. QED

8.2 Henkin construction

The goal of the Henkin construction is to show that every consistent set of formulas has a model. In order to be able to prove such a result, the signature must first be augmented with new constant symbols. Indeed, according to Henkin's idea, the domain of the envisaged canonical model should be the set of closed terms. But, as the following example shows, it may be the case that the closed terms of the given signature are not enough.

Let Σ be the signature with just one constant symbol c and one unary predicate symbol p. Observe that $cT_\Sigma = \{c\}$. Consider the following subset of L_Σ:

$$\Gamma = \{(\neg p(c)), (\exists x\, p(x))\}.$$

Clearly, Γ is consistent. Note that any model of Γ must have at least two elements, one denoted by c and another one allowing the satisfaction of formula $(\exists x\, p(x))$. Therefore, it is impossible to set up a model of Γ with cT_Σ as domain.

The difficulty is easily overcome by providing additional constant symbols that can be used as witnesses for the existential formulas that are to be satisfied.

Given a signature $\Sigma = (F, P, \tau)$, the signature

$$\Sigma^+ = (F^+, P, \tau^+)$$

is defined as follows (assuming that $A_\Sigma \cap \{b_k : k \in \mathbb{N}\} = \emptyset$):

- $F^+ = F \cup \{b_k : k \in \mathbb{N}\}$;

- $\tau^+ = \lambda s \,.\, \begin{cases} \tau(s) & \text{if } s \in F \cup P \\ 0 & \text{otherwise} \end{cases}$.

Given a set $\Gamma \subseteq L_\Sigma$ consistent w.r.t. Σ, the question arises of whether it is still consistent w.r.t. Σ^+. The answer is affirmative, but proving it requires the following technical lemma allowing derivation sequences over Σ^+ to be translated into derivation sequences over Σ. The idea is to replace the new constant symbols used in the derivation by fresh variables.

Proposition 6 Let:

- $\Gamma \subseteq L_\Sigma$;

- $w = (\psi_1, J_1) \ldots (\psi_n, J_n)$ be a derivation sequence for $\Gamma \vdash_{\Sigma^+} \varphi$;

- $\vec{b}$ be the vector $b_{i_1} \ldots b_{i_m}$ (with $i_j < i_{j+1}$ for $j = 1, \ldots, m-1$) of the new constants occurring in w;

- $\vec{y}$ be a vector $y_1 \ldots y_m$ of distinct variables not occurring in w;

- $[w]_{\vec{y}}^{\vec{b}}$ be the sequence obtained from w by replacing each ψ_k by formula $[\psi_k]_{\vec{y}}^{\vec{b}}$ obtained from ψ_k by replacing each b_{i_j} by y_j.

Then, $[w]_{\vec{y}}^{\vec{b}}$ is a derivation sequence for $\Gamma \vdash_\Sigma [\varphi]_{\vec{y}}^{\vec{b}}$.

Proof: Observe that sequence $[w]_{\vec{y}}^{\vec{b}}$ contains only formulas over Σ, with $[\varphi]_{\vec{y}}^{\vec{b}}$ being the last formula. Therefore, it only remains to verify that $[w]_{\vec{y}}^{\vec{b}}$ is a derivation sequence, in other words, that the justifications are still legitimate, which is achieved by case analysis:

If J_k is Hyp, then, since $\Gamma \subseteq L_\Sigma$, $[\psi_k]_{\vec{y}}^{\vec{b}}$ is ψ_k, and therefore $[\psi_k]_{\vec{y}}^{\vec{b}} \in \Gamma$.

If J_k is Axi, then $[\psi_k]_{\vec{y}}^{\vec{b}}$ is also an instance of Axi — clearly, only in the cases of Ax4 and Ax5 one needs to invoke the hypothesis that vector $\vec{y}$ be made up of fresh variables.

If J_k is MP k_1, k_2 and, for example, ψ_{k_2} is $(\psi_{k_1} \Rightarrow \psi_k)$, then $[\psi_k]_{\vec{y}}^{\vec{b}}$ also follows by MP from $[\psi_{k_1}]_{\vec{y}}^{\vec{b}}$ and $[\psi_{k_2}]_{\vec{y}}^{\vec{b}}$, since the latter formula is $([\psi_{k_1}]_{\vec{y}}^{\vec{b}} \Rightarrow [\psi_k]_{\vec{y}}^{\vec{b}})$.

If J_k is Gen k' over variable x, then $[\psi_k]_{\vec{y}}^{\vec{b}}$ also follows by Gen from $[\psi_{k'}]_{\vec{y}}^{\vec{b}}$, since $[\psi_k]_{\vec{y}}^{\vec{b}}$ is $[(\forall x\, \psi_{k'})]_{\vec{y}}^{\vec{b}}$, that is, $(\forall x [\psi_{k'}]_{\vec{y}}^{\vec{b}})$, again taking into account that $\vec{y}$ is composed of fresh variables.   QED

Proposition 7 Let $\Gamma \subseteq L_\Sigma$. Then, Γ is consistent w.r.t. Σ if and only if Γ is consistent w.r.t. Σ^+.

Proof:

($\rightarrow$) By contraposition, assume that Γ is not consistent w.r.t. Σ^+. Then, by Proposition 1, $\Gamma^{\vdash_{\Sigma^+}} = L_{\Sigma^+}$. In particular, since $L_\Sigma \subset L_{\Sigma^+}$, $\Gamma \vdash_{\Sigma^+} \varphi$ for every $\varphi \in L_\Sigma$. Then, from the equivalence between notions of derivations, for every $\varphi \in L_\Sigma$ there exists a derivation sequence w^φ for $\Gamma \vdash_{\Sigma^+} \varphi$. Hence, by the previous lemma, for every $\varphi \in L_\Sigma$, $[w^\varphi]_{\vec{y}}^{\vec{b}}$ is a derivation sequence for $\Gamma \vdash_\Sigma \varphi$. Hence $\Gamma^{\vdash_\Sigma} = L_\Sigma$, whence, again by Proposition 1, it follows that Γ is not consistent w.r.t. Σ.

($\leftarrow$) By contraposition, assume that Γ is not consistent w.r.t. Σ. Then, by Proposition 2, there exists $\psi \in L_\Sigma$ such that $\Gamma \vdash_\Sigma \psi$ and $\Gamma \vdash_\Sigma (\neg \psi)$. Then, by the equivalence of notions of derivation, there exist derivation sequences w and w' for $\Gamma \vdash_\Sigma \psi$ and $\Gamma \vdash_\Sigma (\neg \psi)$, respectively. Note that, since $L_\Sigma \subset L_{\Sigma^+}$, such w and w' are also derivation sequences for $\Gamma \vdash_{\Sigma^+} \psi$ and $\Gamma \vdash_{\Sigma^+} (\neg \psi)$, respectively. Therefore, again using Proposition 2, it follows that Γ is not consistent w.r.t. Σ^+. QED

Let
$$L_{\Sigma^+}^1 = \{\varphi \in L_{\Sigma^+} : \#\mathrm{fv}_{\Sigma^+}(\varphi) = 1\}.$$

That is, $L_{\Sigma^+}^1$ is the set of formulas with only one free variable (known as unary formulas).

Chosen an injective enumeration $\lambda\,k\,.\,\pi_k$ of $L_{\Sigma^+}^1$, let $\lambda\,k\,.\,y_k$ be the sequence of variables such that $\mathrm{fv}_{\Sigma^+}(\pi_k) = \{y_k\}$.

Exercise 8 Define by induction the map $\mathrm{cns}_\Sigma : L_\Sigma \to \wp F_0$ assigning to each formula the set of constant symbols occurring in it.

Consider a sequence $\lambda\,k\,.\,c_k$ such that:

- $c_k \in \{b_i : i \in \mathbb{N}\}$ for each $k \in \mathbb{N}$;

- $c_k \notin \mathrm{cns}_{\Sigma^+}(\pi_k)$ for each $k \in \mathbb{N}$;

- $c_k \notin \{c_0, \ldots, c_{k-1}\}$ for each $k \in \mathbb{N}^+$.

Consider now the sequence $\lambda\,k\,.\,\eta_k$ where each η_k is the closed formula

$$((\neg(\forall y_k\,\pi_k)) \Rightarrow (\neg[\pi_k]_{c_k}^{y_k}))$$

over Σ^+. Formula $(\neg[\pi_k]_{c_k}^{y_k})$ provides a witness for the formula $(\neg(\forall y_k\,\pi_k))$. Clearly, if $(\neg(\forall y_k\,\pi_k))$ is derivable from a set containing η_k, then so is $(\neg[\pi_k]_{c_k}^{y_k})$.

Finally, let

$$\Gamma^\exists = \bigcup_{k \in \mathbb{N}} \Gamma_k^\exists$$

where $\lambda\, k\,.\, \Gamma_k^\exists$ is the sequence defined as follows:

- $\Gamma_0^\exists = \Gamma$;

- $\Gamma_{k+1}^\exists = \Gamma_k^\exists \cup \{\eta_k\}$.

Observe that, by construction, each $\Gamma_k^\exists$ contains all the previous sets in the sequence.

Proposition 9 If Γ is consistent w.r.t. Σ, then its extension $\Gamma^\exists$ is consistent w.r.t. Σ^+.

Proof: First it is proved by induction that $\Gamma_k^\exists$ is consistent w.r.t. Σ^+ for every $k \in \mathbb{N}$:

(Basis) $\Gamma_0^\exists = \Gamma$, hence consistent w.r.t. Σ^+ by Proposition 7, since it is assumed to be consistent w.r.t Σ.

(Step) $\Gamma_{k+1}^\exists = \Gamma_k^\exists \cup \{\eta_k\}$. The proof proceeds by contradiction. Suppose that $\Gamma_{k+1}^\exists$ is not consistent w.r.t. Σ^+. Then, by Proposition 1, $(\Gamma_{k+1}^\exists)^{\vdash_{\Sigma^+}} = L_{\Sigma^+}$. Hence, in particular,

$$\Gamma_{k+1}^\exists \vdash_{\Sigma^+} (\neg\, \eta_k),$$

that is,

$$\Gamma_k^\exists \cup \{\eta_k\} \vdash_{\Sigma^+} (\neg\, \eta_k).$$

Then, applying the MTD (since η_k is closed),

$$\Gamma_k^\exists \vdash_{\Sigma^+} (\eta_k \Rightarrow (\neg\, \eta_k)),$$

whence, since $\vdash_\Sigma ((\eta_k \Rightarrow (\neg\, \eta_k)) \Rightarrow (\neg\, \eta_k))$ as already mentioned in the proof of Proposition 3,

$$\Gamma_k^\exists \vdash_{\Sigma^+} (\neg\, \eta_k).$$

That is,

$$\Gamma_k^\exists \vdash_{\Sigma^+} (\neg((\neg(\forall y_k\, \pi_k)) \Rightarrow (\neg[\pi_k]_{c_k}^{y_k}))).$$

Since $\vdash_{\Sigma^+} ((\neg(\psi \Rightarrow \delta)) \Rightarrow \psi)$ and $\vdash_{\Sigma^+} ((\neg(\psi \Rightarrow \delta)) \Rightarrow (\neg\, \delta))$, as the interested reader should verify, and also taking into account that $\vdash_{\Sigma^+} ((\neg(\neg\, \varphi)) \Rightarrow \varphi)$, one concludes that

$$\begin{cases} (1) & \Gamma_k^\exists \vdash_{\Sigma^+} (\neg(\forall y_k\, \pi_k)) \\ (2) & \Gamma_k^\exists \vdash_{\Sigma^+} [\pi_k]_{c_k}^{y_k}. \end{cases}$$

Applying the equivalence of the two notions of derivation to (2), it follows that there is a derivation sequence $w = (\psi_1, J_1) \dots (\psi_n, J_n)$ for

$$\Gamma_k^\exists \vdash_{\Sigma^+} [\pi_k]_{c_k}^{y_k}.$$

Let $[w]_x^{c_k}$ be the sequence obtained from w by replacing each ψ_i by $[\psi_i]_x^{c_k}$, where x is a variable not occurring in w. Since c_k does not occur in $\Gamma_k^\exists$, it is easily shown, using the technique used in the proof of Proposition 6, that $[w]_x^{c_k}$ is a derivation sequence for

$$\Gamma_k^\exists \vdash_{\Sigma^+} [[\pi_k]_{c_k}^{y_k}]_x^{c_k}.$$

Hence,

$$\Gamma_k^\exists \vdash_{\Sigma^+} [\pi_k]_x^{y_k}$$

because $[[\pi_k]_{c_k}^{y_k}]_x^{c_k}$ is $[\pi_k]_x^{y_k}$, since c_k does not occur in π_k. Thus, by generalization,

$$\Gamma_k^\exists \vdash_{\Sigma^+} (\forall x [\pi_k]_x^{y_k}).$$

On the other hand, by Ax4, one has

$$\vdash_{\Sigma^+} ((\forall x [\pi_k]_x^{y_k}) \Rightarrow [[\pi_k]_x^{y_k}]_{y_k}^x)$$

since $y_k \rhd_{\Sigma^+} x : [\pi_k]_x^{y_k}$, because the new variable x only occurs in $[\pi_k]_x^{y_k}$ where y_k occurs free in π_k. Observe also that $[[\pi_k]_x^{y_k}]_{y_k}^x$ is π_k, since x does not occur in π_k. Therefore,

$$\Gamma_k^\exists \vdash_{\Sigma^+} \pi_k$$

and, so, by generalization,

$$(2') \qquad \Gamma_k^\exists \vdash_{\Sigma^+} (\forall y_k \, \pi_k).$$

Using Proposition 2, from (1) and (2') one concludes that $\Gamma_k^\exists$ is not consistent w.r.t. Σ^+, in contradiction with the induction hypothesis.

Once it has been shown that $\Gamma_k^\exists$ is consistent w.r.t. Σ^+ for every $k \in \mathbb{N}$, it is easy to prove by contradiction that $\Gamma^\exists$ is also consistent w.r.t. Σ^+, a task left as an exercise to the reader. Hint: Follow the technique used in the proof of Lindenbaum's lemma. QED

Choose an injective enumeration e of cL_{Σ^+}. Given $\Gamma \subseteq L_\Sigma$, let

$$\widehat{\Gamma} = (\Gamma^\exists)^e.$$

Clearly, by Proposition 9 and Lindenbaum's lemma, if Γ is consistent w.r.t. Σ, then $\widehat{\Gamma}$ is a consistent and exhaustive w.r.t. Σ^+ extension of $\Gamma^\exists$ (and, hence, also of Γ).

Consider the interpretation structure over Σ^+:

$$I_\Gamma = (D, \_^{\mathsf{F}}, \_^{\mathsf{P}})$$

defined as follows:

- $D = cT_{\Sigma^+}$;

- $c_0^{\mathsf{F}} = c$;

- $f_n^{\mathsf{F}} = \lambda d_1, \ldots, d_n . \, f(d_1, \ldots, d_n)$ for each $n \in \mathbb{N}^+$;

- $p_n^{\mathsf{P}} = \lambda d_1, \ldots, d_n . \begin{cases} 1 & \text{if } \widehat{\Gamma} \vdash_{\Sigma^+} p(d_1, \ldots, d_n) \\ 0 & \text{otherwise} \end{cases}$ for each $n \in \mathbb{N}^+$.

It should be emphasized that, for each tuple $d_1, \ldots, d_n \in cT_{\Sigma^+}$, map f_n^{F} returns the closed term $f[d_1, \ldots, d_n]$, usually written $f(d_1, \ldots, d_n)$, as was mentioned before.

Observe also that the fact that I_Γ is well defined does not depend on the consistency and the exhaustiveness of $\widehat{\Gamma}$. But these properties are essential for establishing the following relationship between satisfaction by I_Γ and derivability from $\widehat{\Gamma}$.

Proposition 10 (Fundamental lemma of the Henkin construction)
If Γ is consistent over Σ, then, for every closed formula φ over Σ^+, the following holds:
$$\widehat{\Gamma} \vdash_{\Sigma^+} \varphi \text{ if and only if } I_\Gamma \Vdash_{\Sigma^+} \varphi.$$

Proof: Recall that $\widehat{\Gamma}$ is consistent and exhaustive since Γ is assumed to be consistent. The result is obtained as follows by structural induction over $\varphi \in cL_{\Sigma^+}$:

(Basis) φ is $p(d_1, \ldots, d_n)$:

$\widehat{\Gamma} \vdash_{\Sigma^+} p(d_1, \ldots, d_n)$ if and only if (construction of I_Γ)

$p_n^{\mathsf{P}}(d_1, \ldots, d_n) = 1$ if and only if (*)

for any ρ,
$$p_n^{\mathsf{P}}(\llbracket d_1 \rrbracket_{\Sigma^+}^{I_\Gamma \rho}, \ldots, \llbracket d_n \rrbracket_{\Sigma^+}^{I_\Gamma \rho}) = 1 \quad \text{if and only if}$$

for any ρ,
$$I_\Gamma \rho \Vdash_{\Sigma^+} p(d_1, \ldots, d_n) \qquad \text{if and only if}$$

$I_\Gamma \Vdash_{\Sigma^+} p(d_1, \ldots, d_n)$.

(*) Taking into account that

$$\llbracket d \rrbracket_{\Sigma^+}^{I_\Gamma \rho} = d \text{ for every } d \in cT_{\Sigma^+} \text{ and assignment } \rho \text{ into } I_\Gamma,$$

a result that the interested reader should prove by induction on the structure of d.

(Step) There are three cases to consider:

(a) φ is $(\neg\psi)$:

($\rightarrow$):

$$\text{if} \quad \widehat{\Gamma} \vdash_{\Sigma+} (\neg\psi), \quad \text{then}$$
$$\widehat{\Gamma} \nvdash_{\Sigma+} \psi \qquad \text{(since } \widehat{\Gamma} \text{ is consistent)}$$
$$I_\Gamma \nVdash_{\Sigma+} \psi \qquad \text{(by the induction hypothesis)}$$
$$I_\Gamma \Vdash_{\Sigma+} (\neg\psi) \qquad \text{(by the lemma of closed formula)}.$$

($\leftarrow$):

$$\text{if} \quad I_\Gamma \Vdash_{\Sigma+} (\neg\psi), \quad \text{then}$$
$$I_\Gamma \nVdash_{\Sigma+} \psi \qquad \text{(by the lemma of closed formula)}$$
$$\widehat{\Gamma} \nvdash_{\Sigma+} \psi \qquad \text{(by the induction hypothesis)}$$
$$\widehat{\Gamma} \vdash_{\Sigma+} (\neg\psi) \qquad \text{(since } \widehat{\Gamma} \text{ is exhaustive)}.$$

(b) φ is $(\psi \Rightarrow \delta)$:

($\rightarrow$) by contraposition:

$$\text{if} \quad I_\Gamma \nVdash_{\Sigma+} (\psi \Rightarrow \delta), \qquad \text{then}$$
$$I_\Gamma \Vdash_{\Sigma+} \psi \text{ and } I_\Gamma \nVdash_{\Sigma+} \delta \quad \text{(by the lemma of closed formula)}$$
$$\widehat{\Gamma} \vdash_{\Sigma+} \psi \text{ and } \widehat{\Gamma} \nvdash_{\Sigma+} \delta \quad \text{(by the induction hypothesis)}$$
$$\widehat{\Gamma} \nvdash_{\Sigma+} (\psi \Rightarrow \delta) \qquad \text{(by MP)}$$

($\leftarrow$) again by contraposition:

$$\text{if} \quad \widehat{\Gamma} \nvdash_{\Sigma+} (\psi \Rightarrow \delta), \qquad \text{then}$$
$$\widehat{\Gamma} \vdash_{\Sigma+} (\neg(\psi \Rightarrow \delta)) \qquad \text{(since } \widehat{\Gamma} \text{ is exhaustive)}$$
$$\widehat{\Gamma} \vdash_{\Sigma+} \psi \text{ and } \widehat{\Gamma} \vdash_{\Sigma+} (\neg\delta) \quad (*)$$
$$\widehat{\Gamma} \vdash_{\Sigma+} \psi \text{ and } \widehat{\Gamma} \nvdash_{\Sigma+} \delta \quad \text{(since } \widehat{\Gamma} \text{ is consistent)}$$
$$I_\Gamma \Vdash_{\Sigma+} \psi \text{ and } I_\Gamma \nVdash_{\Sigma+} \delta \quad \text{(by the induction hypothesis)}$$
$$I_\Gamma \nVdash_{\Sigma+} (\psi \Rightarrow \delta) \qquad \text{(by the lemma of closed formula)}.$$

(*) Since $\vdash_{\Sigma+} ((\neg(\psi \Rightarrow \delta)) \Rightarrow \psi)$ and $\vdash_{\Sigma+} ((\neg(\psi \Rightarrow \delta)) \Rightarrow (\neg\delta))$ as already mentioned in the proof of Proposition 9.

(c) φ is $(\forall x\,\psi)$. There are two cases to consider:

(i) $x \notin \text{fv}_{\Sigma+}(\psi)$, in which case formula ψ is closed:

$$\widehat{\Gamma} \vdash_{\Sigma+} (\forall x\,\psi) \qquad \text{if and only if} \quad \text{(Chapter 5, Exercise 2)}$$
$$\widehat{\Gamma} \vdash_{\Sigma+} \psi \qquad \text{if and only if} \quad \text{(by the induction hypothesis)}$$
$$I_\Gamma \Vdash_{\Sigma+} \psi \qquad \text{if and only if} \quad \text{(Chapter 7, Exercise 25)}$$
$$I_\Gamma \Vdash_{\Sigma+} (\forall x\,\psi)$$

(ii) $x \in \text{fv}_{\Sigma+}(\psi)$, in which case $\psi \in L^1_{\Sigma+}$:

Let $k \in \mathbb{N}$ be such that π_k is ψ, and hence x is y_k; in other words, φ is $(\forall y_k\,\pi_k)$.

Then:

($\rightarrow$) by contradiction:

Assume

$$\begin{cases} (1) & \widehat{\Gamma} \vdash_{\Sigma^+} (\forall y_k \, \pi_k) \\ (2) & I_\Gamma \nVdash_{\Sigma^+} (\forall y_k \, \pi_k). \end{cases}$$

From (2) it follows that there exists an assignment ρ such that

$$I_\Gamma \rho \nVdash_{\Sigma^+} (\forall y_k \, \pi_k).$$

Hence, there is an assignment σ that is y_k-equivalent to ρ such that

$$I_\Gamma \sigma \nVdash_{\Sigma^+} \pi_k.$$

Let $d = \sigma(y_k)$. Then, using the lemma of substitution in formula (Proposition 19 of Chapter 7), since $d \rhd_{\Sigma^+} y_k : \pi_k$ and $\sigma(y_k) = d = [\![d]\!]_{\Sigma^+}^{I_\Gamma \rho}$, one concludes

$$I_\Gamma \rho \nVdash_{\Sigma^+} [\pi_k]_d^{y_k}.$$

Therefore,

$$(2') \quad I_\Gamma \nVdash_{\Sigma^+} [\pi_k]_d^{y_k}.$$

On the other hand, from (1) it follows

$$\widehat{\Gamma} \vdash_{\Sigma^+} [\pi_k]_d^{y_k}$$

and, so, by the induction hypothesis,

$$I_\Gamma \Vdash_{\Sigma^+} [\pi_k]_d^{y_k},$$

in contradiction with $(2')$.

($\leftarrow$) again by contradiction:

Assume

$$\begin{cases} (1) & I_\Gamma \Vdash_{\Sigma^+} (\forall y_k \, \pi_k) \\ (2) & \widehat{\Gamma} \nvdash_{\Sigma^+} (\forall y_k \, \pi_k). \end{cases}$$

Using soundness of Ax4, (1) yields

$$I_\Gamma \Vdash_{\Sigma^+} [\pi_k]_{c_k}^{y_k}.$$

and, so, by the induction hypothesis,

$$(1') \quad \widehat{\Gamma} \vdash_{\Sigma^+} [\pi_k]_{c_k}^{y_k}.$$

On the other hand, since $\widehat{\Gamma}$ is exhaustive, (2) yields

$$\widehat{\Gamma} \vdash_{\Sigma^+} (\neg(\forall y_k \, \pi_k)).$$

and, so, since $\eta_k \in \widehat{\Gamma}$, by MP one gets

$$\widehat{\Gamma} \vdash_{\Sigma^+} (\neg[\pi_k]_{c_k}^{y_k}),$$

in contradiction with (1′). QED

Notice that $\forall \Gamma = \{(\forall \gamma) : \gamma \in \Gamma\} \subseteq \widehat{\Gamma}^{\vdash_{\Sigma^+}}$. Therefore, assuming that Γ is consistent, by the fundamental lemma of the Henkin construction, the interpretation structure I_Γ over Σ^+ is a model of $\forall \Gamma$ and, so, by the instantiation lemma (Exercise 25 of Chapter 7), I_Γ is a model of Γ.

All that remains is to obtain a model of Γ over the original signature Σ, which is easily achieved by forgetting the interpretation of the additional constants in Σ^+.

Recall from Section 7.6 that, given a signature $\Sigma = (F, P, \tau)$ and an interpretation structure $J = (D, \_^F, \_^P)$ over $\Sigma^+ = (F^+, P, \tau^+)$, the reduct of J along the inclusion $\Sigma \hookrightarrow \Sigma^+$ is the interpretation structure

$$J|_\Sigma = (D, \_^{F|_F}, \_^P)$$

over Σ where:

$$\_^{F|_F} = \lambda n \in \mathbb{N} . \_{-n}^{F|_F}$$

with each $\_{-n}^{F|_F} : F_n \to (D^n \to D)$ such that

$$f_n^{F|_F} = f_n^{F} \text{ for each } f \in F_n.$$

Exercise 11 Establish a cryptomorphism from $J|_\Sigma$ to J.

Proposition 12 (Lemma of model existence)
If $\Gamma \subseteq L_\Sigma$ is consistent, then the interpretation structure $(I_\Gamma)|_\Sigma$ over Σ is a model of Γ.

Proof: The result is a consequence of the fact that I_Γ is a model of Γ assumed to be consistent, as observed above, and of the following lemma:

For every $\psi \in L_\Sigma$ and $\rho : X \to D$, $I_\Gamma \, \rho \Vdash_{\Sigma^+} \psi$ if and only if $(I_\Gamma)|_\Sigma \, \rho \Vdash_\Sigma \psi$.

The proof of this lemma by structural induction over ψ is straightforward and, so, it is left as an exercise to the reader. Hint: Begin by showing the following auxiliary result:

For every $t \in T_\Sigma$ and $\rho : X \to D$, $[\![t]\!]_{\Sigma^+}^{I_\Gamma \rho} = [\![t]\!]_\Sigma^{(I_\Gamma)|_\Sigma \, \rho}$.

It is worth mentioning that these two results are particular cases of those in Exercise 48 of Section 7.6. The signature morphism at work here is the inclusion $\Sigma \hookrightarrow \Sigma^+$. QED

8.3   Gödel's completeness theorem

The completeness of first-order logic was first shown by Kurt Gödel in 1929, but using a very different technique from the one proposed by Henkin. Gödel's method is specific to first-order logic while Henkin's method is quite universal and has been adapted with great success for proving the completeness of many other logics.

Following Henkin's method, the completeness of the Hilbert calculus is easily obtained using the lemma of model existence.

Proposition 13 (Gödel's completeness theorem)

$$\text{If } \Gamma \vDash_\Sigma \varphi, \text{ then } \Gamma \vdash_\Sigma \varphi.$$

Proof: Towards a proof by contraposition, assume that $\Gamma \nvdash_\Sigma \varphi$. Then:

$\Gamma \nvdash_\Sigma (\forall \varphi)$	(Chapter 5, Exercise 2)
$\Gamma \nvdash_\Sigma (\neg(\neg(\forall \varphi)))$	(using (a) in Section 5.4)
$\Gamma \cup \{(\neg(\forall \varphi))\}$ is consistent	(Proposition 3)

Hence, by the lemma of model existence, there exists an interpretation structure I over Σ that is a model of $\Gamma \cup \{(\neg(\forall \varphi))\}$. That is, there is I such that:

$$\begin{cases} (1) & I \Vdash_\Sigma \Gamma \\ (2) & I \Vdash_\Sigma (\neg(\forall \varphi)). \end{cases}$$

From (2) and the lemma of closed formula it follows that $I \nVdash_\Sigma (\forall \varphi)$, whence $I \nVdash_\Sigma \varphi$, using the counter-reciprocal of the generalization lemma (Exercise 25 of Chapter 7).

In other words, there is an interpretation structure I such that

$$\begin{cases} (1) & I \Vdash_\Sigma \Gamma \\ (2') & I \nVdash_\Sigma \varphi \end{cases}$$

and, so, $\Gamma \nvDash_\Sigma \varphi$.   QED

It should be pointed out again that this proof is not constructive, since it does not provide a derivation for $\Gamma \vdash_\Sigma \varphi$ when $\Gamma \vDash_\Sigma \varphi$. Note also that the original proof by Gödel is also non constructive and only infers theoremhood from validity.

The following implication is a direct corollary of the completeness of the Hilbert calculus:

$$\text{if } \varphi \text{ is a valid formula, then } \varphi \text{ is a theorem.}$$

This weaker result is known as weak completeness of the Hilbert calculus. For this reason, some authors refer to Proposition 13 as the strong completeness of the Hilbert calculus.

The results below are further consequences of the completeness theorem and the technique used for establishing it.

Proposition 14 (Derivability of tautological formulas)
Every tautological formula is a theorem.[1]

This result allows the use of justification Taut (tautological formula) in derivation sequences when using the Hilbert calculus.

Example 15 Recall Proposition 30 of Chapter 5 where x be a variable that is not free in ψ. The sequence

1	$(\forall x(\varphi \Rightarrow \psi))$	Hyp
2	$((\forall x((\neg\psi) \Rightarrow (\neg\varphi))) \Rightarrow ((\neg\psi) \Rightarrow (\forall x(\neg\varphi))))$	Ax5
3	$((\forall x(\varphi \Rightarrow \psi)) \Rightarrow (\varphi \Rightarrow \psi))$	Ax4
4	$(\varphi \Rightarrow \psi)$	MP 1,3
5	$((\varphi \Rightarrow \psi) \Rightarrow ((\neg\psi) \Rightarrow (\neg\varphi)))$	Taut
6	$((\neg\psi) \Rightarrow (\neg\varphi))$	MP 4,5
7	$(\forall x((\neg\psi) \Rightarrow (\neg\varphi)))$	Gen 6
8	$((\neg\psi) \Rightarrow (\forall x(\neg\varphi)))$	MP 7,2
9	$(((\neg\psi) \Rightarrow (\forall x(\neg\varphi))) \Rightarrow ((\neg(\forall x(\neg\varphi))) \Rightarrow (\neg(\neg\psi))))$	Taut
10	$((\neg(\forall x(\neg\varphi))) \Rightarrow (\neg(\neg\psi)))$	MP 8,9
11	$((\neg(\neg\psi)) \Rightarrow \psi)$	Taut
12	$((\neg(\forall x(\neg\varphi))) \Rightarrow \psi)$	HS 10,11

can now be taken as a derivation of $(\forall x (\varphi \Rightarrow \psi)) \vdash_\Sigma ((\exists x\, \varphi) \Rightarrow \psi)$.

Proposition 16 (Skolem–Löwenheim theorem)
If a set of formulas has a model, then it has a model with denumerable domain.

Proof: Let I be a model of Γ. Then, Γ is consistent (which is left as an exercise). Therefore, by the lemma of model existence in the Henkin construction, $(I_\Gamma)|_\Sigma$ is a model of Γ with a denumerable domain, since the latter is the set of closed terms over Σ^+. QED

Proposition 17 (Cardinality theorem)
If a set Γ of formulas has a model whose domain has cardinality ξ, then Γ has a model whose domain has cardinality ξ' for every cardinal $\xi' \geq \xi$.

---

[1]Although this result is obtained here as a simple corollary of the completeness theorem for first-order logic, it can be shown that any tautological formula can be derived just from axioms 1 to 3 and MP.

Proof: Let $I = (D, \_^{\text{F}}, \_^{\text{P}})$ be a model of set of formulas Γ such that $\#D = \xi$. Let $D' \supseteq D$ be such that $\#D' = \xi'$.

Recall the interpretation structure

$$I' = (D', \_^{\text{F}'}, \_^{\text{P}'})$$

built in Exercise 46 of Chapter 7. Recall also that, choosing $a \in D$, for every assignment ρ' into I', one has:

$$I\rho'_a \Vdash_\Sigma \varphi \text{ if and only if } I'\rho' \Vdash_\Sigma \varphi$$

where

$$\rho'_a = \lambda x . \begin{cases} \rho'(x) & \text{if } \rho'(x) \in D \\ a & \text{otherwise.} \end{cases}$$

It can then be shown by contraposition that if I is a model of γ, then so is I'. Indeed, suppose that $I' \nVdash_\Sigma \gamma$. Then, there is an assignment ρ' into I' such that $I'\rho' \nVdash_\Sigma \gamma$. Hence $I\rho'_a \nVdash_\Sigma \gamma$, and therefore $I \nVdash_\Sigma \gamma$.

Thus, as required, if $I \Vdash_\Sigma \Gamma$, then $I' \Vdash_\Sigma \Gamma$.   QED

The next exercise provides a sufficient condition for the validity of a formula and illustrates the application of the Skolem–Löwenheim and cardinality theorems.

Exercise 18 Show that if a formula is satisfied by every interpretation structure whose domain has cardinality greater than $\aleph_0 = \#\mathbb{N}$, then that formula is valid.

As the reader should expect, completeness (together with correctness) provides a means for proving compactness of model existence from compactness of derivation. Other compactness results at the semantic level can be obtained in the same way.

Proposition 19 (Compactness theorem)
If every finite subset of a set Γ of formulas has a model, then Γ has a model.

Proof: Towards a proof by contradiction, assume that Γ does not have a model. Choose a formula φ. Then:

$$\begin{cases} \Gamma \vDash_\Sigma \varphi \\ \Gamma \vDash_\Sigma (\neg\varphi). \end{cases}$$

Hence, by the completeness of the Hilbert calculus, it can be deduced that:

$$\begin{cases} \Gamma \vdash_\Sigma \varphi \\ \Gamma \vdash_\Sigma (\neg\varphi). \end{cases}$$

Furthermore, using the compactness of $\vdash_\Sigma$, there is a finite set $\Psi \subseteq \Gamma$ such that:

$$\begin{cases} \Psi \vdash_\Sigma \varphi \\ \Psi \vdash_\Sigma (\neg\,\varphi). \end{cases}$$

Therefore, using the soundness of the Hilbert calculus, one can infer:

$$\begin{cases} \Psi \vDash_\Sigma \varphi \\ \Psi \vDash_\Sigma (\neg\,\varphi). \end{cases}$$

One may thus conclude that Ψ does not have a model, in contradiction with the hypothesis of the theorem. QED

It is worthwhile to mention that the compactness theorem for the propositional logic is a result of Tychonoff's theorem (which says that the product of compact spaces is compact, see [31]) applied to Stone spaces (see [30]), hence the name traditionally given by logicians to the result in each logic where it holds.

Proposition 20 A set is consistent if and only if all its finite subsets are consistent.

Proof:

($\rightarrow$) Assume that Γ is consistent. Then, by Proposition 12, Γ has a model I. Then, I is also a model for any finite subset Φ of Γ.

($\leftarrow$) Assume that each finite subset Φ of Γ is consistent. Then, each Φ has a model using Proposition 12. Therefore, by the compactness theorem, Γ has a model and, so, it is consistent. QED

Exercise 21 Show that the operator $\vDash_\Sigma$ is compact. That is,

$$\Gamma^{\vDash_\Sigma} = \bigcup_{\Phi \in \wp_{\mathrm{fin}}\Gamma} \Phi^{\vDash_\Sigma}.$$

Adapting the technique used in this chapter for proving the completeness of the Hilbert (global) calculus presented in Chapter 5 with respect to the (global) semantic entailment provided in Chapter 7, the interested reader should be able to prove the completeness of the local calculus (Exercise 36 in Section 7.4) with respect to local semantic entailment, as it is asked in the next exercise.

Exercise 22 Show that if $\Gamma \vDash_\Sigma^\ell \varphi$ then $\Gamma \vdash_\Sigma^\ell \varphi$.

Chapter 9

Gentzen calculus

This chapter serves two aims. First, proving the consistency of first-order logic by purely syntactical means, that is, without resorting to semantic techniques, following Gerhard Gentzen's original proposal in 1936 (see [18]), thus giving a positive answer to one of the key questions raised by David Hilbert. Second, introducing the sequent calculus, which is more practical than the Hilbert calculus, at least having in mind automatic theorem proving. In fact, the former objective is served by the latter.

In the Hilbert calculus (as seen in Chapter 5), the unit of reasoning is the formula. Axioms are formulas; inference rules are pairs composed by a set of formulas (the premises of the rule) and a formula (the conclusion of the rule); and derivations are finite sequences of formulas.

In the Gentzen calculus (as seen below), the unit of reasoning is the sequent, a pair composed by two sequences of formulas. The first component of the pair is known as the antecedent and the second component is called the consequent. Intuitively, a sequent $\vec{\gamma} \to \vec{\eta}$ states that if all the formulas of the antecedent $\vec{\gamma}$ are true, then at least a formula of the consequent $\vec{\eta}$ should be true.

In the Gentzen calculus, axioms are sequents; rules are pairs composed by a set of sequents (the premises of the rule) and a sequent (the conclusion of the rule); and derivations are sequences of sequents.

In order to show in the Gentzen calculus that a formula φ is derivable from a set Γ of hypotheses, one has to derive the sequent $\varepsilon \to \varphi$ (with empty antecedent ε and singleton consequent composed of φ) from the following set of sequents $\{\varepsilon \to \gamma : \gamma \in \Gamma\}$. That is, in order to make inferences on formulas in the Gentzen calculus one has to restate the problem in terms of inferences on sequents.

The counterpart in the Hilbert calculus of the reasoning on sequents can be recognized to some extent in the metatheorems.

In a Gentzen calculus, there are general rules like cut, permutation, weakening and contraction that do not depend of the logic at hand, and specific rules that depend on the constructors of the formulas of the logic.

When defining a sequent calculus for a logic, one has to decide which general rules to use. This is an important decision that constrains the nature of the antecedent and the consequent of a sequent. For example, including permutation as a rule means that the order of formulas in the antecedent and in the consequent of a sequent is not relevant.

Specific rules should be provided for each operator in the language. In the case of first-order logic it is necessary to define specific rules for $\neg$, $\Rightarrow$, $\forall x$ and so on. Moreover, specific rules for each operator c of arity n should include a left rule where a formula $c(\varphi_1, \ldots, \varphi_n)$ appears in the antecedent of the conclusion of the rule and a right rule where the formula appears in the consequent of the conclusion of the rule. Atomic formulas (the basic constructors of the language) are used for choosing the axioms.

Gentzen calculi have been used extensively in automatic theorem proving. All the first-order rules are suitable for this purpose, with the exception of the cut rule. Fortunately, the cut rule is not essential as shown below. This fact is also essential in the proof of the consistency of the Gentzen calculus.

Soundness and completeness of the Gentzen calculus with respect to the Tarskian semantics can also be stated at the level of the sequents and at the level of formulas. For the main purpose of this book, soundness and completeness at the level of formulas is of particular interest.

In Section 9.1, the axioms and rules of the Gentzen calculus are presented, as well as the notion of derivation sequence. Section 9.2 is dedicated to proving the consistency of the Gentzen calculus by purely symbolic means. The section starts with some technical concepts and includes as the main result that cuts can be eliminated from derivation sequences. Consistency is then proved as a simple corollary of the cut elimination result. Section 9.3 starts by relating the Hilbert calculus and the Gentzen calculus on formulas. This purely symbolic relationship allows the proof of consistency of the Hilbert calculus from the consistency of the Gentzen calculus. Section 9.4 discusses the soundness and completeness of the Gentzen calculus with respect to the Tarskian semantics. Soundness is established for reasoning on both sequents and formulas. Completeness for reasoning on formulas is proved via the Hilbert calculus. Finally, Gentzen formula derivability is proved to be equivalent to Hilbert derivability.

It is worthwhile to mention that there are other calculi for first-order logic, namely the tableau calculus (proposed by Raymond Smullyan [43]) and natural deduction (proposed by Gerhard Gentzen [18]). The former is very suitable to automatic theorem proving. The latter was introduced as a formalization of the proof methods used by mathematicians in their every day work.

9.1 Rules and derivations

Let Σ be a first-order signature. A (finite[1]) *sequent* over Σ is a pair

$$(\gamma_1 \ldots \gamma_e, \eta_1 \ldots \eta_d) \in L_\Sigma^* \times L_\Sigma^*$$

commonly written

$$\gamma_1, \ldots, \gamma_e \to \eta_1, \ldots, \eta_d$$

or even

$$\gamma_1 \ldots \gamma_e \to \eta_1 \ldots \eta_d.$$

The set of sequents over Σ is denoted by S_Σ.

It helps at this point to give the notion of satisfaction of a sequent. An interpretation structure I over Σ *satisfies* sequent $\gamma_1 \ldots \gamma_e \to \eta_1 \ldots \eta_d$, written

$$I \Vdash_\Sigma \gamma_1 \ldots \gamma_e \to \eta_1 \ldots \eta_d,$$

if, for every assignment ρ in I, there is some $j = 1, \ldots, d$ for which $I\rho \Vdash_\Sigma \eta_j$ whenever $I\rho \Vdash_\Sigma \gamma_i$ for every $i = 1, \ldots, e$. Hence, $I \Vdash_\Sigma \gamma_1 \ldots \gamma_e \to \eta_1 \ldots \eta_d$ if and only if, for every ρ,

$$I\rho \Vdash_\Sigma \bigvee_{j=1}^{d} \eta_j \quad \text{whenever} \quad I\rho \Vdash_\Sigma \bigwedge_{i=1}^{e} \gamma_i.$$

Moreover a sequent s is said to *valid*, written $\models_\Sigma s$, if $I \Vdash_\Sigma s$ for every interpretation structure I.

Recall that B_Σ stands for the set of *basic* (or *atomic*) formulas over Σ:

$$B_\Sigma = \bigcup_{n \in \mathbb{N}^+} \{p(t_1, \ldots, t_n) : p \in P_n \,\&\, t_1, \ldots, t_n \in T_\Sigma\}.$$

One may write $\vec{\delta}$ in place of $\delta_1 \ldots \delta_{|\vec{\delta}|}$ whenever the length of the sequence is not relevant.

The main notion of derivability in Gentzen calculi is to say when a sequent is derivable from a set of sequents. In the case at hand, rules are provided for each operator independently of the fact that all the operators can be defined as abbreviations of $\neg$, $\Rightarrow$ and $\forall$. Herein, we provide the $\exists$ rules. Let $\mathfrak{S} \subseteq S_\Sigma$. The set $\mathfrak{S}^{\vdash_\Sigma}$ of sequents *derivable* from $\mathfrak{S}$ is inductively defined as follows:

- $\mathfrak{S}^{\vdash_\Sigma}$ includes the *hypotheses*:

    - $\mathfrak{S} \subseteq \mathfrak{S}^{\vdash_\Sigma}$.

---

[1]Although it is possible to define a Gentzen calculus working with infinite sequents, herein only finite sequents are considered since they are enough for the objectives of the chapter.

- $\mathfrak{S}^{\vdash_\Sigma}$ includes the *axioms*:

  - $\beta \to \beta \in \mathfrak{S}^{\vdash_\Sigma}$ for every $\beta \in B_\Sigma$.

- $\mathfrak{S}^{\vdash_\Sigma}$ is closed under the *inference rules*:

  - *Permutation on the right* (PR):
    $\vec{\gamma} \to \vec{\eta}\,\vec{\nu}\,\psi \in \mathfrak{S}^{\vdash_\Sigma}$ if $\vec{\gamma} \to \vec{\eta}\,\psi\,\vec{\nu} \in \mathfrak{S}^{\vdash_\Sigma}$;

  - *Permutation on the left* (PL):
    $\psi\,\vec{\gamma}\,\vec{\nu} \to \vec{\eta} \in \mathfrak{S}^{\vdash_\Sigma}$ if $\vec{\gamma}\,\psi\,\vec{\nu} \to \vec{\eta} \in \mathfrak{S}^{\vdash_\Sigma}$;

  - *Contraction on the right* (CR):
    $\vec{\gamma} \to \vec{\eta}\,\psi \in \mathfrak{S}^{\vdash_\Sigma}$ if $\vec{\gamma} \to \vec{\eta}\,\psi\,\psi \in \mathfrak{S}^{\vdash_\Sigma}$;

  - *Contraction on the left* (CL):
    $\psi\,\vec{\gamma} \to \vec{\eta} \in \mathfrak{S}^{\vdash_\Sigma}$ if $\psi\,\psi\,\vec{\gamma} \to \vec{\eta} \in \mathfrak{S}^{\vdash_\Sigma}$;

  - *Weakening on the right* (WR):
    $\vec{\gamma} \to \vec{\eta}\,\psi \in \mathfrak{S}^{\vdash_\Sigma}$ if $\vec{\gamma} \to \vec{\eta} \in \mathfrak{S}^{\vdash_\Sigma}$;

  - *Weakening on the left* (WL):
    $\psi\,\vec{\gamma} \to \vec{\eta} \in \mathfrak{S}^{\vdash_\Sigma}$ if $\vec{\gamma} \to \vec{\eta} \in \mathfrak{S}^{\vdash_\Sigma}$;

  - *Cut rule* (Cut):
    $\vec{\gamma} \to \vec{\eta} \in \mathfrak{S}^{\vdash_\Sigma}$ if $\vec{\gamma} \to \vec{\eta}\,\psi \in \mathfrak{S}^{\vdash_\Sigma}$ and $\psi\,\vec{\gamma} \to \vec{\eta} \in \mathfrak{S}^{\vdash_\Sigma}$;

  - *Negation on the right* ($\neg$R):
    $\vec{\gamma} \to \vec{\eta}\,(\neg\,\varphi) \in \mathfrak{S}^{\vdash_\Sigma}$ if $\varphi\,\vec{\gamma} \to \vec{\eta} \in \mathfrak{S}^{\vdash_\Sigma}$;

  - *Negation on the left* ($\neg$L):
    $(\neg\,\varphi)\,\vec{\gamma} \to \vec{\eta} \in \mathfrak{S}^{\vdash_\Sigma}$ if $\vec{\gamma} \to \vec{\eta}\,\varphi \in \mathfrak{S}^{\vdash_\Sigma}$;

  - *Implication on the right* ($\Rightarrow$R):
    $\vec{\gamma} \to \vec{\eta}\,(\varphi \Rightarrow \psi) \in \mathfrak{S}^{\vdash_\Sigma}$ if $\varphi\,\vec{\gamma} \to \vec{\eta}\,\psi \in \mathfrak{S}^{\vdash_\Sigma}$;

  - *Implication on the left* ($\Rightarrow$L):
    $(\varphi \Rightarrow \psi)\,\vec{\gamma} \to \vec{\eta} \in \mathfrak{S}^{\vdash_\Sigma}$ if $\psi\,\vec{\gamma} \to \vec{\eta} \in \mathfrak{S}^{\vdash_\Sigma}$ and $\vec{\gamma} \to \vec{\eta}\,\varphi \in \mathfrak{S}^{\vdash_\Sigma}$;

  - *Universal quantification on the right* ($\forall$R):
    $\vec{\gamma} \to \vec{\eta}\,(\forall x\,\varphi) \in \mathfrak{S}^{\vdash_\Sigma}$ if $\vec{\gamma} \to \vec{\eta}\,[\varphi]_y^x \in \mathfrak{S}^{\vdash_\Sigma}$ and $y \notin \mathrm{fv}_\Sigma(\vec{\gamma}) \cup \mathrm{fv}_\Sigma(\vec{\eta})$;

  - *Universal quantification on the left* ($\forall$L):
    $(\forall x\,\varphi)\,\vec{\gamma} \to \vec{\eta} \in \mathfrak{S}^{\vdash_\Sigma}$ if $[\varphi]_t^x\,\vec{\gamma} \to \vec{\eta} \in \mathfrak{S}^{\vdash_\Sigma}$ and $t \triangleright_\Sigma x : \varphi$;

  - *Existential quantification on the right* ($\exists$R):
    $\vec{\gamma} \to \vec{\eta}\,(\exists x\,\varphi) \in \mathfrak{S}^{\vdash_\Sigma}$ if $\vec{\gamma} \to \vec{\eta}\,[\varphi]_t^x \in \mathfrak{S}^{\vdash_\Sigma}$ and $t \triangleright_\Sigma x : \varphi$;

    – *Existential quantification on the left* ($\exists$L):

$$(\exists x\,\varphi)\,\vec{\gamma} \to \vec{\eta} \in \mathfrak{S}^{\vdash_\Sigma} \text{ if } [\varphi]^x_y\,\vec{\gamma} \to \vec{\eta} \in \mathfrak{S}^{\vdash_\Sigma} \text{ and } y \notin \mathrm{fv}_\Sigma(\vec{\gamma}) \cup \mathrm{fv}_\Sigma(\vec{\eta});$$

for every $\vec{\gamma}, \vec{\eta}, \vec{\nu} \in L^*_\Sigma$, $\varphi, \psi \in L_\Sigma$, $x, y \in X$ and $t \in T_\Sigma$.

The weakening, contraction, permutation and cut rules are *general rules*. All the other rules are *specific* to first-order logic. For each connective and each quantifier there are a left and a right rules. The rules of $\forall$R, $\forall$L, $\exists$R and $\exists$L have provisos. They can only be applied when the provisos are fulfilled. Contrarily to the Hilbert calculus where the provisos appeared in axioms (4 and 5), here provisos appear on inference rules.

One usually writes $\mathfrak{S} \vdash_\Sigma s$ instead of $s \in \mathfrak{S}^{\vdash_\Sigma}$. Sequent s is said to be a *theorem* if $\emptyset \vdash_\Sigma s$, written $\vdash_\Sigma s$.

A *derivation sequence* for $s \in S_\Sigma$ from $\mathfrak{S} \subseteq S_\Sigma$ is a sequence

$$(s_1, J_1) \ldots (s_n, J_n)$$

such that:

- each $s_i \in S_\Sigma$;

- s_1 is s;

- each J_i is the justification for s_i:

    – if J_i is Hyp, then $s_i \in \mathfrak{S}$;

    – if J_i is Ax, then s_i is an axiom;

    – if J_i is $R\,k$ with

$$R \in \{\mathrm{PR}, \mathrm{PL}, \mathrm{CR}, \mathrm{CL}, \mathrm{WR}, \mathrm{WL}, \neg\mathrm{R}, \neg\mathrm{L}, \Rightarrow\mathrm{R}, \forall\mathrm{R}, \forall\mathrm{L}\}$$

    and $n \geq k > i$, then s_i is obtained by applying rule R to s_k;

    – if J_i is $R\,k_1, k_2$ with $R \in \{\mathrm{Cut}, \Rightarrow\mathrm{L}\}$ and $n \geq k_1, k_2 > i$, then s_i is obtained by applying rule R to s_{k_1} and s_{k_2}.

Contrarily to the bottom up nature of derivations in the Hilbert calculus introduced in Chapter 5, following the tradition of Gentzen calculi, herein derivations are built, top down, from the target sequent to the hypotheses and axioms. It should also be mentioned that derivations in Gentzen calculi are usually presented as inverted trees. Using sequences instead of trees is simpler, brings no theoretical problems, and shows that the difference between Hilbert calculi and Gentzen calculi is the unit of reasoning (formula versus sequent) and not the nature of a derivation (sequence versus tree). In fact, it is easy to define Hilbert calculi where derivations are presented by trees.

Exercise 1 Show that $\mathfrak{S} \vdash_\Sigma s$ if and only if there exists a derivation sequence for s from $\mathfrak{S}$.

As a first example,

$$
\begin{array}{lll}
1 & \varepsilon \to (p(x) \Rightarrow (q(x) \Rightarrow p(x))) & \Rightarrow\!\text{R } 2 \\
2 & p(x) \to (q(x) \Rightarrow p(x)) & \Rightarrow\!\text{R } 3 \\
3 & q(x)p(x) \to p(x) & \text{WL } 4 \\
4 & p(x) \to p(x) & \text{Ax}
\end{array}
$$

is a derivation sequence for $\vdash_\Sigma \varepsilon \to (p(x) \Rightarrow (q(x) \Rightarrow p(x)))$. This simple derivation illustrates that proofs in the Gentzen calculus are carried out by following the structure of the formulas in the sequent to be proved.

Exercise 2 Derive the *generalized axioms* (Axg):

$$\vdash_\Sigma \vec{\gamma}\psi\vec{\nu} \to \vec{\eta}\psi\vec{\mu}\,.$$

Hence, it is legitimate to allow these generalized axioms in derivations. For instance,

$$
\begin{array}{lll}
1 & \varepsilon \to (\varphi \Rightarrow (\psi \Rightarrow \varphi)) & \Rightarrow\!\text{R } 2 \\
2 & \varphi \to (\psi \Rightarrow \varphi) & \Rightarrow\!\text{R } 3 \\
3 & \psi\varphi \to \varphi & \text{Axg}
\end{array}
$$

is a derivation sequence for $\vdash_\Sigma \varepsilon \to (\varphi \Rightarrow (\psi \Rightarrow \varphi))$ that the reader should compare with Ax1 of the Hilbert calculus.

The following example illustrates the use of inference rules with provisos in order to derive a sequent corresponding to axiom Ax5 of the Hilbert calculus.

Example 3 Assume that $x \notin \mathrm{fv}_\Sigma(\varphi)$. Then,

$$\vdash_\Sigma \varepsilon \to ((\forall x(\varphi \Rightarrow \psi)) \Rightarrow (\varphi \Rightarrow (\forall x\,\psi)))$$

is established by the following derivation sequence:

$$
\begin{array}{lll}
1 & \varepsilon \to ((\forall x(\varphi \Rightarrow \psi)) \Rightarrow (\varphi \Rightarrow (\forall x\,\psi))) & \Rightarrow\!\text{R } 2 \\
2 & (\forall x(\varphi \Rightarrow \psi)) \to (\varphi \Rightarrow (\forall x\,\psi)) & \Rightarrow\!\text{R } 3 \\
3 & \varphi, (\forall x(\varphi \Rightarrow \psi)) \to (\forall x\,\psi) & \forall\text{R } 4\ (*) \\
4 & \varphi, (\forall x(\varphi \Rightarrow \psi)) \to \psi & \text{PL } 5 \\
5 & (\forall x(\varphi \Rightarrow \psi)), \varphi \to \psi & \forall\text{L } 6 \\
6 & (\varphi \Rightarrow \psi), \varphi \to \psi & \Rightarrow\!\text{L } 7{,}8 \\
7 & \psi, \varphi \to \psi & \text{PL } 9 \\
8 & \varphi \to \psi, \varphi & \text{PR } 10 \\
9 & \varphi, \psi \to \psi & \text{WL } 11 \\
10 & \varphi \to \varphi, \psi & \text{WR } 12 \\
11 & \psi \to \psi & \text{Axg} \\
12 & \varphi \to \varphi & \text{Axg}
\end{array}
$$

(∗) The ∀R rule can be used because $x \notin \mathrm{fv}_\Sigma(\varphi)$, by hypothesis, and $x \notin \mathrm{fv}_\Sigma((\forall x(\varphi \Rightarrow \psi)))$, by definition of free variable in a formula.

As a first illustration of the use of the cut rule, see the derivation in the next example, where, as usual, permutation steps are omitted.

Example 4 The sequence

$$
\begin{array}{lll}
1 & (\forall x\,(\forall y\,p(y,x))) \to p(t,t) & \text{Cut 2,3} \\
2 & (\forall z\,p(z,z)), (\forall x\,(\forall y\,p(y,x))) \to p(t,t) & \forall\text{L } 4 \\
3 & (\forall x\,(\forall y\,p(y,x))) \to p(t,t), (\forall z\,p(z,z)) & \forall\text{R } 6 \\
4 & p(t,t), (\forall x\,(\forall y\,p(y,x))) \to p(t,t) & \text{WR } 5 \\
5 & p(t,t) \to p(t,t) & \text{Ax} \\
6 & (\forall x\,(\forall y\,p(y,x))) \to p(t,t), p(z,z) & \forall\text{L } 7 \\
7 & (\forall y\,p(y,z)) \to p(t,t), p(z,z) & \forall\text{L } 8 \\
8 & p(z,z) \to p(t,t), p(z,z) & \text{WR } 9 \\
9 & p(z,z) \to p(z,z) & \text{Ax}
\end{array}
$$

is a derivation of sequent $(\forall x\,(\forall y\,p(y,x))) \to p(t,t)$ provided that $z \notin \mathrm{var}_\Sigma(t)$. Compare with Example 34 of Chapter 5.

The following exercise shows that the (left and right) rules for negation, implication and universal quantification are enough to obtain the (left and right) rules for the other connectives and the existential quantification.

Exercise 5 State and prove admissibility of rules for conjunction, disjunction, equivalence and existential quantification.

Observe that the contraction rules are essential for being able to use the quantification rules more than once. That is, for instance

$$(\forall x\,\varphi), \vec{\gamma} \to \vec{\eta} \in \mathfrak{S}^{\vdash_\Sigma} \quad \text{whenever} \quad [\varphi]^x_t, (\forall x\,\varphi), \vec{\gamma} \to \vec{\eta} \in \mathfrak{S}^{\vdash_\Sigma} \text{ and } t \rhd_\Sigma x : \varphi$$

is a derivable rule. Indeed consider the derivation sequence

$$
\begin{array}{lll}
1 & (\forall x\,\varphi), \vec{\gamma} \to \vec{\eta} & \text{CL 2} \\
2 & (\forall x\,\varphi), (\forall x\,\varphi), \vec{\gamma} \to \vec{\eta} & \forall\text{L 3} \\
3 & [\varphi]^x_t, (\forall x\,\varphi), \vec{\gamma} \to \vec{\eta} & \text{Hyp}
\end{array}
$$

Note also that if w is a derivation sequence and s_i is the sequent in position i, then there is a subsequence v of w that proves $\mathfrak{S} \vdash_\Sigma s_i$.

9.2   Cut elimination and consistency

A set $\mathfrak{S}$ of sequents is said to be *consistent* if $\mathfrak{S}^{\vdash_\Sigma} \subsetneq S_\Sigma$. It is easily verified that $\mathfrak{S}$ is consistent if and only if $\varepsilon \to \varepsilon \notin \mathfrak{S}^{\vdash_\Sigma}$. Sequent $\varepsilon \to \varepsilon$ is known as the *empty* or *contradictory sequent*.

The goal of this section is to establish the consistency of the Gentzen calculus by purely symbolic means, that is, to show, without resorting to semantics, that sequent $\varepsilon \to \varepsilon$ is not derivable from the axioms.

Some technical concepts and results are necessary. Recall that if $\vec{\gamma} \in L_\Sigma^*$, then $|\vec{\gamma}|$ denotes the length of sequence $\vec{\gamma}$. The *length of sequent* $s = \vec{\gamma} \to \vec{\eta}$ is $|s| = |\vec{\gamma}| + |\vec{\eta}|$.

Proposition 6 (Lemma of length of sequent)
Let $(s_1, J_1) \ldots (s_n, J_n)$ be a derivation sequence for $\vdash_\Sigma s_1$ that does not use the cut rule. Then, for each $i = 1, \ldots, n$, $|s_i| > 0$.

Proof: For each $i = 1, \ldots, n$, a case analysis establishes that the length of s_i is greater than 0:

- J_i is Ax: in this case $|s_i| = 2 > 0$;

- J_i is $R\ k$ with $R \in \{\text{PR}, \text{PL}, \text{CR}, \text{CL}, \text{WR}, \text{WL}, \neg\text{R}, \neg\text{L}, \Rightarrow\text{R}, \forall\text{R}, \forall\text{L}\}$ or J_i is $\Rightarrow\text{L } k_1, k_2$: in these cases $|s_i| \geq 1 > 0$.

QED

Therefore, without resorting to the cut rule, it is not possible to derive the empty sequent solely from the axioms. This fact led Gerhard Gentzen to attempt and succeed to prove that the cut rule is not essential. This result is established herein following a technique proposed by Samuel Buss [7].

The *depth of a formula* is inductively defined as follows:

- $\mathrm{dp}_\Sigma(\beta) = 0$ for each $\beta \in B_\Sigma$;

- $\mathrm{dp}_\Sigma(\neg\,\varphi) = 1 + \mathrm{dp}_\Sigma(\varphi)$;

- $\mathrm{dp}_\Sigma(\varphi \Rightarrow \psi) = 1 + \max\{\mathrm{dp}_\Sigma(\varphi), \mathrm{dp}_\Sigma(\psi)\}$;

- $\mathrm{dp}_\Sigma(\forall x\,\varphi) = 1 + \mathrm{dp}_\Sigma(\varphi)$.

The *depth of an application of the cut rule* in a derivation sequence is the depth of the formula to which the cut is applied.

Let s be the sequent $\gamma_1 \ldots \gamma_e \to \eta_1 \ldots \eta_d$. Recall that $|s| = e + d$. In the sequel, $\mathrm{ant}(s)$ stands for the antecedent $\gamma_1 \ldots \gamma_e$ and $\mathrm{con}(s)$ for the consequent

$\eta_1 \ldots \eta_d$. Hence, $|\text{ant}(s)| = e$ and $|\text{con}(s)| = d$. For $i = 1, \ldots, |s|$, the i-th formula of s is:

$$\begin{cases} \gamma_i & \text{if } i \leq |\text{ant}(s)|; \\ \eta_{i-|\text{ant}(s)|} & \text{if } i > |\text{ant}(s)|. \end{cases}$$

Let w be the derivation sequent $(s_1, J_1) \ldots (s_n, J_n)$. The j-th formula of s_k is a *(direct) ancestor* in w of the i-th formula of s_m if

$$\begin{cases} 1 \leq j \leq |s_k| \\ 1 \leq i \leq |s_m| \end{cases}$$

and at least one of the following alternatives holds:

- J_m is PR k, the j'-th formula of s_k is the other target of the permutation, $j = j'$ and $i = |s_m|$;

- J_m is PR k, the j'-th formula of s_k is the other target of the permutation, $j = |s_k|$ and $i = j'$;

- J_m is PR k, the j'-th formula of s_k is the other target of the permutation, $j = i$, $j \neq |s_k|$, $j \neq j'$, $i \neq |s_m|$ and $i \neq j'$;

- J_m is PL k, the j'-th formula of s_k is the other target of the permutation, $j = j'$ and $i = 1$;

- J_m is PL k, the j'-th formula of s_k is the other target of the permutation, $j = 1$ and $i = j'$;

- J_m is PL k, the j'-th formula of s_k is the other target of the permutation, $j = i$, $j \neq 1$, $j \neq j'$, $i \neq 1$ and $i \neq j'$;

- J_m is CR k and $j = i$;

- J_m is CR k, $j = |s_k|$ and $i = |s_m|$;

- J_m is CL k and $j = i + 1$;

- J_m is CL k, $j = 1$ and $i = 1$;

- J_m is WR k and $j = i$;

- J_m is WL k and $j = i - 1$ and $i > 1$;

- J_m is Cut k_1, k_2, $k = k_1$ and $j = i$;

- J_m is Cut k_1, k_2, $k = k_2$ and $j = i + 1$;

- J_m is $\neg R\, k$, $j = i + 1$ and $i < |s_m|$;

- J_m is $\neg R\, k$, $j = 1$ and $i = |s_m|$;

- J_m is $\neg L\, k$, $j = i - 1$ and $i > 1$;

- J_m is $\neg L\, k$, $j = |s_k|$ and $i = 1$;

- J_m is $\Rightarrow R\, k$, $j = i + 1$ and $i < |s_m|$;

- J_m is $\Rightarrow R\, k$, $j = 1$ and $i = |s_m|$;

- J_m is $\Rightarrow R\, k$, $j = |s_k|$ and $i = |s_m|$;

- J_m is $\Rightarrow L\, k_1, k_2$, $k = k_1$ and $j = i$;

- J_m is $\Rightarrow L\, k_1, k_2$, $k = k_2$, $j = i - 1$ and $i > 1$;

- J_m is $\Rightarrow L\, k_1, k_2$, $k = k_2$, $j = |s_k|$ and $i = 1$;

- J_m is $\forall R\, k$ and $j = i$;

- J_m is $\forall L\, k$ and $j = i$.

The next result shows that if a cut rule is used in a derivation sequence on some formula ψ, then there is an equivalent derivation sequence (that is, with the same hypotheses and the same goal) where that cut is eliminated in favor of cuts on subformulas of ψ.

Proposition 7 (Lemma of reduction of cut depth)
Let w be a derivation sequence for $\vdash_\Sigma s$ where the justification of the first step is a cut with depth r and such that all other applications of the cut rule have depth smaller than r. Then, there exists a derivation sequence w' for $\vdash_\Sigma s$ where all applications of the cut rule have depth smaller than r.

Proof: Let s be the sequent $\vec{\gamma} \to \vec{\eta}$, ψ the target formula (of depth r) of the cut in the first step of sequence w, Cut k_1, k_2 the justification for this step, w'' the subsequence[2] of w that provides a derivation sequence for $\vdash_\Sigma \vec{\gamma} \to \vec{\eta}\psi$ and w''' the subsequence of w that provides a derivation sequence for $\vdash_\Sigma \psi\vec{\gamma} \to \vec{\eta}$. The proof proceeds by a case analysis:

(1) $\psi \in B_\Sigma$:
Consider the sequence u''' obtained from w''' by replacing each sequent

$$\vec{\gamma}_k''' \to \vec{\eta}_k''' \quad \text{by} \quad (\vec{\gamma}_k''' \setminus \psi)\vec{\gamma} \to \vec{\eta}\vec{\eta}_k''',$$

---

[2]With the references in the justifications adequately changed.

where $(\vec{\gamma}_k''' \setminus \psi)$ is obtained from $\vec{\gamma}_k'''$ by removing the ancestors (in the k-th sequent of w''') of the first formula in the first sequent of w''', whenever the latter exist. Note that u''' is not, in general, a derivation sequence.

- If the justification at step k is Ax over an ancestor of the first formula in the first sequent of w''', then $\vec{\gamma}_k'''$ and $\vec{\eta}_k'''$ are both the singleton sequence ψ and $(\vec{\gamma}_k''' \setminus \psi)$ is the sequence ε, whence step k in u''' is $\vec{\gamma} \to \vec{\eta}\psi$, which is not justifiable by Ax. However, derivation sequence w'' establishes this sequent. Therefore, it suffices to combine u''' with w'' in order to justify the relevant step in u''' legitimately.

- If the justification at step k is Ax over some β that is not an ancestor of the first formula in the first sequent of w''', then $\vec{\gamma}_k'''$ and $\vec{\eta}_k'''$ are both the singleton sequence β and $(\vec{\gamma}_k''' \setminus \psi)$ is still the singleton sequence β, whence step k of u''' is $\beta\vec{\gamma} \to \vec{\eta}\beta$, which is not justifiable by Ax. However, it suffices to extend u''' with enough weakenings in order to justify the relevant step in u''' legitimately.

Let v''' be the sequence obtained from u''' as detailed above. This is a derivation sequence for $\vdash_\Sigma \vec{\gamma}, \vec{\gamma} \to \vec{\eta}\vec{\eta}$. Thus, the intended derivation sequence w' for $\vdash_\Sigma \vec{\gamma} \to \vec{\eta}$ is obtained by adding the necessary contractions to the beginning of v'''.

(2) ψ is $(\neg \delta)$:

Consider the sequence u_1'' obtained from w'' by replacing each sequent

$$\vec{\gamma}_k'' \to \vec{\eta}_k'' \quad \text{by} \quad \delta, \vec{\gamma}_k'' \to (\vec{\eta}_k'' \setminus \psi),$$

where $(\vec{\eta}_k'' \setminus \psi)$ is obtained from $\vec{\eta}_k''$ by removing the ancestors (in the k-th sequent of w'') of the last formula in the first sequent of w'', whenever the latter exist. The first sequent of u_1'' is

$$\delta, \vec{\gamma} \to \vec{\eta}.$$

Note that u_1'' is not, in general, a derivation sequence.

- Rule $\neg$R applied to ψ.
  Assume that at a certain point k' of w'' rule $\neg$ was applied to ψ. This means that in w'' the corresponding sequent was $\vec{\gamma}_{k'}'' \to \eta_{k'}'', \psi$ and, with the application of the rule, sequent $\delta, \vec{\gamma}_{k'}'' \to \eta_{k'}''$ was obtained. In u_1'' the sequents above are as follows:

$$
\begin{array}{ll}
k' & \delta, \vec{\gamma}_{k'}'' \to \eta_{k'}'' \\
k' + 1 & \delta, \delta, \gamma_{k'}'' \to \eta_{k'}''
\end{array}
$$

The justification should be as follows:

$$
\begin{array}{lll}
k' & \delta, \vec{\gamma}''_{k'} \to \eta''_{k'} & \text{CL } k'+1 \\
k'+1 & \delta, \delta, \gamma''_{k'} \to \eta''_{k'} &
\end{array}
$$

- Axioms.
  Assume that the k-th sequent of w'' is $\beta \to \beta$ where β is an atomic formula. Then, in u''_1 the corresponding sequent is $\delta, \beta \to \beta$ which is not justifiable by Ax. However the sequence can be extended with a weakening so that an axiom is obtained.

Let u''_2 be the sequence obtained from u''_1 as detailed above and with the relevant $\neg$R justification changed. Then, u''_2 is a derivation sequence for $\vdash_\Sigma \delta, \vec{\gamma} \to \vec{\eta}$ where the required reordering is assumed. In a similar way it is possible to obtain a derivation sequence u''' for $\vdash_\Sigma \vec{\gamma} \to \vec{\eta}, \delta$. Consider the sequence

$$
\begin{array}{lll}
1 & \vec{\gamma} \to \vec{\eta} & \text{Cut 2,3} \\
2 & \vec{\gamma} \to \vec{\eta}, \delta & J_1 \\
3 & \delta, \vec{\gamma} \to \vec{\eta} & J_2 \\
& \vdots &
\end{array}
$$

where J_1 and J_2 are the justifications of the first sequents in u''' and u''_2, respectively and the dots indicate the sequences u''' and u''_2 without the first sequents. Thus, the intended derivation sequence w' for $\vdash_\Sigma \vec{\gamma} \to \vec{\eta}$ is obtained by reordering the steps and eliminating possible repeated steps.

(3) ψ is $(\delta_1 \Rightarrow \delta_2)$:

Consider the sequence u''_1 obtained from w'' by replacing each sequent

$$
\vec{\gamma}''_k \to \vec{\eta}''_k \quad \text{by} \quad \delta_1, \vec{\gamma}'' \to (\vec{\eta}''_k \setminus \psi), \delta_2
$$

where $(\vec{\eta}''_k \setminus \psi)$ is obtained from $\vec{\eta}''_k$ by removing the ancestors (in the k-th sequent of w'') of the last formula in the first sequent of w'', whenever the latter exist. Thus the first sequent of u''_1 is

$$
\delta_1, \vec{\gamma} \to \vec{\eta}, \delta_2.
$$

Note that u''_1 is not, in general, a derivation sequence.

- Rule $\Rightarrow$R applied to ψ.
  Assume that at a certain point k' of w'' rule $\Rightarrow$R was applied to ψ. This means that the corresponding sequent was $\vec{\gamma}''_{k'} \to \eta''_{k'}, \psi$ and, with the application of the rule, sequent $\delta_1, \vec{\gamma}''_{k'} \to \eta''_{k'}, \delta_2$ was obtained. In u''_1 the sequents above are transformed as follows:

$$\begin{array}{ll} k' & \delta_1, \vec{\gamma}''_{k'} \to \eta''_{k'}, \delta_2 \\ k'+1 & \delta_1, \delta_1, \gamma''_{k'} \to \eta''_{k'}, \delta_2 \end{array}$$

where no justifications are provided. The sequence above should be replaced by the sequence

$$\begin{array}{lll} k' & \delta_1, \vec{\gamma}''_{k'} \to \eta''_{k'}, \delta_2 & \text{CL } k'+1 \\ k'+1 & \delta_1, \delta_1, \gamma''_{k'} \to \eta''_{k'}, \delta_2 & \end{array}$$

where the proper justifications were already inserted.

- Axiom.
  An axiom $\alpha \to \alpha$ in w'' was transformed in u''_1 into $\delta_1, \alpha \to \alpha, \delta_2$. Hence two new steps justified by WL and WR have to be added to u''_1 so that the sequence also ends with an axiom.

Let u''_2 be the sequence obtained from u''_1 as detailed above and with the relevant $\Rightarrow$R justification changed. This is a derivation sequence for $\vdash_\Sigma \delta_1, \vec{\gamma} \to \vec{\eta}, \delta_2$ doing the necessary reordering. In a similar way it is possible to obtain derivation sequences u''' and v''' for $\vdash_\Sigma \vec{\gamma} \to \vec{\eta}, \delta_1$ and $\vdash_\Sigma \delta_2, \vec{\gamma} \to \vec{\eta}$, respectively. Thus, the intended derivation sequence w' for $\vdash_\Sigma \vec{\gamma} \to \vec{\eta}$ is obtained as follows:

$$\begin{array}{lll} 1 & \vec{\gamma} \to \vec{\eta} & \text{Cut } 2,3 \\ 2 & \vec{\gamma} \to \vec{\eta}, \delta_2 & \text{Cut } 4,5 \\ 3 & \delta_2, \vec{\gamma} \to \vec{\eta} & J_1 \\ 4 & \delta_1, \vec{\gamma} \to \vec{\eta}, \delta_2 & J_2 \\ 5 & \vec{\gamma} \to \vec{\eta}, \delta_2, \delta_1 & \text{PR } 6 \\ 6 & \vec{\gamma} \to \vec{\eta}, \delta_1, \delta_2 & \text{WR } 7 \\ 7 & \vec{\gamma} \to \vec{\eta}, \delta_1 & J_3 \\ & \vdots & \end{array}$$

where J_1, J_2 and J_3 are the justifications of the first sequents in v''', u''_2 and u''', respectively and the dots indicate the sequences v''', u''_2 and u''' without the first sequents.

(4) ψ is $(\forall x \delta)$:

For the sake of simplicity, assume that rule $\forall$L was applied once at step k' to a formula $(\forall x \delta)$ using term t. That is, by application of $\forall$L, if the sequent $w_{k'}$ is $(\forall x \delta), \vec{\gamma}_{k'} \to \vec{\eta}_{k'}$, then the sequent $w_{k'+1}$ is $[\delta]^x_t, \vec{\gamma}_{k'} \to \vec{\eta}_{k'}$.

Consider the sequence u'''_1 obtained from w''' by replacing each sequent

$$\vec{\omega}'''_k \to \vec{\delta}'''_k \text{ by } (\vec{\omega}'''_k \setminus \psi), \vec{\gamma} \to \vec{\eta}, \vec{\delta}'''_k$$

where $(\vec{\omega}_k''' \setminus \psi)$ is obtained from $\vec{\omega}_k'''$ by removing the ancestors (in the k-th sequent of w'') of the first formula in the first sequent of w''', whenever the latter exist. Observe that u_1''' is not a derivation sequence. Indeed:

- An axiom $\alpha \to \alpha$ in w''' will become a sequent $\alpha, \vec{\gamma} \to \vec{\eta}, \alpha$ in u_1'''.

- Sequents $w_{k'}$ and $w_{k'+1}$ in w''' become

$$\vec{\gamma}_{k'}, \vec{\gamma} \to \vec{\eta}, \vec{\eta}_{k'} \quad \text{and} \quad [\delta]_t^x, \vec{\gamma}_{k'}, \vec{\gamma} \to \vec{\eta}, \vec{\eta}_{k'}.$$

Hence, axioms in w''' are no longer axioms in u_1''' and no justification can be given for the sequents that replace $w_{k'}$ and $w_{k'+1}$. Consider the sequence u_2''' obtained from u_1''' by replacing steps k' and $k'+1$ by the following sequence:

$$\vec{\gamma}, \vec{\gamma} \to \vec{\eta}, \vec{\eta}$$
$$\vdots$$

$\vec{\gamma}_{k'}, \vec{\gamma} \to \vec{\eta}, \vec{\eta}_{k'}$	Cut 2, 3
$\vec{\gamma}_{k'}, \vec{\gamma} \to \vec{\eta}, \vec{\eta}_{k'}, [\delta]_t^x$	(∗)
$[\delta]_t^x, \vec{\gamma}_{k'}, \vec{\gamma} \to \vec{\eta}, \vec{\eta}_{k'}$	(∗∗)

$$\vdots$$

where (∗∗) is justified in the sequence u_1''' (there is in u_1''' a derivation of this sequent) but (∗) is not justified in sequence u_2'''. On the other hand, consider the sequence u'' obtained from w'' by deleting $(\forall x \delta)$, adding $[\delta]_t^x$, substituting $[\delta]_y^x$ by $[\delta]_t^x$ and changing some justifications accordingly. The sequence u'' is a derivation for $\vec{\gamma} \to \vec{\eta}, [\delta]_t^x$. Then, (∗) can be justified by several weakenings, a permutation and u''. Let u_3''' be obtained from u_2''' by doing the insertions of the u'', that is, the sequence above becomes

$$\vec{\gamma}, \vec{\gamma} \to \vec{\eta}, \vec{\eta}$$
$$\vdots$$

$\vec{\gamma}_{k'}, \vec{\gamma} \to \vec{\eta}, \vec{\eta}_{k'}$	Cut 2, 3
$\vec{\gamma}_{k'}, \vec{\gamma} \to \vec{\eta}, \vec{\eta}_{k'}, [\delta]_t^x$	J_1
$[\delta]_t^x, \vec{\gamma}_{k'}, \vec{\gamma} \to \vec{\eta}, \vec{\eta}_{k'}$	J_2

$$\vdots$$

where J_1 and J_2 are the justifications of the first sequents in u_k'' and u_1''', respectively and the dots indicate the sequences u_k'' with the referred weakenings and permutation and u_1''' without the first sequents.

Let u_4''' be the sequence obtained from u_3''' by inserting the permutations and weakenings over sequents $\alpha, \vec{\gamma} \to \vec{\eta}, \alpha$ so that they are transformed into axioms. Let w' be the sequence u_4''' by adding in the beginning the left and right contractions and permutations so that $\vec{\gamma}$ is the left hand side and $\vec{\eta}$ is the right hand side of the initial sequent. Moreover, in w' repeated steps are to be eliminated and the ordering corrected. Then, w' is a derivation sequence for $\vdash_\Sigma \vec{\gamma} \to \vec{\eta}$ where the applications of the cut are over $[\delta]_t^x$.

In the case where the rule $\forall$L was applied more than once to an ancestor of the cut formula, the procedure described above has to be successively applied to the sub-derivations ending in an application of the rule $\forall$L, starting from the topmost ones. QED

Finally, it is now possible to prove that the sequent calculus for first-order logic has cut elimination.

Proposition 8 (Cut elimination)
If $\vdash_\Sigma s$, then there exists a derivation sequence for s from the empty set of hypotheses without any application of the cut rule.

Proof: Let w be a derivation sequence of s. The result is true when there are no applications of the cut rule in w. Assume that there is at least an application of the cut rule in w. Assume that all the applications of the cut rule are of depth at most r. Then, it is possible to get a derivation sequence for s where all the applications of the cut rule are of depth at most $r - 1$ which can easily be proved using Proposition 7. Applying the reduction several times the result follows. QED

The following is a concrete example of cut elimination where, for instance, WR^n is used to indicate that the weakening on the right rule was used n times and $+$ is used to combine two rules.

Example 9 Consider Example 4. It is possible to use the relevant steps of Proposition 7 to eliminate the cut rule in the second derivation. The cut has depth 1. The sequence w''' is

$$
\begin{array}{clc}
2 & (\forall z\, p(z, z)), (\forall x\, (\forall y\, p(y, x))) \to p(t, t) & \forall\text{L } 4 \\
4 & p(t, t), (\forall x\, (\forall y\, p(y, x))) \to p(t, t) & \text{WR } 5 \\
5 & p(t, t) \to p(t, t) & \text{Ax}
\end{array}
$$

and the sequence w'' is

$$
\begin{array}{clc}
3 & (\forall x\, (\forall y\, p(y, x))) \to p(t, t), (\forall z\, p(z, z)) & \forall\text{L } 6 \\
6 & (\forall x\, (\forall y\, p(y, x))) \Rightarrow p(t, t), p(z, z) & \forall\text{L } 7 \\
7 & (\forall y\, p(y, z)) \to p(t, t), p(z, z) & \forall\text{L } 8 \\
9 & p(z, z) \to p(t, t), p(z, z) & \text{WR } 10 \\
10 & p(z, z) \to p(z, z) & \text{Ax}
\end{array}
$$

We start by using step (4) of the proof of the referred proposition. Sequence u_1''' is as follows:

$$
\begin{array}{lll}
2 & (\forall x\,(\forall y\,p(y,x)))^2 \to p(t,t)^2 & \\
4 & p(t,t), (\forall x\,(\forall y\,p(y,x)))^2 \to p(t,t)^2 & \text{WR } 5 \\
5 & p(t,t), (\forall x\,(\forall y\,p(y,x))) \to p(t,t)^2 &
\end{array}
$$

where φ^2 is an abbreviation of φ, φ. Sequence u_2''' is as follows:

$$
\begin{array}{lll}
2 & (\forall x\,(\forall y\,p(y,x)))^2 \to p(t,t)^2 & \text{Cut 3,4} \\
3 & (\forall x\,(\forall y\,p(y,x)))^2 \to p(t,t)^3 & \\
4 & p(t,t), (\forall x\,(\forall y\,p(y,x)))^2 \to p(t,t)^2 & \text{WR } 5 \\
5 & p(t,t), (\forall x\,(\forall y\,p(y,x))) \to p(t,t)^2 &
\end{array}
$$

Sequence u'' is as follows:

$$
\begin{array}{lll}
6 & (\forall x\,(\forall y\,p(y,x))) \Rightarrow p(t,t)^2 & \forall\text{L } 7 \\
7 & (\forall y\,p(y,z)) \to p(t,t)^2 & \forall\text{L } 8 \\
9 & p(t,t) \to p(t,t)^2 & \text{WR } 10 \\
10 & p(t,t) \to p(t,t) & \text{Ax}
\end{array}
$$

Sequence u_4''' is the following:

$$
\begin{array}{lll}
2 & (\forall x\,(\forall y\,p(y,x)))^2 \to p(t,t)^2 & \text{Cut 3,4} \\
3 & (\forall x\,(\forall y\,p(y,x)))^2 \to p(t,t)^3 & \\
4 & p(t,t), (\forall x\,(\forall y\,p(y,x)))^2 \to p(t,t)^2 & \text{WR}^2\ 5 \\
5 & p(t,t), (\forall x\,(\forall y\,p(y,x))) \to p(t,t)^2 & \\
6 & (\forall x\,(\forall y\,p(y,x))) \Rightarrow p(t,t)^2 & \forall\text{L } 7 \\
7 & (\forall y\,p(y,z)) \to p(t,t)^2 & \forall\text{L } 8 \\
9 & p(t,t) \to p(t,t)^2 & \text{WR } 10 \\
10 & p(t,t) \to p(t,t) & \text{Ax}
\end{array}
$$

Finally sequence w' (already ordered and eliminating repeated steps)

$$
\begin{array}{lll}
1 & (\forall x\,(\forall y\,p(y,x))) \to p(t,t) & \text{CR+CL 2} \\
2 & (\forall x\,(\forall y\,p(y,x)))^2 \to p(t,t)^2 & \text{Cut 3,4} \\
3 & (\forall x\,(\forall y\,p(y,x)))^2 \to p(t,t)^3 & \text{WL+WR 6} \\
4 & p(t,t), (\forall x\,(\forall y\,p(y,x)))^2 \to p(t,t)^2 & \text{WL}^2\ 5 \\
5 & p(t,t) \to p(t,t)^2 & \text{WR 9} \\
6 & (\forall x\,(\forall y\,p(y,x))) \to p(t,t)^2 & \forall\text{L } 7 \\
7 & (\forall y\,p(y,z)) \to p(t,t)^2 & \forall\text{L } 8 \\
8 & p(t,t) \to p(t,t)^2 & \text{WR 9} \\
9 & p(t,t) \to p(t,t) & \text{Ax}
\end{array}
$$

has one application of the cut rule to the formula $p(t,t)$ of lower depth. The application of the cut rule over the atomic formula $p(t,t)$ has to be eliminated by using step (1) of the proof of Proposition 7 and is left as an exercise.

Cut elimination is a key ingredient of the proof by purely symbolic means of the consistency of the Gentzen calculus. It is also very helpful in automatic theorem proving, since the cut rule is not analytical in the sense that, in general, the formulas in the premises are not all subformulas of the formulas in the conclusion. Clearly, the use of analytical rules is much easier to automate.

Proposition 10 (Consistency of the Gentzen calculus)

$$\varepsilon \to \varepsilon \notin \emptyset^{\vdash_\Sigma}.$$

Proof: The proof is by contradiction. Assume that $\vdash_\Sigma \varepsilon \to \varepsilon$. Then, by the cut elimination theorem, there exists a cut-free derivation sequence for $\emptyset \vdash_\Sigma \varepsilon \to \varepsilon$. Hence, by the lemma of length of sequent, $|\varepsilon \to \varepsilon| > 0$, which is absurd. QED

Observe that Proposition 10 states only the weak consistency of the Gentzen calculus. Indeed, strong consistency (if $\mathfrak{G}$ is consistent, then $\mathfrak{G}^{\vdash_\Sigma}$ is consistent) was not established since the key cut elimination result was proved only in the absence of hypotheses. In fact, such strong consistency is out of question in full generality.

9.3 Consistency of first-order logic

In order to establish, still by purely symbolic means, that Hilbertian first-order logic is consistent, it is necessary to relate the Gentzen calculus with the Hilbert calculus. Hence, it is mandatory to work simultaneously with the notions of Hilbert- and Gentzen-derivations, which requires some additional notation. Let $\Gamma \subseteq L_\Sigma$. By

$$\Gamma \vdash_\Sigma^{\mathsf{H}} \varphi$$

it is meant that formula φ is derivable in the Hilbert calculus from the set of hypotheses Γ. By

$$\Gamma \vdash_\Sigma^{\mathsf{G}} \varphi$$

it is meant that there exists a finite subset $\{\gamma_1, \ldots, \gamma_e\}$ of Γ such that $\vdash_\Sigma^{\mathsf{G}} \varepsilon \to \varphi$ whenever $\vdash_\Sigma^{\mathsf{G}} \varepsilon \to \gamma_i$ for every $i = 1, \ldots, e$. Note that, due to the rule of permutation on the left, the order in which the sequence $\gamma_1 \ldots \gamma_e$ is extracted from $\{\gamma_1, \ldots, \gamma_e\}$ is not relevant. By

$$\Gamma \Vdash_\Sigma^{\mathsf{G}} \varphi$$

it is meant that there is a finite subset $\{\gamma_1, \ldots, \gamma_e\}$ of Γ such that

$$\vdash_\Sigma^{\mathsf{G}} \gamma_1 \ldots \gamma_e \to \varphi.$$

Observe that it is not always the case that

$$\vdash_{\Sigma}^{\mathsf{G}} \varphi \to (\forall x \varphi)$$

since the $\forall$R rule cannot be applied when $x \in \mathrm{fv}_{\Sigma}(\varphi)$. Hence,

$$\varphi \not\Vdash_{\Sigma}^{\mathsf{G}} (\forall x\, \varphi).$$

But, it is the case that

$$\varphi \vdash_{\Sigma}^{\mathsf{G}} (\forall x\, \varphi).$$

This fact helps in understanding the definition of $\vdash_{\Sigma}^{\mathsf{G}}$ and $\Vdash_{\Sigma}^{\mathsf{G}}$.

At this point, it is convenient to introduce the following two operators over the complete lattice $\langle \wp L_{\Sigma}, \subseteq \rangle$:

- (global operator) $\lambda\,\Gamma\,.\,\Gamma^{\vdash_{\Sigma}^{\mathsf{G}}}$ where $\Gamma^{\vdash_{\Sigma}^{\mathsf{G}}} = \{\varphi : \Gamma \vdash_{\Sigma}^{\mathsf{G}} \varphi\}$;

- (local operator) $\lambda\,\Gamma\,.\,\Gamma^{\Vdash_{\Sigma}^{\mathsf{G}}}$ where $\Gamma^{\Vdash_{\Sigma}^{\mathsf{G}}} = \{\varphi : \Gamma \Vdash_{\Sigma}^{\mathsf{G}} \varphi\}$.

Proposition 11 Operator $\Vdash_{\Sigma}^{\mathsf{G}}$ is a compact closure operator.

Proof: (a) Extensitivity. Let $\gamma \in \Gamma$. Then, $\gamma \to \gamma$ is an axiom and so there is a finite set $\{\gamma\}$ such that $\vdash_{\Sigma} \gamma \to \gamma$.
(b) Monotonicity. Assume that $\Gamma_1 \subseteq \Gamma_2$ and that $\Gamma_1 \Vdash_{\Sigma}^{\mathsf{G}} \varphi$. Then, there is a finite set $\Psi \subseteq \Gamma_1$ such that $\vdash_{\Sigma} \Psi \to \varphi$. But Ψ is also a subset of Γ_2 and so $\Gamma_2 \Vdash_{\Sigma}^{\mathsf{G}} \varphi$.
(c) Idempotence. Assume that $\Gamma^{\Vdash_{\Sigma}^{\mathsf{G}}} \Vdash_{\Sigma}^{\mathsf{G}} \varphi$. Then, there is a finite subset Ψ of $\Gamma^{\Vdash_{\Sigma}^{\mathsf{G}}}$ such that $\vdash_{\Sigma} \Psi \to \varphi$. Moreover, there is a finite set $\Omega_{\psi} \subseteq \Gamma$ such that $\vdash_{\Sigma} \Omega_{\psi} \to \psi$, for each $\psi \in \Psi$. It is proved below that

$$\vdash_{\Sigma} \bigcup_{\psi \in \Psi} \Omega_{\psi} \to \varphi.$$

Assume, without loss of generality that $\Psi = \{\psi_1, \psi_2\}$ and $\Omega = \bigcup_{\psi \in \Psi} \Omega_{\psi}$. The sequence

1	$\Omega \to \varphi$	Cut 2,3
2	$\Omega \to \varphi, \psi_1$	WL$^+$+WR 4
3	$\psi_1, \Omega \to \varphi$	Cut 5,6
4	$\Omega_{\psi_1} \to \psi_1$	Thm
5	$\psi_1, \Omega \to \varphi, \psi_2$	WL$^+$+WR 7
6	$\psi_2, \psi_1, \Omega \to \varphi$	WL$^+$ 8
7	$\Omega_{\psi_2} \to \psi_2$	Thm
8	$\psi_2, \psi_1 \to \varphi$	Thm

shows the thesis. The proof of compactness is straightforward from the definition and is left as an exercise. QED

Exercise 12 Analyze the properties of operator $\vdash_{\Sigma}^{G}$.

The following result states an important relationship between the local and the global Gentzen operators on sets of formulas.

Proposition 13 If $\Gamma \Vdash_{\Sigma}^{G} \varphi$, then $\Gamma \vdash_{\Sigma}^{G} \varphi$.

Proof: Assume that $\Gamma \Vdash_{\Sigma}^{G} \varphi$. Then, there is $\{\gamma_1, \ldots, \gamma_n\} \subseteq \Gamma$ such that $\gamma_1, \ldots, \gamma_n \Vdash_{\Sigma}^{G} \varphi$. That is, $\gamma_1, \ldots, \gamma_n \to \varphi$ is a theorem. The sequence

1	$\varepsilon \to \varphi$	Cut 2,3
2	$\varepsilon \to \varphi, \gamma_n$	WR 4
3	$\gamma_n \to \varphi$	Cut 5,6
4	$\varepsilon \to \gamma_n$	Hyp
	...	
3n-2	$\gamma_2, \ldots, \gamma_n \to \varphi$	Cut 3n-1,3n
3n-1	$\gamma_2, \ldots, \gamma_n \to \varphi, \gamma_1$	WR+WL 3n+1
3n	$\gamma_1, \ldots, \gamma_n \to \varphi$	Thm
3n+1	$\varepsilon \to \gamma_1$	Hyp

shows that if $\varepsilon \to \gamma_i$ is a theorem for $i = 1, \ldots, n$, then $\varepsilon \to \varphi$ is also a theorem and so $\Gamma \vdash_{\Sigma}^{G} \varphi$. QED

For instance $\{\varphi, (\varphi \Rightarrow \psi)\} \Vdash_{\Sigma}^{G} \psi$ and, therefore, $\{\varphi, (\varphi \Rightarrow \psi)\} \vdash_{\Sigma}^{G} \psi$. The following two results relate derivability in the Hilbert and the Gentzen calculi.

Proposition 14 Let w be a derivation sequence for $\Gamma \vdash_{\Sigma}^{H} \varphi$ where no generalizations were applied to variables that occur free in Γ. Then, $\Gamma \Vdash_{\Sigma}^{G} \varphi$.

Proof: The proof is by induction on the length n of w.
Base: There are two cases.
(i) $w_n = \langle \varphi, \text{Hyp} \rangle$. Then, $\varphi \to \varphi$ is an axiom.
(ii) $w_n = \langle \varphi, \text{Ax} \rangle$. Then, $\varepsilon \to \varphi$ is a theorem.
Step: Only one case is considered and the others are left to the reader.
(iii) $w_n = \langle \varphi, \text{Gen } k \rangle$. Then, by the induction hypothesis, $\Gamma \Vdash_{\Sigma}^{G} \psi_k$ and so there is a finite set $\Phi \subseteq \Gamma$ such that $\Phi \to \psi_k$ is a theorem. Moreover

1	$\Phi \to \varphi$	$\forall$R 2
2	$\Phi \to \psi_k$	Thm

is a derivation sequence where the $\forall$R rule can be applied because of the hypothesis. QED

Proposition 15

$$\text{If } \Gamma \vdash^{\mathsf{H}}_{\Sigma} \varphi, \text{ then } \Gamma \vdash^{\mathsf{G}}_{\Sigma} \varphi.$$

Proof: One begins by checking that $\vdash^{\mathsf{G}}_{\Sigma} \alpha$, that is, $\vdash_{\Sigma} \varepsilon \to \alpha$ for each axiom α of the Hilbert calculus.

One then checks that if $\vdash^{\mathsf{G}}_{\Sigma} \varphi$ and $\vdash^{\mathsf{G}}_{\Sigma} (\varphi \Rightarrow \psi)$, that is, if $\vdash_{\Sigma} \varepsilon \to \varphi$ and $\vdash_{\Sigma} \varepsilon \to (\varphi \Rightarrow \psi)$, then $\vdash_{\Sigma} \varepsilon \to \psi$, that is, $\vdash^{\mathsf{G}}_{\Sigma} \psi$.

Next one checks that if $\vdash^{\mathsf{G}}_{\Sigma} \varphi$, that is, if $\vdash_{\Sigma} \varepsilon \to \varphi$, then $\vdash_{\Sigma} \varepsilon \to (\forall x\, \varphi)$, that is, $\vdash^{\mathsf{G}}_{\Sigma} (\forall x\, \varphi)$.

Finally, the result is obtained by induction on the length of the derivation sequence for φ from Γ in the Hilbert calculus.   QED

Using this result and the consistency of the Gentzen calculus, it is possible, as David Hilbert intended, to establish, by purely symbolic means, the (weak) consistency of first-order logic (as defined in Chapter 5).

Proposition 16 (Proof-theoretic consistency of Hilbert calculus)

$$\emptyset^{\vdash_{\Sigma}} \neq L_{\Sigma}.$$

Proof: The proof is by contradiction. Assume that $\emptyset^{\vdash_{\Sigma}} = L_{\Sigma}$. In particular, for every $\psi \in L_{\Sigma}$, one would have $\vdash^{\mathsf{H}}_{\Sigma} \psi$ and $\vdash^{\mathsf{H}}_{\Sigma} (\neg \psi)$. So, by Proposition 15, it would follow that $\vdash^{\mathsf{G}}_{\Sigma} \psi$ and $\vdash^{\mathsf{G}}_{\Sigma} (\neg \psi)$. That is,

$$(a) \quad \vdash_{\Sigma} \varepsilon \to \psi$$
$$(b) \quad \vdash_{\Sigma} \varepsilon \to (\neg \psi)$$

which would allow the empty sequent to be derived, contradicting consistency of the Gentzen calculus, in the following way:

$$
\begin{array}{llll}
1 & \varepsilon \to \varepsilon & \text{Cut 2,3} \\
2 & \varepsilon \to (\neg \psi) & (b) \\
3 & (\neg \psi) \to \varepsilon & \neg\text{L 4} \\
4 & \varepsilon \to \psi & (a)
\end{array}
$$

Therefore, the Hilbert calculus is consistent as well.   QED

9.4   Soundness and completeness

The (strong) soundness of the calculus consists in proving that if $\mathfrak{S} \vdash^{\mathsf{G}}_{\Sigma} s$, then $\mathfrak{S}$ semantically entails s. As expected, a sequent s is said to be *entailed* by a set of sequents $\mathfrak{S}$, written $\mathfrak{S} \vDash_{\Sigma} s$, if, for every interpretation structure I, $I \Vdash_{\Sigma} s$ whenever $I \Vdash_{\Sigma} s'$ for every $s' \in \mathfrak{S}$.

The following results are useful to relate satisfaction of sequents with satisfaction of formulas.

Proposition 17 Let $\varphi \in L_\Sigma$ be a formula and I an interpretation structure over Σ. Then,

$$I \Vdash_\Sigma \varphi \text{ if and only if } I \Vdash_\Sigma \varepsilon \to \varphi.$$

As a corollary, it follows:

$$\varphi \text{ is valid if and only if } \varepsilon \to \varphi \text{ is valid.}$$

Moreover, it is possible to relate semantic entailment between formulas and semantic entailment between sequents.

Proposition 18 Let $\Gamma \cup \varphi \subseteq L_\Sigma$. Then,

$$\Gamma \vDash_\Sigma \varphi \text{ if and only if } \{\varepsilon \to \gamma : \gamma \in \Gamma\} \vDash_\Sigma \varepsilon \to \varphi.$$

The proof of the proposition above is also a corollary of Proposition 17. It is possible to define a local notion of entailment between formulas. A formula φ is said to be *locally entailed* by a set of formulas Γ, written $\Gamma \rhd_\Sigma \varphi$, if, for every interpretation structure I and every assignment ρ, $I\rho \Vdash_\Sigma \varphi$ whenever $I\rho \Vdash_\Sigma \gamma$ for every $\gamma \in \Gamma$.

Proposition 19 Let $\Gamma \cup \varphi \subseteq L_\Sigma$ and Γ be a finite set. Then,

$$\Gamma \rhd_\Sigma \varphi \quad \text{if and only if} \quad \bigwedge_{\gamma \in \Gamma} \gamma \to \varphi \text{ is valid.}$$

The two notions of entailment can be related as shown below.

Proposition 20 Let $\Gamma \cup \varphi \subseteq L_\Sigma$. Then,

$$\Gamma \rhd_\Sigma \varphi \quad \text{implies} \quad \Gamma \vDash_\Sigma \varphi.$$

The technique that was provided for proving the soundness of the Hilbert calculus, see Chapter 7, can also be applied to prove soundness of the Gentzen calculus: it is enough to prove that the axioms are valid sequents (soundness of the axioms) and that the set of the premises of a rule semantically entails the conclusion of the rule (soundness of the rules).

Proposition 21 The axioms are sound.

Proof: Straightforward from the definition of satisfaction of a sequent by an interpretation structure. QED

Proposition 22 The inference rules are sound.

Proof: Only the proof of soundness of the cut rule and the right and left rules for the universal quantifier is provided. The soundness of the other rules is left as an exercise.

(i) Soundness of the cut rule. It is necessary to prove

$$\vec{\gamma} \to \vec{\eta}\,\psi, \ \psi\,\vec{\gamma} \to \vec{\eta} \models_\Sigma \vec{\gamma} \to \vec{\eta}.$$

Let I be an interpretation structure for Σ. Assume that

(†) $I \Vdash_\Sigma \vec{\gamma} \to \vec{\eta}, \psi$ and (††) $I \Vdash_\Sigma \psi, \vec{\gamma} \to \vec{\eta}$.

Let ρ be an assignment in I. Assume that

(†††) $I\rho \Vdash_\Sigma \vec{\gamma}.$

There are two cases to consider:
(a) $I\rho \Vdash_\Sigma \psi$. Then, by (††) one concludes that $I\rho \Vdash_\Sigma \eta_j$ for some $j = 1, \ldots, e$ and so $I\rho \Vdash_\Sigma \vec{\eta}$.
(b) $I\rho \not\Vdash_\Sigma \psi$. By (†), either $I\rho \Vdash_\Sigma \eta_j$ for some $j = 1, \ldots, e$ or $I\rho \Vdash_\Sigma \psi$. The latter one is impossible by hypothesis. Hence, $I\rho \Vdash_\Sigma \eta_j$ for some $j = 1, \ldots, e$. In both cases, $I \Vdash_\Sigma \vec{\gamma} \to \vec{\eta}$.

(ii) Soundness of the $\forall$R rule. It is necessary to prove

$$\vec{\gamma} \to \vec{\eta}, [\varphi]_y^x \models_\Sigma \vec{\gamma} \to \vec{\eta}, (\forall x \varphi)$$

whenever $y \notin \mathrm{fv}_\Sigma(\vec{\gamma} \cup \vec{\eta})$. Let I be an interpretation structure for Σ. Assume that

(†) $I \Vdash_\Sigma \vec{\gamma} \to \vec{\eta}, [\varphi]_y^x.$

Let ρ be an assignment in I. Assume that

(††) $I\rho \Vdash_\Sigma \vec{\gamma}.$

Then, either (a) $I\rho \Vdash_\Sigma \eta_i$ for some $i = 1, \ldots, d$ or (b) $I\rho \Vdash_\Sigma [\varphi]_y^x$.
(a) It follows immediately that $I\rho \Vdash_\Sigma \vec{\gamma} \to \vec{\eta}, (\forall x\varphi)$.
(b) Let σ be an assignment x-equivalent to ρ. Take σ' as the assignment y-equivalent to σ such that $\sigma'(y) = \sigma(x)$. Then, by Proposition 19 of Chapter 7, $I\sigma' \Vdash_\Sigma [\varphi]_y^x$ iff $I\sigma \Vdash_\Sigma \varphi$, because φ is $[[\varphi]_y^x]_x^y$. Since, $y \notin \mathrm{fv}_\Sigma(\vec{\gamma} \cup \vec{\eta})$, then, by the lemma of the absent free variables (see Proposition16 of Chapter 7), $I\sigma' \Vdash_\Sigma \gamma_i$ for $i = 1, \ldots, d$ and $I\sigma' \not\Vdash_\Sigma \eta_j$ for $j = 1, \ldots, e$. Hence $I\sigma' \Vdash_\Sigma [\varphi]_y^x$ and so $I\sigma \Vdash_\Sigma \varphi$.

(iii) Soundness of the $\forall$L rule. It is necessary to prove

$$[\varphi]_t^x\vec{\gamma} \to \vec{\eta} \models_\Sigma (\forall x\,\varphi)\vec{\gamma} \to \vec{\eta}$$

whenever $t \rhd_\Sigma x : \varphi$. Let I be an interpretation structure for Σ. Assume that

$$(\dagger) \quad I \Vdash_\Sigma [\varphi]_t^x \vec{\gamma} \to \vec{\eta}.$$

Let ρ be an assignment in I. Assume that

$$(\dagger\dagger) \quad I\rho \Vdash_\Sigma (\forall x \, \varphi)\vec{\gamma}.$$

In particular, $I\rho' \Vdash_\Sigma \varphi$ where ρ' is x-equivalent to ρ and $\rho'(x) = [\![t]\!]_\Sigma^{I\rho}$. Since $t \rhd_\Sigma x : \varphi$, by Proposition 19 of Chapter 7,

$$(\dagger\dagger\dagger) \quad I\rho \Vdash_\Sigma [\varphi]_t^x.$$

From ($\dagger\dagger$) and ($\dagger\dagger\dagger$),

$$I\rho \Vdash_\Sigma [\varphi]_t^x \vec{\gamma}$$

and so, by ($\dagger$), it can be concluded that there is j such that

$$I\rho \Vdash_\Sigma \eta_j.$$

Therefore, $I\rho \Vdash_\Sigma (\forall x \, \varphi)\vec{\gamma} \to \vec{\eta}$. $\hfill$ QED

Proposition 23 (Soundness of Gentzen calculus)

$$\text{If } \mathfrak{S} \vdash_\Sigma^{\mathsf{G}} s, \text{ then } \mathfrak{S} \vDash_\Sigma s.$$

Proof: The result is obtained by induction on the length of the derivation sequence for $\mathfrak{S} \vdash_\Sigma^{\mathsf{G}} s$. $\hfill$ QED

The soundness more appropriate for the purpose of the book follows as a corollary.

Proposition 24 (Soundness of Gentzen calculus on formulas)

$$\text{If } \Gamma \vdash_\Sigma^{\mathsf{G}} \varphi, \text{ then } \Gamma \vDash_\Sigma \varphi.$$

This amounts to global soundness. However, there is a local counterpart:

Proposition 25 Assume that $\Gamma \subseteq L_\Sigma$ is a finite set.

$$\text{If } \Gamma \Vvdash_\Sigma^{\mathsf{G}} \varphi, \text{ then } \Gamma \rhd_\Sigma \varphi.$$

The Gentzen calculus is also complete with respect to the Tarskian semantics. Herein, only completeness of the Gentzen calculus on formulas (as opposed to sequents) is proved. The result is achieved in an indirect way via the Hilbert calculus.

Proposition 26 (Completeness of Gentzen calculus on formulas)

$$\text{If } \Gamma \vDash_\Sigma \varphi, \text{ then } \Gamma \vdash^{\mathsf{G}}_\Sigma \varphi.$$

Proof: The result follows from Proposition 15 by completeness of the Hilbert calculus.   QED

Finally, it is possible to prove that the Hilbert calculus and the Gentzen calculus on formulas are inter-derivable.

Proposition 27 (Equivalence of Hilbert and Gentzen calculi)

$$\Gamma \vdash^{\mathsf{H}}_\Sigma \varphi \text{ if and only if } \Gamma \vdash^{\mathsf{G}}_\Sigma \varphi.$$

Proof: Indeed,

$$
\begin{array}{ll}
\text{if} & \Gamma \vdash^{\mathsf{H}}_\Sigma \varphi \\[4pt]
\text{then} & \Gamma \vdash^{\mathsf{G}}_\Sigma \varphi \quad \text{(Proposition 15)} \\[4pt]
& \Gamma \vDash_\Sigma \varphi \quad \text{(Proposition 23)} \\[4pt]
& \Gamma \vdash^{\mathsf{H}}_\Sigma \varphi \quad \text{(Completeness of the Hilbert calculus)}
\end{array}
$$

QED

Chapter 10

Equality

This chapter is dedicated to equality, first by looking at theories with equality and, afterward, by mentioning first-order logic with equality built-in.

As expected, a theory with equality should include as theorems the characteristic properties of the equality binary relation: reflexivity, symmetry, transitivity and congruence. The latter should be taken to mean that if two terms are equal, then they are interchangeable (in every term and atomic formula).

A particular example of theory with equality is the minimal theory of equality containing only the strict knowledge about equality. It has no function symbols and no predicate symbols besides equality.

The equality predicate symbol can be interpreted by any congruence relation (not necessarily the diagonal relation =) while still satisfying its characteristic properties. However, one of the main results of the chapter is precisely to show that if a theory with equality has a model, then it also has a normal model, that is, a model where the equality symbol is interpreted by the diagonal relation.

Theories with equality are the focus of Section 10.1. Normal models are addressed in Section 10.2. Finally, Section 10.3 contains a very brief reference to first-order predicate logic with equality.

10.1 Theories with equality

A *signature with equality* is a first-order signature where $\cong \in P_2$. It is usual to write $(t_1 \cong t_2)$ instead of $\cong [t_1, t_2]$. One uses $\cong$ in place of $=$ in order to stress that the predicate symbol $\cong$ is not always interpreted as equality. This option has the additional advantage of distinguishing between equality in the formal language (of first-order logic) and equality in the meta-language of Mathematics.

Recall that a first-order theory over a signature Σ is a set of formulas over Σ that is closed under $\vdash_\Sigma$. A theory Θ over Σ is said to be a *theory with equality* if:

- Σ is a signature with equality;

- Θ contains the following theorems:

  E1 $(\forall x_1(x_1 \cong x_1))$;

  E2 $(\forall x_1(\ldots(\forall x_{n+n}((x_1 \cong x_{n+1}) \Rightarrow (\ldots \Rightarrow ((x_n \cong x_{n+n}) \Rightarrow$
  $(f(x_1,\ldots,x_n) \cong f(x_{n+1},\ldots,x_{n+n})))\ldots)))\ldots))$ for each $f \in F_n$;

  E3 $(\forall x_1(\ldots(\forall x_{n+n}((x_1 \cong x_{n+1}) \Rightarrow (\ldots \Rightarrow ((x_n \cong x_{n+n}) \Rightarrow$
  $(p(x_1,\ldots,x_n) \Rightarrow p(x_{n+1},\ldots,x_{n+n})))\ldots)))\ldots))$ for each $p \in P_n$.

Theorem E1 states the reflexivity of equality. Theorems E2 and E3 state that equality is congruently compatible with functions and predicates, respectively.

The reader is asked to show in the following exercise that the minimal theory of equality (the theory that only contains knowledge about equality) is a theory with equality.

Exercise 1 Consider the signature $\Sigma_\cong$ such that:

- $F_n = \emptyset$ for every $n \in \mathbb{N}$;

- $P_2 = \{\cong\}$;

- $P_n = \emptyset$ for every $n \neq 2$.

Let the *theory of equality* be the theory over $\Sigma_\cong$ with the following specific axioms:

Ref $(\forall x_1(x_1 \cong x_1))$;

Sym $(\forall x_1(\forall x_2((x_1 \cong x_2) \Rightarrow (x_2 \cong x_1))))$;

Trans $(\forall x_1(\forall x_2(\forall x_3((x_1 \cong x_2) \Rightarrow ((x_2 \cong x_3) \Rightarrow (x_1 \cong x_3))))))$.

Show that the theory of equality is a theory with equality.

The next result shows that, in every theory with equality, $\cong$ fulfills the requirements of an equivalence relation over the set of terms.

Proposition 2 In every theory with equality, for all terms t, t_1, t_2 and t_3, the following are theorems:

RE1 $(t \cong t)$;

RE2 $((t_1 \cong t_2) \Rightarrow (t_2 \cong t_1))$;

RE3 $((t_1 \cong t_2) \Rightarrow ((t_2 \cong t_3) \Rightarrow (t_1 \cong t_3)))$.

Proof:

RE1: Immediate by E1 using Ax4.

RE2: Observe first that, for all distinct variables y_1, y_2, y_3 and y_4, the formula

gE3 $(\forall y_1(\forall y_2(\forall y_3(\forall y_4((y_1 \cong y_3) \Rightarrow ((y_2 \cong y_4) \Rightarrow ((y_1 \cong y_2) \Rightarrow (y_3 \cong y_4))))))))$

belongs to every theory with equality. Indeed, consider the following derivation sequence where the auxiliary variables z_1, z_2 and z_3 are chosen distinct from each other and from the elements of $\{x_1, \ldots, x_4, y_1, \ldots, y_4\}$:

1	$(\forall x_1(\forall x_2(\forall x_3(\forall x_4$ $((x_1 \cong x_3) \Rightarrow ((x_2 \cong x_4) \Rightarrow ((x_1 \cong x_2) \Rightarrow (x_3 \cong x_4))))))))$	E3
2	$((\forall x_1(\forall x_2(\forall x_3(\forall x_4$ $((x_1 \cong x_3) \Rightarrow ((x_2 \cong x_4) \Rightarrow ((x_1 \cong x_2) \Rightarrow (x_3 \cong x_4))))))))$ $\Rightarrow$ $(\forall x_2(\forall x_3(\forall x_4$ $((z_1 \cong x_3) \Rightarrow ((x_2 \cong x_4) \Rightarrow ((z_1 \cong x_2) \Rightarrow (x_3 \cong x_4))))))))$	Ax4
3	$(\forall x_2(\forall x_3(\forall x_4((z_1 \cong x_3) \Rightarrow$ $((x_2 \cong x_4) \Rightarrow ((z_1 \cong x_2) \Rightarrow (x_3 \cong x_4))))))$	MP 1,2
	$\ldots$	
7	$((\forall x_4((z_1 \cong z_3) \Rightarrow$ $((z_2 \cong x_4) \Rightarrow ((z_1 \cong z_2) \Rightarrow (z_3 \cong x_4)))))$	MP 5,6
8	$(\forall x_4((z_1 \cong z_3) \Rightarrow ((z_2 \cong x_4) \Rightarrow ((z_1 \cong z_2) \Rightarrow (z_3 \cong x_4)))))$ $\Rightarrow$ $((z_1 \cong z_3) \Rightarrow ((z_2 \cong y_4) \Rightarrow ((z_1 \cong z_2) \Rightarrow (z_3 \cong y_4))))$	Ax4
9	$((z_1 \cong z_3) \Rightarrow ((z_2 \cong y_4) \Rightarrow ((z_1 \cong z_2) \Rightarrow (z_3 \cong y_4))))$	MP 7,8
10	$(\forall z_1((z_1 \cong z_3) \Rightarrow$ $((z_2 \cong y_4) \Rightarrow ((z_1 \cong z_2) \Rightarrow (z_3 \cong y_4)))))$	Gen 9
11	$((\forall z_1((z_1 \cong z_3) \Rightarrow$ $((z_2 \cong y_4) \Rightarrow ((z_1 \cong z_2) \Rightarrow (z_3 \cong y_4)))))$ $\Rightarrow$ $((y_1 \cong z_3) \Rightarrow ((z_2 \cong y_4) \Rightarrow ((y_1 \cong z_2) \Rightarrow (z_3 \cong y_4)))))$	Ax4

12    $((y_1 \cong z_3) \Rightarrow ((z_2 \cong y_4) \Rightarrow ((y_1 \cong z_2) \Rightarrow (z_3 \cong y_4))))$ MP 10,11

$\dots$

18    $((y_1 \cong y_3) \Rightarrow ((y_2 \cong y_4) \Rightarrow ((y_1 \cong y_2) \Rightarrow (y_3 \cong y_4))))$ MP 16,17

$\dots$

22    $(\forall y_1 (\forall y_2 (\forall y_3 (\forall y_4 ((y_1 \cong y_3) \Rightarrow$
$((y_2 \cong y_4) \Rightarrow ((y_1 \cong y_2) \Rightarrow (y_3 \cong y_4)))))))$ Gen 21

With gE3 at hand, consider now the following derivation sequence, where variables $y_1, \dots, y_4$ are chosen distinct from each other and from those in $\mathrm{var}_\Sigma(t_1) \cup \mathrm{var}_\Sigma(t_2)$:

1    $(t_1 \cong t_2)$ Hyp
2    $(\forall y_1 (\forall y_2 (\forall y_3 (\forall y_4$
       $((y_1 \cong y_3) \Rightarrow ((y_2 \cong y_4) \Rightarrow ((y_1 \cong y_2) \Rightarrow (y_3 \cong y_4)))))))$ gE3

$\dots$ (*)

10    $((t_1 \cong t_2) \Rightarrow ((t_1 \cong t_1) \Rightarrow ((t_1 \cong t_1) \Rightarrow (t_2 \cong t_1))))$ MP 8,9
11    $((t_1 \cong t_1) \Rightarrow ((t_1 \cong t_1) \Rightarrow (t_2 \cong t_1)))$ MP 1,10
12    $(t_1 \cong t_1)$ RE1
13    $((t_1 \cong t_1) \Rightarrow (t_2 \cong t_1))$ MP 11,12
14    $(t_2 \cong t_1)$ MP 12,13

(*) Using Ax4 and MP in order to achieve the following instantiations: $y_1 \mapsto t_1$, $y_2 \mapsto t_1$, $y_3 \mapsto t_2$ and $y_4 \mapsto t_1$.
Hence, applying the MTD, it follows that RE2 belongs to every theory with equality.

RE3: A similar technique establishes RE3, using again RE1 and gE3, but now with the following instantiations: $y_1 \mapsto t_1$, $y_2 \mapsto t_2$, $y_3 \mapsto t_1$ and $y_4 \mapsto t_3$. QED

Exercise 3 Simplify the proof of Proposition 2 using the rule of substitution (Proposition 33 of Section 5.4).

From now on, in the context of an interpretation structure $I = (D, \_^{\mathsf{F}}, \_^{\mathsf{P}})$ over a signature with equality, one may confuse the map $\cong_2^{\mathsf{P}} \colon D^2 \to \{0, 1\}$ with the binary relation

$$\{(d_1, d_2) \in D^2 : \cong_2^{\mathsf{P}}(d_1, d_2) = 1\}.$$

Furthermore, $d_1 \cong_2^{\mathsf{P}} d_2$ may be used instead of $\cong_2^{\mathsf{P}}(d_1, d_2) = 1$.
The next result shows that $E1$, $E2$ and $E3$ constrain as envisaged the interpretation of the predicate symbol $\cong$.

Proposition 4 In every model of a theory with equality, $\cong_2^P$ is a congruence relation. In other words, $\cong_2^P$ is an equivalence relation such that:

- for each $f \in F_n$,

  if $d_1 \cong_2^P d_1'$, ..., $d_n \cong_2^P d_n'$, then $f_n^F(d_1, \ldots, d_n) \cong_2^P f_n^F(d_1', \ldots, d_n')$;

- for each $p \in P_n$,

  if $d_1 \cong_2^P d_1'$, ..., $d_n \cong_2^P d_n'$, then $p_n^P(d_1, \ldots, d_n) = p_n^P(d_1', \ldots, d_n')$.

Proof: First note that, according to the previous proposition, the theory contains the following formulas:

(1) $(x_1 \cong x_1)$;

(2) $((x_1 \cong x_2) \Rightarrow (x_2 \cong x_1))$;

(3) $((x_1 \cong x_2) \Rightarrow ((x_2 \cong x_3) \Rightarrow (x_1 \cong x_3)))$.

Let $I = (D, \_^F, \_^P)$ be a model of the theory. Then:

(a) The relation $\cong_2^P$ is an equivalence relation. Indeed:

The relation $\cong_2^P$ is reflexive thanks to (1). In fact, $I \Vdash_\Sigma (x_1 \cong x_1)$, whence, for every assignment ρ into I, $I\rho \Vdash_\Sigma (x_1 \cong x_1)$, that is, $\rho(x_1) \cong_2^P \rho(x_1)$. In particular, for every $d \in D$, choosing ρ such that $\rho(x_1) = d$, it follows that $d \cong_2^P d$. Symmetry and transitivity of relation $\cong_2^P$ are established in a similar way, using the fact that the theory contains (2) and (3), respectively.

(b) For each $f \in F_n$,

  if $d_1 \cong_2^P d_1', \ldots, d_n \cong_2^P d_n'$, then $f_n^F(d_1, \ldots, d_n) \cong_2^P f_n^F(d_1', \ldots, d_n')$.

Indeed,

$$I \Vdash_\Sigma ((x_1 \cong x_{n+1}) \Rightarrow (\ldots \Rightarrow ((x_n \cong x_{n+n}) \Rightarrow \\ (f(x_1, \ldots, x_n) \cong f(x_{n+1}, \ldots, x_{n+n}))) \ldots))$$

and, so, for every assignment ρ into I,

(1) $\quad I\rho \Vdash_\Sigma ((x_1 \cong x_{n+1}) \Rightarrow (\ldots \Rightarrow ((x_n \cong x_{n+n}) \Rightarrow \\ (f(x_1, \ldots, x_n) \cong f(x_{n+1}, \ldots, x_{n+n}))) \ldots)).$

Choosing ρ such that $\rho(x_i) = d_i$ and $\rho(x_{n+i}) = d_i'$ for $i = 1, \ldots, n$, from the hypotheses

$$\begin{cases} d_1 \cong_2^P d_1' \\ \ldots \\ d_n \cong_2^P d_n' \end{cases}$$

one is able to obtain:

$$(2) \quad \begin{cases} I\rho \Vdash_{\Sigma} (x_1 \cong x_{n+1}) \\ \ldots \\ I\rho \Vdash_{\Sigma} (x_n \cong x_{n+n}). \end{cases}$$

From (1) and (2), it follows that

$$I\rho \Vdash_{\Sigma} (f(x_1, \ldots, x_n) \cong f(x_{n+1}, \ldots, x_{n+n})),$$

whence

$$f_n^{\mathsf{F}}(d_1, \ldots, d_n) \cong_2^{\mathsf{P}} f_n^{\mathsf{F}}(d_1', \ldots, d_n').$$

(c) For each $p \in P_n$,

if $d_1 \cong_2^{\mathsf{P}} d_1', \ldots, d_n \cong_2^{\mathsf{P}} d_n'$, then $p_n^{\mathsf{P}}(d_1, \ldots, d_n) = p_n^{\mathsf{P}}(d_1', \ldots, d_n')$.

Indeed, the reader should not have any difficulty in proving this result. Hint: Begin by showing that, for each $p \in P_n$, the formula

E3 $\Leftrightarrow (\forall x_1(\ldots (\forall x_{n+n} ((x_1 \cong x_{n+1}) \Rightarrow (\ldots \Rightarrow ((x_n \cong x_{n+n}) \Rightarrow$
$$(p(x_1, \ldots, x_n) \Leftrightarrow p(x_{n+1}, \ldots, x_{n+n}))) \ldots))) \ldots))$$

belongs to every theory with equality.   QED

The next exercise asks the reader to prove that, in any theory with equality, replacing terms by equal terms leads to equivalent formulas, provided that collateral capturing of variables by quantifiers is avoided.

Exercise 5 (Principle of substitution by equal – PSE)
Show that the following holds in any theory Θ with equality:

$$((x \cong y) \Rightarrow (\varphi \Rightarrow \psi)) \in \Theta,$$

where ψ is obtained from φ by replacing, outside the scope of quantifications over y, zero, one or more free occurrences of x by y.

When equality is available, it is possible to define the useful quantifier "exists one and only one", as shown in the following exercise.

Exercise 6 ($\exists^1$ quantifier) Let $(\exists^1 x \, \varphi)$ stand for:

$$((\exists x \, \varphi) \wedge (\forall x \, (\forall x_i((\varphi \wedge [\varphi]_{x_i}^x) \Rightarrow (x \cong x_i)))))$$

where i is the least j such that variable x_j is not x and it does not occur in φ. State and show properties of this quantifier in a theory with equality.

10.2   Normal models

Among the interpretation structures over a signature with equality it is interesting to identify those that interpret $\cong$ by the diagonal relation.

Let $I = (D, \_^{\mathsf{F}}, \_^{\mathsf{P}})$ be an interpretation structure over a signature with equality. Then, I is said to be *normal* if the interpretation of $\cong$ is the characteristic map of equality, that is, if:

$$\cong_2^{\mathsf{P}} = \lambda\, d_1, d_2 \, . \begin{cases} 1 & \text{if } d_1 = d_2 \\ 0 & \text{otherwise.} \end{cases}$$

Proposition 7 (Existence of normal model)
Every consistent theory with equality has a normal model with a countable domain.

Proof: Let Θ be a consistent theory with equality over Σ. The Henkin construction yields an interpretation structure

$$I = (D, \_^{\mathsf{F}}, \_^{\mathsf{P}})$$

with denumerable domain that is a model of Θ. Recall that, thanks to Proposition 4, $\cong_2^{\mathsf{P}}$ is a congruence relation. Consider

$$I/{\cong} \; = \; (D/{\cong}, \_^{\mathsf{F}/\cong}, \_^{\mathsf{P}/\cong})$$

(the quotient of I by $\cong_2^{\mathsf{P}}$) defined as expected:

- $D/{\cong} \; = \; D/{\cong_2^{\mathsf{P}}} \; = \; \{ [d]_{\cong_2^{\mathsf{P}}} : d \in D \}$
  where, as usual, $[d]_{\cong_2^{\mathsf{P}}}$ denotes the equivalence class of d with respect to the equivalence relation $\cong_2^{\mathsf{P}}$;

- $f_n^{\mathsf{F}/\cong}([d_1]_{\cong_2^{\mathsf{P}}}, \ldots, [d_n]_{\cong_2^{\mathsf{P}}}) = [f_n^{\mathsf{F}}(d_1, \ldots, d_n)]_{\cong_2^{\mathsf{P}}}$,
  for each $n \in \mathbb{N}$, $f \in F_n$ and $(d_1, \ldots, d_n) \in D^n$;

- $p_n^{\mathsf{P}/\cong}([d_1]_{\cong_2^{\mathsf{P}}}, \ldots, [d_n]_{\cong_2^{\mathsf{P}}}) = p_n^{\mathsf{P}}(d_1, \ldots, d_n)$,
  for each $n \in \mathbb{N}^+$, $p \in P_n$ and $(d_1, \ldots, d_n) \in D^n$.

Observe that if

$$\begin{cases} [d_1]_{\cong_2^{\mathsf{P}}} = [d_1']_{\cong_2^{\mathsf{P}}} \\ \ldots \\ [d_n]_{\cong_2^{\mathsf{P}}} = [d_n']_{\cong_2^{\mathsf{P}}}, \end{cases}$$

then

$$\begin{cases} d_1 \cong_2^{\mathsf{P}} d_1' \\ \ldots \\ d_n \cong_2^{\mathsf{P}} d_n' \end{cases}$$

and, thus, by Proposition 4, it follows that:

- $f_n^{\mathsf{F}}(d_1, \ldots, d_n) \cong_2^{\mathsf{P}} f_n^{\mathsf{F}}(d_1', \ldots, d_n')$
  and, so, $[f_n^{\mathsf{F}}(d_1, \ldots, d_n)]_{\cong_2^{\mathsf{P}}} = [f_n^{\mathsf{F}}(d_1', \ldots, d_n')]_{\cong_2^{\mathsf{P}}}$;

- $p_n^{\mathsf{P}}(d_1, \ldots, d_n) = p_n^{\mathsf{P}}(d_1', \ldots, d_n')$.

Therefore, each equation

$$f_n^{\mathsf{F}/\cong}([d_1]_{\cong_2^{\mathsf{P}}}, \ldots, [d_n]_{\cong_2^{\mathsf{P}}}) = [f_n^{\mathsf{F}}(d_1, \ldots, d_n)]_{\cong_2^{\mathsf{P}}}$$

yields a well-defined map

$$f_n^{\mathsf{F}/\cong} : D/\cong^n \to D/\cong$$

and each equation

$$p_n^{\mathsf{P}/\cong}([d_1]_{\cong_2^{\mathsf{P}}}, \ldots, [d_n]_{\cong_2^{\mathsf{P}}}) = p_n^{\mathsf{P}}(d_1, \ldots, d_n)$$

yields a well-defined map

$$p_n^{\mathsf{P}/\cong} : D/\cong^n \to \{0, 1\}.$$

Hence, the triple $I/\cong$ is an interpretation structure over Σ. Note that it has a countable domain. It remains to check that this is a normal model of Θ.

First, observe that the interpretation given by $I/\cong$ to $\cong$ is the intended one:

$$\cong_2^{\mathsf{P}/\cong}([d_1]_{\cong_2^{\mathsf{P}}}, [d_2]_{\cong_2^{\mathsf{P}}}) \quad = \quad \cong_2^{\mathsf{P}}(d_1, d_2)$$

$$= \quad \begin{cases} 1 & \text{if } d_1 \cong_2^{\mathsf{P}} d_2 \\ 0 & \text{otherwise.} \end{cases} \quad = \quad \begin{cases} 1 & \text{if } [d_1]_{\cong_2^{\mathsf{P}}} = [d_2]_{\cong_2^{\mathsf{P}}} \\ 0 & \text{otherwise.} \end{cases}$$

Furthermore, for every assignment ρ into I, denoting $([\_]_{\cong_2^{\mathsf{P}}} \circ \rho)$ by $[\rho]_{\cong_2^{\mathsf{P}}}$, it can be shown by induction, which is left as an exercise, that:

- for every $t \in T_\Sigma$,

$$[\![t]\!]_\Sigma^{I/\cong \, [\rho]_{\cong_2^{\mathsf{P}}}} \quad = \quad [[\![t]\!]_\Sigma^{I\rho}]_{\cong_2^{\mathsf{P}}};$$

- for every $\varphi \in L_\Sigma$,

$$I/\cong \, [\rho]_{\cong_2^{\mathsf{P}}} \Vdash_\Sigma \varphi \quad \text{if and only if} \quad I\rho \Vdash_\Sigma \varphi.$$

Hence, for every $\theta \in \Theta$, since $I \Vdash_\Sigma \theta$, that is, since $I\rho \Vdash_\Sigma \theta$ for every assignment ρ into I, using the result above one concludes

$$I/\cong [\rho]_{\cong_2^P} \Vdash_\Sigma \theta \text{ for any } \rho \text{ over } I,$$

whence, since the map $\lambda d . [d]_{\cong_2^P}$ is surjective, it follows easily by contraposition

$$I/\cong \sigma \Vdash_\Sigma \theta \text{ for any } \sigma \text{ over } I/\cong$$

and, so,

$$I/\cong \Vdash_\Sigma \theta .$$

QED

Let Θ be a theory with equality over Σ. Recall the category Int_Σ of interpretation structures over signature Σ introduced in Section 7.6. Consider the subcategory Mod_Θ of Int_Σ composed of the models of Θ together with all their homomorphisms.[1] Let nMod_Σ be the subcategory of Mod_Θ composed of the models of Θ where equality is interpreted by $=$ together with all their homomorphisms.[2]

In the proof of Proposition 7 it is shown that map

$$h = \lambda d . [d]_{\cong_2^P} : D \to D/\cong$$

is an homomorphism in Mod_Θ and, moreover, that it is elementary. In fact, more can be said in categorical terms:

Exercise 8 Show that[3] given $I \in \text{Mod}_\Theta$ there is $h' : I \to I'$ in Mod_Θ such that:

- I' is an object of nMod_Θ;

- for every $g : I \to I''$ with I'' an object of nMod_Θ, there is a unique $g' : I' \to I''$ such that $g' \circ h' = g$.

Show that h' is elementary. Conclude that if g is elementary, then so is g'.

The following exercises show that equality provides to some extent the means for imposing the cardinality of models.

Exercise 9 Show that, for every $k \in \mathbb{N}$, there exists a theory with equality that only admits normal models whose domains have cardinality:

---

[1]Category theoreticians would say that Mod_Θ is a full subcategory of Int_Σ.

[2]Again, nMod_Θ is a full subcategory of Mod_Θ.

[3]Category theoreticians would say that nMod_Θ is a reflective subcategory of Mod_Θ.

1. less than or equal to k;

2. equal to k.

Exercise 10 Show that there is no theory with equality that admits normal models with domains of every finite cardinality and no normal models with infinite domains.

10.3 First-order logic with equality

Some authors would rather work with first-order predicate logic endowed with equality from the start. The next exercise is an invitation to study this alternative.

Exercise 11 (First-order logic with equality)
(a) Define the language, the Hilbert calculus and the semantics of first-order logic with equality (FOLE), imposing the normal interpretation for the predicate symbol $\cong$.
(b) Let FOL be the first-order logic introduced in earlier chapters. For each signature Σ with equality, show the following results:

- $\Gamma \vDash_{\Sigma}^{\mathrm{FOLE}} \varphi$ if and only if $\Gamma, \mathrm{E}1, \mathrm{E}2, \mathrm{E}3 \vDash_{\Sigma}^{\mathrm{FOL}} \varphi$;

- $\Gamma \vdash_{\Sigma}^{\mathrm{FOLE}} \varphi$ if and only if $\Gamma, \mathrm{E}1, \mathrm{E}2, \mathrm{E}3 \vdash_{\Sigma}^{\mathrm{FOL}} \varphi$.

(c) Establish soundness and completeness of FOLE.

For more results on equality in first-order logic, the reader is referred to Section 2.8 of Elliott Mendelson's textbook [35].

Part III

Arithmetic and incompleteness theorems

Chapter 11

Arithmetic

Consider the set of natural numbers endowed with three operations (successor, addition and multiplication), as well as two binary relations (equal to and less than). Consider a signature, denoted herein by $\Sigma_{\mathbb{N}}$, with function and predicate symbols to represent these operations and relations, respectively, plus a constant symbol for zero. Let $\mathbb{N}$ denote the interpretation structure over $\Sigma_{\mathbb{N}}$ with set $\mathbb{N}$ as domain and interpreting the function and predicate symbols as expected.[1]

Following Hilbert's programme, after first-order logic was perfected, logicians attempted to formalize, as an axiomatizable first-order theory, the knowledge about this structure $\mathbb{N}$. In due course, the task failed when Kurt Gödel proved that $\mathrm{Th}(\mathbb{N})$ is not axiomatizable (the main result established in Chapter 13).

Nevertheless, it is possible to axiomatize fragments of the arithmetical knowledge. For instance, it is possible to state that

$$\langle \mathbb{N}, +, 0, \times, 1 \rangle$$

is a commutative semiring. That is, $\langle \mathbb{N}, + \rangle$ is a commutative monoid (+ is associative and commutative) with identity 0 and $\langle \mathbb{N}, \times \rangle$ is also a commutative monoid ($\times$ is associative and commutative) with identity 1. Moreover, $\times$ distributes over $+$ on the left and on the right and 0 is absorbent with respect to $\times$. The injectivity of successor is also easily stated. Finally, with respect to $<$, it raises no difficulty to impose that $<$ is stable for addition (if $n < m$, then $n + k < m + k$). This partial body of knowledge about arithmetic is captured by the axiomatized theory **P** introduced in this chapter. This theory includes the first-order version of the induction principle proposed by Richard Dedekind and Giuseppe Peano.

---

[1]No harm should result from this notation overload.

Theory **N** is a much weaker theory of arithmetic (with no induction principle) also introduced in this chapter. Contrarily to **P**, theory **N** is finitely axiomatizable, but, nevertheless, still rich enough to represent all computable maps. This fact will be used at the end of Chapter 13 to establish the undecidability of first-order logic.

It should be stressed that Gödel's incompleteness results require a sufficiently rich language where theory $\text{Th}(\mathbb{N})$ allows the representation of all computable maps. Indeed, arithmetic may even be decidable when a less rich language is used. That is the case of the theory of arithmetic proposed in 1929 by Mojżesz Presburger (see [37]) using a language only with 0, 1, + and $\cong$. Clearly, in Presburger theory not all computable maps are representable. The decidability of Presburger theory of arithmetic is proved in Chapter 15.

The signature and the standard semantics of arithmetic are presented in Section 11.1. Along the way, the concept of numeral is introduced. Section 11.2 is dedicated to theories of arithmetic, including the standard theory $\text{Th}(\mathbb{N})$ and the axiomatized theories **N** and **P**. Finally, a very brief reference is made in Section 11.3 to first-order predicate logic with arithmetic built in.

The issue of representing computable maps in theories of arithmetic, a key ingredient of Gödel's incompleteness proof, is left to Chapter 12.

11.1   Standard language and semantics

The *signature of arithmetic* used throughout this book is the first-order signature $\Sigma_{\mathbb{N}}$ such that:

- $F_0 = \{\mathbf{0}\}$;

- $F_1 = \{'\}$;

- $F_2 = \{+, \times\}$;

- $F_n = \emptyset$ for $n > 2$;

- $P_2 = \{\cong, <\}$;

- $P_n = \emptyset$ for $n \neq 2$.

This signature is sufficiently rich for establishing theories that allow the representability of computable maps, as required by Gödel's proofs of the incompleteness theorems.

It is common practice to write t' in place of $'[t]$ and to use infix notation for the binary function and predicate symbols.

The set of *numerals* is the subset of $cT_{\Sigma_{\mathbb{N}}}$ inductively defined as follows:

- **0** is a numeral;

- **n**′ is a numeral whenever **n** is a numeral.

It is convenient to use the following abbreviations: **1** stands for **0**′, **2** for **0**″, **3** for **0**‴, etc. Note that the natural number n should not be confused with the numeral **n**. The former is an element of the domain of the interpretation structure one has in mind when speaking about arithmetic, whereas the latter is a term of the formal language of arithmetic.

One may write $\overline{0}$ for **0**, $\overline{1}$ for **1**, $\overline{2}$ for **2**, etc. Accordingly, given $k \in \mathbb{N}$, we may denote the corresponding numeral **k** by $\overline{k}$. This alternative notation is more convenient for handwriting and also when one needs to refer to the sequence $\overline{a_1}, \ldots, \overline{a_n}$ of numerals corresponding to a given sequence $a_1, \ldots, a_n$ of natural numbers or the numeral corresponding to a natural number given by an expression. At this stage, the reader should ponder on the computability of the map

$$\lambda k . \overline{k} : \mathbb{N} \to cT_{\Sigma_{\mathbb{N}}}.$$

The *standard structure for arithmetic* is the interpretation structure over $\Sigma_{\mathbb{N}}$

$$\mathbb{N} = (\mathbb{N}, \_^{F_{\mathbb{N}}}, \_^{P_{\mathbb{N}}})$$

such that:[2]

- $\mathbf{0}_0^{F_{\mathbb{N}}} = 0$;

- $\prime_1^{F_{\mathbb{N}}} = \lambda d . d + 1$;

- $+_2^{F_{\mathbb{N}}} = \lambda d_1, d_2 . d_1 + d_2$;

- $\times_2^{F_{\mathbb{N}}} = \lambda d_1, d_2 . d_1 \times d_2$;

- $\cong_2^{P_{\mathbb{N}}} = \lambda d_1, d_2 . \begin{cases} 1 & \text{if } d_1 = d_2 \\ 0 & \text{otherwise;} \end{cases}$

- $<_2^{P_{\mathbb{N}}} = \lambda d_1, d_2 . \begin{cases} 1 & \text{if } d_1 < d_2 \\ 0 & \text{otherwise.} \end{cases}$

It is worth mentioning that $\Sigma_{\mathbb{N}}$ is a signature with equality and that the interpretation structure $\mathbb{N} = (\mathbb{N}, \_^{F_{\mathbb{N}}}, \_^{P_{\mathbb{N}}})$ is normal.

Exercise 1 Show that $[\![\mathbf{k}]\!]_{\Sigma_{\mathbb{N}}}^{\mathbb{N}\rho} = k$ for every $k \in \mathbb{N}$.

---

[2]No harm should result from using $\mathbb{N}$ for the interpretation structure and also for its domain, as well as by the traditional notation overload concerning $+$, $\times$ and $<$. The reader should be able to infer from the context what is at hand. For instance, $+$ can be either a binary function symbol in the language or a binary function in the interpretation structure.

11.2   Theories of arithmetic

A *theory of arithmetic* is a theory over $\Sigma_\mathbb{N}$ with equality and containing

$$(\neg(\mathbf{1} \cong \mathbf{0}))$$

as one of its theorems.

Among the theories of arithmetic, the *standard theory of arithmetic*

$$\mathrm{Th}_{\Sigma_\mathbb{N}}(\mathbb{N}) = \{\varphi \in L_{\Sigma_\mathbb{N}} : \mathbb{N} \Vdash_{\Sigma_\mathbb{N}} \varphi\},$$

also known simply as *arithmetic* and denoted shortly by $\mathrm{Th}(\mathbb{N})$, is of particular importance. It follows from Exercise 37 in Chapter 7 that arithmetic is consistent and exhaustive.[3] Later on, it will be seen that $\mathrm{Th}(\mathbb{N})$ is not axiomatizable, a consequence of the fact that it is possible to represent all computable maps in that theory.

Interestingly, computable maps are already representable in theories of arithmetic significantly weaker than $\mathrm{Th}(\mathbb{N})$, in particular in *theory* $\mathbf{N}$, the theory over $\Sigma_\mathbb{N}$ presented by the following fifteen specific axioms:

E1 $(\forall x_1(x_1 \cong x_1))$;

E2a $(\forall x_1(\forall x_2((x_1 \cong x_2) \Rightarrow (x_1' \cong x_2'))))$;

E2b $(\forall x_1(\forall x_2(\forall x_3(\forall x_4((x_1 \cong x_3) \Rightarrow ((x_2 \cong x_4) \Rightarrow$
$((x_1 + x_2) \cong (x_3 + x_4)))))))))$;

E2c $(\forall x_1(\forall x_2(\forall x_3(\forall x_4((x_1 \cong x_3) \Rightarrow ((x_2 \cong x_4) \Rightarrow$
$((x_1 \times x_2) \cong (x_3 \times x_4)))))))))$;

E3a $(\forall x_1(\forall x_2(\forall x_3(\forall x_4((x_1 \cong x_3) \Rightarrow ((x_2 \cong x_4) \Rightarrow$
$((x_1 \cong x_2) \Rightarrow (x_3 \cong x_4)))))))))$;

E3b $(\forall x_1(\forall x_2(\forall x_3(\forall x_4((x_1 \cong x_3) \Rightarrow ((x_2 \cong x_4) \Rightarrow$
$((x_1 < x_2) \Rightarrow (x_3 < x_4)))))))))$;

N1 $(\forall x_1(\neg(x_1' \cong \mathbf{0})))$;

N2 $(\forall x_1(\forall x_2((x_1' \cong x_2') \Rightarrow (x_1 \cong x_2))))$;

N3 $(\forall x_1((x_1 + \mathbf{0}) \cong x_1))$;

N4 $(\forall x_1(\forall x_2((x_1 + x_2') \cong (x_1 + x_2)')))$;

N5 $(\forall x_1((x_1 \times \mathbf{0}) \cong \mathbf{0}))$;

N6 $(\forall x_1(\forall x_2((x_1 \times x_2') \cong ((x_1 \times x_2) + x_1))))$;

[3]Recall that consistency of $\mathrm{Th}_\Sigma(I)$ was established semantically, and hence it depends upon the consistency of the naive set theory in which this semantics was defined.

N7 $(\forall x_1(\neg(x_1 < \mathbf{0})));$

N8 $(\forall x_1(\forall x_2((x_1 < x_2') \Leftrightarrow ((x_1 < x_2) \vee (x_1 \cong x_2))))));$

N9 $(\forall x_1(\forall x_2((x_1 < x_2) \vee (x_1 \cong x_2) \vee (x_2 < x_1)))).$

Axioms **E** express that equality is reflexive and congruent with respect to the function and predicate symbols. Axiom **N1** imposes that no successor is zero. Axiom **N2** states that the successor map is injective. Axiom **N3** imposes zero as the identity element for addition. Axiom **N4** completes the definition of addition using successor. Axiom **N5** expresses that zero is absorbent for multiplication. Axiom **N6** completes the definition of multiplication using addition and successor. Axiom **N7** states that zero is minimal with respect to $<$. Axiom **N8** relates $<$ with the successor map. Finally, **N9** expresses that $\leq$ is a total order relation. The set of specific axioms of **N** is henceforth denoted by $\mathrm{Ax_N}$.

Observe that **N** is a theory of arithmetic, since it is a theory with equality, due to the first six axioms, and $(\neg(\mathbf{1} \cong \mathbf{0})) \in \mathbf{N}$, thanks to axiom **N1**.

Exercise 2 Check that $\mathbb{N}$ is a model of theory **N**. Go on to show that $\mathrm{Th}(\mathbb{N})$ is a theory of arithmetic.

Theory **N** is very weak (as the following exercise illustrates). However, as it was already mentioned, it will be established in due course that **N** is strong enough to represent all computable maps, a fact that will be used in Chapter 13 to prove the undecidability of first-order logic.

Exercise 3 Show that $(\neg(x_1 \cong x_1'))$ is not a theorem of **N**.

The *theory* **P** over $\Sigma_\mathbb{N}$ is obtained by adding the following first-order induction principle to the specific axioms of theory **N**:

Ind $([\varphi]_{\mathbf{0}}^x \Rightarrow ((\forall x(\varphi \Rightarrow [\varphi]_{x'}^x)) \Rightarrow (\forall x\,\varphi))).$

Observe that the set $\mathrm{Ax_P} = \mathrm{Ax_N} \cup \mathbf{Ind}$ of specific axioms of **P** is not finite. In fact, the induction principle contains a denumerable set of axioms. Note also that **P** is a theory of arithmetic since it contains **N**.

Exercise 4 Show that $\mathbb{N}$ is a model of **P**.

The induction principle is a very powerful tool for deriving universally quantified formulas. Indeed, in order to show that $(\forall x\,\varphi)$ is a theorem of **P**, it is enough:

- to derive formula $[\varphi]_{\mathbf{0}}^x$ from $\mathrm{Ax_P}$;

- and to derive, without essential generalizations on dependents of φ, formula $[\varphi]_{x'}^{x}$ from $\mathrm{Ax_P} \cup \{\varphi\}$.

Due to the induction scheme, theory $\mathbf{P}$ is quite stronger than theory $\mathbf{N}$, as it is illustrated in the following exercise.

Exercise 5 Show that the following are theorems of $\mathbf{P}$:

1. $(\neg(x_1 \cong x_1'))$;

2. associativity and commutativity of addition and multiplication;

3. distributivity of multiplication over addition.

Exercise 6 Show that axiom $\mathbf{N9}$ is derivable from the other axioms in $\mathrm{Ax_P}$.

For more results on theories $\mathbf{N}$ and $\mathbf{P}$, the interested reader is advised to consult Joseph Shoenfield's textbook [42].

The crucial step toward Gödel's first incompleteness theorem is the proof of the undecidability of any consistent theory of arithmetic where computable maps are representable (Church's theorem).

The representability of computable maps within $\mathrm{Th}(\mathbb{N})$ is established in Chapter 12. It is also sketched therein the proof of the representability of computable maps within $\mathbf{N}$, and therefore within any of its extensions, which incidentally include $\mathbf{P}$ and $\mathrm{Th}(\mathbb{N})$.

In Chapter 13, following the proof of the aforementioned theorem due to Church, several fundamental theorems are established as corollaries, including non exhaustiveness of any axiomatizable, consistent theory of arithmetic where computable maps are representable (Gödel's first incompleteness theorem, in a formulation due to Rosser) and the impossibility of axiomatizing $\mathrm{Th}(\mathbb{N})$. Finally, undecidability of first-order logic is established, using the representability of computable maps within theory $\mathbf{N}$ and capitalizing on the finite cardinality of $\mathrm{Ax_N}$.

In Chapter 14, after extending (in a weak sense) representability to computably enumerable sets, in order to allow a (weak) representation of derivability, it is shown that, whenever Θ is a computably enumerable, consistent and sufficiently strong (able to represent computable maps, *inter alia*) theory of arithmetic, it is not possible within Θ to derive the formula stating the consistency of Θ (Gödel's second incompleteness theorem).

11.3   Arithmetic as a logic

As in the case of equality (recall Section 10.3) it is possible to introduce arithmetic as a logic instead of working with theories.

The *first-order logic with arithmetic* (FOLA) is defined as follows:

- every signature contains $\Sigma_{\mathbb{N}}$;

- for each signature, the language is defined as for FOL;

- the reduct of any interpretation structure to $\Sigma_{\mathbb{N}}$ is the standard structure $\mathbb{N}$;

- derivation is defined as in FOL, but enriching the set of axioms with the specific axioms of **P**.

Exercise 7 (Incompleteness of FOLA)

1. Show that FOLA is not (strongly) complete. Hint: Start by showing that entailment in FOLA is not compact.

2. Is it possible, maintaining consistency, to enrich the set of axioms of FOLA in order to make it (strongly) complete?

In fact, there is no computable set $Ax_{\mathbf{P}}^+$ of sound axioms that would guarantee even weak completeness[4] of FOLA$^+$, the logic obtained by replacing in FOLA the specific axioms of **P** by those in $Ax_{\mathbf{P}}^+$. Indeed, since

$$Ax_{\mathbf{P}}^+ \vdash_{\Sigma_{\mathbb{N}}}^{\text{FOL}} \varphi \text{ if and only if } \vdash_{\Sigma_{\mathbb{N}}}^{\text{FOLA}^+} \varphi,$$

the weak soundness and completeness of FOLA$^+$

$$\vdash_{\Sigma_{\mathbb{N}}}^{\text{FOLA}^+} \varphi \text{ if and only if } \vDash_{\Sigma_{\mathbb{N}}}^{\text{FOLA}^+} \varphi$$

would allow to conclude

$$Ax_{\mathbf{P}}^+ \vdash_{\Sigma_{\mathbb{N}}}^{\text{FOL}} \varphi \text{ if and only if } \vDash_{\Sigma_{\mathbb{N}}}^{\text{FOLA}^+} \varphi$$

and, therefore,

$$Ax_{\mathbf{P}}^+ \vdash_{\Sigma_{\mathbb{N}}}^{\text{FOL}} \varphi \text{ if and only if } \varphi \in \text{Th}(\mathbb{N}),$$

in contradiction with the results of Chapter 13.

However, proving (strong) incompleteness of (any variant of) FOLA only requires a compactness argument, as the reader was asked to show in Exercise 7, and, so, it does not require the use of Gödel's first incompleteness theorem.

---

[4]Recall that a weakly complete logic is one where $\vDash \varphi$ implies $\vdash \varphi$. Observe that (strong) completeness implies weak completeness. The converse does not hold. For example, propositional temporal linear logic is weakly complete, but not (strongly) complete.

Chapter 12

Representability

As mentioned before, representing computable maps in theories of arithmetic is a crucial ingredient of the technique used by Kurt Gödel for establishing the incompleteness theorems.

Section 12.1 introduces and illustrates the notions of representable map and of representable set in a given theory of arithmetic. The section ends with the proof of the equivalence between the representability of a set and the representability of its characteristic map.

Section 12.2 is dedicated to characterizing the set of computable maps. To this end, it is necessary to introduce first the notion of primitive recursive map and, afterward, prove Kleene's normal form theorem.

Section 12.3 presents the notion of computable map à la Gödel. This notion does not involve recursion in order to facilitate the representation of computable maps in theories of arithmetic. The section ends with the proof of the equivalence of Kleene's and Gödel's formalizations of the notion of computable map.

Finally, Section 12.4 and Section 12.5 show that all computable maps are representable in theory $\mathrm{Th}(\mathbb{N})$ and in theory $\mathbf{N}$, respectively. The former is essential to the results in the following chapters. The latter is briefly sketched only because it is needed in Chapter 13 for proving the undecidability of first-order logic as a corollary of the first incompleteness theorem.

12.1   Notion of representability

Let Θ be a theory of arithmetic and $h : \mathbb{N}^n \to \mathbb{N}^m$. Then, the map h is said to be *represented* in Θ by formula $\varphi \in L_{\Sigma_{\mathbb{N}}}$ if:

- $\mathrm{fv}_{\Sigma_{\mathbb{N}}}(\varphi) = \{x_1, \ldots, x_{n+m}\}$;

- for every $a_1, \ldots, a_n, b_1, \ldots, b_m \in \mathbb{N}$,

$$\text{if } h(a_1, \ldots, a_n) = (b_1, \ldots, b_m) \text{ then } ([\varphi]_{a_1, \ldots, a_n}^{x_1, \ldots, x_n} \Leftrightarrow (\bigwedge_{i=1}^{m} (x_{n+i} \cong \overline{b_i}))) \in \Theta.$$

A map is said to be *representable* in Θ if it is represented in Θ by some formula.

Exercise 1 Check the following statements:

1. The map $\lambda k.\, k+1$ is represented in $\text{Th}(\mathbb{N})$ by $(x_2 \cong x_1')$.

2. The map P_1^2 is represented in $\text{Th}(\mathbb{N})$ by $((x_3 \cong x_1) \wedge (x_2 \cong x_2))$.

The concept of representability is extended to sets as follows. Let Θ be a theory of arithmetic and $C \subseteq \mathbb{N}^n$. Then, the set C is said to be *represented* in Θ by formula $\varphi \in L_{\Sigma_\mathbb{N}}$ if:

- $\text{fv}_{\Sigma_\mathbb{N}}(\varphi) = \{x_1, \ldots, x_n\}$;

- for every $a_1, \ldots, a_n \in \mathbb{N}$,

    - if $(a_1, \ldots, a_n) \in C$ then $[\varphi]_{a_1, \ldots, a_n}^{x_1, \ldots, x_n} \in \Theta$;

    - otherwise, $(\neg[\varphi]_{a_1, \ldots, a_n}^{x_1, \ldots, x_n}) \in \Theta$.

A set is said to be *representable* in Θ if it is represented in Θ by some formula.

Exercise 2 Show that the set $\{(k_1, k_2) : k_1 \leq k_2\}$ is represented in $\text{Th}(\mathbb{N})$ by the formula $((x_1 < x_2) \vee (x_1 \cong x_2))$.

Although an immediate consequence of the definitions, it is worth mentioning that, given two theories of arithmetic Θ_1 and Θ_2 such that $\Theta_1 \subseteq \Theta_2$, if a map/set is representable in Θ_1 then it is also representable in Θ_2. Observe also that every map/set is representable in the inconsistent theory of arithmetic $L_{\Sigma_\mathbb{N}}$.

Proposition 3 Let Θ be a theory of arithmetic. Then, a set is representable in Θ if and only if its characteristic map is representable in Θ.

Proof: Let $C \subseteq \mathbb{N}^n$.

($\leftarrow$) Suppose φ represents χ_C in Θ. That is, assume that:

- $\text{fv}_{\Sigma_\mathbb{N}}(\varphi) = \{x_1, \ldots, x_{n+1}\}$;

- if $\chi_C(a_1, \ldots, a_n) = b$ then $([\varphi]_{a_1, \ldots, a_n}^{x_1, \ldots, x_n} \Leftrightarrow (x_{n+1} \cong \overline{b})) \in \Theta.$

Then, C is represented by $[\varphi]_1^{x_{n+1}}$. Indeed:

(i)   if $(a_1, \ldots, a_n) \in C$ then

$$\chi_C(a_1, \ldots, a_n) = 1$$

$$([\varphi]_{\overline{a_1},\ldots,\overline{a_n}}^{x_1,\ldots,x_n} \Leftrightarrow (x_{n+1} \cong \mathbf{1})) \in \Theta \quad \text{(representability of } \chi_C)$$

$$[[\varphi]_{\overline{a_1},\ldots,\overline{a_n}}^{x_1,\ldots,x_n}]_1^{x_{n+1}} \in \Theta \qquad (*)$$

$$[[\varphi]_1^{x_{n+1}}]_{\overline{a_1},\ldots,\overline{a_n}}^{x_1,\ldots,x_n} \in \Theta \qquad (**)$$

(*) Recalling that Θ is a theory with equality, consider the following derivation sequence:

1	$([\varphi]_{\overline{a_1},\ldots,\overline{a_n}}^{x_1,\ldots,x_n} \Leftrightarrow (x_{n+1} \cong \mathbf{1}))$	$\in \Theta$
2	$(([\varphi]_{\overline{a_1},\ldots,\overline{a_n}}^{x_1,\ldots,x_n} \Leftrightarrow (x_{n+1} \cong \mathbf{1})) \Rightarrow$	
	$\qquad\qquad\qquad ((x_{n+1} \cong \mathbf{1}) \Rightarrow [\varphi]_{\overline{a_1},\ldots,\overline{a_n}}^{x_1,\ldots,x_n}))$	Taut
3	$((x_{n+1} \cong \mathbf{1}) \Rightarrow [\varphi]_{\overline{a_1},\ldots,\overline{a_n}}^{x_1,\ldots,x_n})$	MP 1,2
4	$(\forall x_{n+1}((x_{n+1} \cong \mathbf{1}) \Rightarrow [\varphi]_{\overline{a_1},\ldots,\overline{a_n}}^{x_1,\ldots,x_n}))$	Gen 3
5	$((\forall x_{n+1}((x_{n+1} \cong \mathbf{1}) \Rightarrow [\varphi]_{\overline{a_1},\ldots,\overline{a_n}}^{x_1,\ldots,x_n})) \Rightarrow$	
	$\qquad\qquad\qquad ((\mathbf{1} \cong \mathbf{1}) \Rightarrow [[\varphi]_{\overline{a_1},\ldots,\overline{a_n}}^{x_1,\ldots,x_n}]_1^{x_{n+1}}))$	Ax4
6	$((\mathbf{1} \cong \mathbf{1}) \Rightarrow [[\varphi]_{\overline{a_1},\ldots,\overline{a_n}}^{x_1,\ldots,x_n}]_1^{x_{n+1}})$	MP 4,5
7	$(\mathbf{1} \cong \mathbf{1})$	RE1
8	$[[\varphi]_{\overline{a_1},\ldots,\overline{a_n}}^{x_1,\ldots,x_n}]_1^{x_{n+1}}$	MP 6,7

(**) Since variables $x_1, \ldots, x_{n+1}$ are distinct and terms $\overline{a_1}, \ldots, \overline{a_n}$ and $\mathbf{1}$ are closed, formula $[[\varphi]_1^{x_{n+1}}]_{\overline{a_1},\ldots,\overline{a_n}}^{x_1,\ldots,x_n}$ coincides with formula $[[\varphi]_{\overline{a_1},\ldots,\overline{a_n}}^{x_1,\ldots,x_n}]_1^{x_{n+1}}$.

(ii)   if $(a_1, \ldots, a_n) \notin C$ then

$$\chi_C(a_1, \ldots, a_n) = 0$$

$$([\varphi]_{\overline{a_1},\ldots,\overline{a_n}}^{x_1,\ldots,x_n} \Leftrightarrow (x_{n+1} \cong \mathbf{0})) \in \Theta \quad \text{(representability of } \chi_C)$$

$$([\varphi]_{\overline{a_1},\ldots,\overline{a_n}}^{x_1,\ldots,x_n} \Rightarrow (x_{n+1} \cong \mathbf{0})) \in \Theta \quad \text{(tautologically)}$$

$$([[\varphi]_{\overline{a_1},\ldots,\overline{a_n}}^{x_1,\ldots,x_n}]_1^{x_{n+1}} \Rightarrow (\mathbf{1} \cong \mathbf{0})) \in \Theta \quad \text{(Gen, Ax4 and MP)}$$

$$(\neg[[\varphi]_{\overline{a_1},\ldots,\overline{a_n}}^{x_1,\ldots,x_n}]_1^{x_{n+1}}) \in \Theta \qquad (\dagger)$$

$$(\neg[[\varphi]_1^{x_{n+1}}]_{\overline{a_1},\ldots,\overline{a_n}}^{x_1,\ldots,x_n}) \in \Theta \qquad (\dagger\dagger)$$

($\dagger$) Recalling that $(\neg(\mathbf{1} \cong \mathbf{0})) \in \Theta$, consider the following derivation sequence:

1	$([[\varphi]_{\overline{a_1},\ldots,\overline{a_n}}^{x_1,\ldots,x_n}]_1^{x_{n+1}} \Rightarrow (\mathbf{1} \cong \mathbf{0}))$	$\in \Theta$
2	$(\neg(\mathbf{1} \cong \mathbf{0}))$	$\in \Theta$

3   $((([[\varphi]_{a_1,\ldots,a_n}^{x_1,\ldots,x_n}]_1^{x_{n+1}} \Rightarrow (\mathbf{1} \cong \mathbf{0})) \Rightarrow$
$((\neg(\mathbf{1} \cong \mathbf{0})) \Rightarrow (\neg[[\varphi]_{a_1,\ldots,a_n}^{x_1,\ldots,x_n}]_1^{x_{n+1}})))$ Thm 5.4(f)

4   $((\neg(\mathbf{1} \cong \mathbf{0})) \Rightarrow (\neg[[\varphi]_{a_1,\ldots,a_n}^{x_1,\ldots,x_n}]_1^{x_{n+1}}))$ MP 1,3

5   $(\neg[[\varphi]_{a_1,\ldots,a_n}^{x_1,\ldots,x_n}]_1^{x_{n+1}})$ MP 2,4

(††) Justification identical to (**) above.

($\rightarrow$) Suppose that ψ represents C in Θ. That is, assume that $\mathrm{fv}_{\Sigma_{\mathrm{N}}}(\psi) = \{x_1,\ldots,x_n\}$ and:

- if $(a_1,\ldots,a_n) \in C$ then $[\psi]_{a_1,\ldots,a_n}^{x_1,\ldots,x_n} \in \Theta$;

- otherwise, $(\neg[\psi]_{a_1,\ldots,a_n}^{x_1,\ldots,x_n}) \in \Theta$.

Then, χ_C is represented by the formula

$$((\psi \wedge (x_{n+1} \cong \mathbf{1})) \vee ((\neg \psi) \wedge (x_{n+1} \cong \mathbf{0})))$$

denoted below by φ. Indeed:

(i)

if $\chi_C(a_1,\ldots,a_n) = 1$ then
$(a_1,\ldots,a_n) \in C$
$[\psi]_{a_1,\ldots,a_n}^{x_1,\ldots,x_n} \in \Theta$ (representability of C)
$([\varphi]_{a_1,\ldots,a_n}^{x_1,\ldots,x_n} \Leftrightarrow (x_{n+1} \cong \mathbf{1})) \in \Theta$ ($\ddagger$)

($\ddagger$) By tautological reasoning,

$$\vdash_{\Sigma} (\alpha \Rightarrow (((\alpha \wedge \beta_1) \vee ((\neg \alpha) \wedge \beta_2)) \Leftrightarrow \beta_1))$$

that the reader should verify.

(ii)

if $\chi_C(a_1,\ldots,a_n) = 0$ then
$(a_1,\ldots,a_n) \notin C$
$(\neg[\psi]_{a_1,\ldots,a_n}^{x_1,\ldots,x_n}) \in \Theta$ (representability of C)
$([\varphi]_{a_1,\ldots,a_n}^{x_1,\ldots,x_n} \Leftrightarrow (x_{n+1} \cong \mathbf{0})) \in \Theta$ ($\ddagger\ddagger$)

($\ddagger\ddagger$) By tautological reasoning,

$$\vdash_{\Sigma} ((\neg \alpha) \Rightarrow (((\alpha \wedge \beta_1) \vee ((\neg \alpha) \wedge \beta_2)) \Leftrightarrow \beta_2))$$

that the reader will have no difficulty in verifying.   QED

12.2 Computable maps

In order to facilitate the proof of the fact that every computable map is representable in theory $\mathrm{Th}(\mathbb{N})$ and in the other theories of arithmetic introduced in Chapter 11, it is convenient to invoke that $\mathcal{C} = \mathcal{R}$ (see Section 3.5). In particular, a map $h : \mathbb{N}^n \to \mathbb{N}^m$ is computable if and only if it is recursive. Thus, it will be shown that every recursive map is representable in each of those theories.

Observe that the basis of the inductive definition of $\mathcal{R}$ contains only maps and that all constructions, with the exception of minimization, generate maps from maps.

However, it is possible to restrict the use of minimization in order to generate only maps: a *guarded minimization* is a minimization that produces a map because there is a zero of the argument map for every tuple of data.

More concretely, given a map $f : \mathbb{N}^{n+1} \to \mathbb{N}$, the minimization of f is said to be *guarded* if for each tuple $(k_1, \ldots, k_n)$ there is k such that $f(k_1, \ldots, k_n, k) = 0$. In this case, the result of the minimization, $\mathsf{min}(f)$, is still a map.

Let $gm\mathcal{R}$ be the least set of maps with arguments and outputs in $\mathbb{N}$ such that:

- it contains the constant k (of arity zero) for every $\mathsf{k} \in \mathbb{N}$;

- it contains maps Z and S (of arity one);

- it contains the projections (of all different positive arities);

- it is closed under aggregation, composition, primitive recursion and guarded minimization.

In face of this definition, one question immediately arises: does $gm\mathcal{R}$ coincide with the set $\mathcal{R}{\downarrow}$ of maps in $\mathcal{R}$?

The answer is positive, but its proof is not trivial and requires some technical preliminaries. It is also necessary to consider the class $p\mathcal{R}$ of *primitive recursive maps* — the least set of maps with arguments and outputs in $\mathbb{N}$ such that:

- it contains the constant k (of arity zero) for every $\mathsf{k} \in \mathbb{N}$;

- it contains maps Z and S (of arity one);

- it contains the projections (of all different positive arities);

- it is closed under aggregation, composition and primitive recursion.

In other words, $p\mathcal{R}$ is built in a similar way to $gm\mathcal{R}$, but without using guarded minimization. Observe, *en passant*, that $p\mathcal{R}$ is strictly contained in $gm\mathcal{R}$: for

example, Ackermann-Péter's map is computable but not primitive recursive
(see, for instance, Section 3 in Chapter 12 of [16]). That is, in synthesis,

$$p\mathcal{R} \subsetneq gm\mathcal{R} = \mathcal{R}\!\downarrow.$$

Before proving that indeed $gm\mathcal{R} = \mathcal{R}\!\downarrow$, it is necessary to show that some
maps are in $p\mathcal{R}$. The reader is asked in the following exercises to carry out this
straightforward verification.

Exercise 4 (Basic predicates)
Show that the following predicates are primitive recursive:[1]

- $\mathsf{eq} = \lambda\, k_1, k_2\, . \begin{cases} 1 & \text{if } k_1 = k_2 \\ 0 & \text{otherwise} \end{cases} : \mathbb{N}^2 \to \{0,1\};$

- $\mathsf{lt} = \lambda\, k_1, k_2\, . \begin{cases} 1 & \text{if } k_1 < k_2 \\ 0 & \text{otherwise} \end{cases} : \mathbb{N}^2 \to \{0,1\};$

- $\mathsf{leq} = \lambda\, k_1, k_2\, . \begin{cases} 1 & \text{if } k_1 \le k_2 \\ 0 & \text{otherwise} \end{cases} : \mathbb{N}^2 \to \{0,1\}.$

Exercise 5 (Numerical operations)
Recall that the map

$$\mathsf{add} = \lambda\, k_1, k_2\, . \, k_1 + k_2 : \mathbb{N}^2 \to \mathbb{N}$$

is primitive recursive. Show that the following maps are also primitive recursive:

- $\mathsf{mult} = \lambda\, k_1, k_2\, . \, k_1 \times k_2 : \mathbb{N}^2 \to \mathbb{N};$

- $\mathsf{pred} = \lambda\, k\, . \begin{cases} k - 1 & \text{if } k \ge 1 \\ 0 & \text{otherwise} \end{cases} : \mathbb{N} \to \mathbb{N}.$

Exercise 6 (Propositional operators)
Assume that $f, f_1, f_2 : \mathbb{N}^n \to \{0,1\} \in p\mathcal{R}$. Let:

- $\mathsf{neg}(f) : \mathbb{N}^n \to \{0,1\}$ be

$$\lambda\, k_1, \ldots, k_n\, . \, 1 - f(k_1, \ldots, k_n);$$

- $\mathsf{conj}(f_1, f_2) : \mathbb{N}^n \to \{0,1\}$ be

$$\lambda\, k_1, \ldots, k_n\, . \, f_1(k_1, \ldots, k_n) \times f_2(k_1, \ldots, k_n);$$

---

[1] In this context, it is usual say that a map with range $\{0, 1\}$ is a predicate.

- disj$(f_1, f_2) : \mathbb{N}^n \to \{0, 1\}$ be

$$\lambda k_1, \ldots, k_n \cdot 1 - ((1 - f_1(k_1, \ldots, k_n)) \times (1 - f_2(k_1, \ldots, k_n))).$$

Show that these predicates are primitive recursive.

Exercise 7 (Finite product and sum operators)

Assume that $f_i : \mathbb{N}^n \to \mathbb{N} \in p\mathcal{R}$ for $i = 1, \ldots, k$. Let:

- prod$(f_1, \ldots, f_k) : \mathbb{N}^n \to \mathbb{N}$ be

$$\lambda k_1, \ldots, k_n \cdot \prod_{i=1}^{k} f_i(k_1, \ldots, k_n);$$

- sum$(f_1, \ldots, f_k) : \mathbb{N}^n \to \mathbb{N}$ be

$$\lambda k_1, \ldots, k_n \cdot \sum_{i=1}^{k} f_i(k_1, \ldots, k_n).$$

Show that these maps are primitive recursive.

Exercise 8 (Bounded quantifiers)

Assume that $b : \mathbb{N}^n \to \mathbb{N}, f : \mathbb{N}^{n+1} \to \{0, 1\} \in p\mathcal{R}$. Let:

- forall$(b, f) : \mathbb{N}^n \to \{0, 1\}$ be

$$\lambda k_1, \ldots, k_n \cdot \prod_{i=0}^{b(k_1, \ldots, k_n)-1} f(k_1, \ldots, k_n, i).$$

- exists$(b, f) : \mathbb{N}^n \to \{0, 1\}$ be

$$\lambda k_1, \ldots, k_n \cdot 1 - \left(\prod_{i=0}^{b(k_1, \ldots, k_n)-1} (1 - f(k_1, \ldots, k_n, i)) \right).$$

Show that these predicates are primitive recursive.

Exercise 9 (Bounded minimization)

Assume that $b : \mathbb{N}^n \to \mathbb{N}, f : \mathbb{N}^{n+1} \to \mathbb{N} \in p\mathcal{R}$. Let

$$\mathsf{bmin}(b, f) = \lambda k_1, \ldots, k_n \cdot \begin{cases} b(k_1, \ldots, k_n) & \text{if } W^{b,f}_{k_1, \ldots, k_n} = \emptyset \\ \text{minimum of } W^{b,f}_{k_1, \ldots, k_n} & \text{otherwise} \end{cases} : \mathbb{N}^n \to \mathbb{N}$$

where

$$W^{b,f}_{k_1, \ldots, k_n} = \{k : k < b(k_1, \ldots, k_n) \ \& \ f(k_1, \ldots, k_n, k) = 0\}.$$

Show that $\mathsf{bmin}(b, f) \in p\mathcal{R}$.

Exercise 10 (Definition by cases)
Assume that:

- $k \in \mathbb{N}^+$;

- $f_i : \mathbb{N}^n \to \mathbb{N} \in p\mathcal{R}$ for $i = 1, \ldots, k+1$;

- $g_j : \mathbb{N}^n \to \{0,1\} \in p\mathcal{R}$ for $j = 1, \ldots, k$;

- for each $(k_1, \ldots, k_n) \in \mathbb{N}^n$ there is at most one $j \in \{1, \ldots, k\}$ such that

$$g_j(k_1, \ldots, k_n) = 1.$$

Let $\mathsf{dbc}(f_1 \leftarrow g_1; \ldots; f_k \leftarrow g_k; f_{k+1}) : \mathbb{N}^n \to \mathbb{N}$ be

$$\lambda k_1, \ldots, k_n \,.\, \begin{cases} f_1(k_1, \ldots, k_n) & \text{if } g_1(k_1, \ldots, k_n) = 1 \\ \ldots \\ f_k(k_1, \ldots, k_n) & \text{if } g_k(k_1, \ldots, k_n) = 1 \\ f_{k+1}(k_1, \ldots, k_n) & \text{otherwise.} \end{cases}$$

Show that $\mathsf{dbc}(f_1 \leftarrow g_1; \ldots; f_k \leftarrow g_k; f_{k+1}) \in p\mathcal{R}$.

Exercise 11 (Joint recursion)
Assume that $f_0, h_0 : \mathbb{N}^n \to \mathbb{N}$ and $f_1, h_1 : \mathbb{N}^{n+3} \to \mathbb{N}$ are in $p\mathcal{R}$. Show that

$$\mathsf{jrec}_1(f_0, f_1, h_0, h_1), \ \mathsf{jrec}_2(f_0, f_1, h_0, h_1) : \mathbb{N}^{n+1} \to \mathbb{N}$$

are both in $p\mathcal{R}$ where

- $\mathsf{jrec}_1(f_0, f_1, h_0, h_1)(k_1, \ldots, k_n, 0) = f_0(k_1, \ldots, k_n)$;

- $\mathsf{jrec}_2(f_0, f_1, h_0, h_1)(k_1, \ldots, k_n, 0) = h_0(k_1, \ldots, k_n)$;

- $\mathsf{jrec}_1(f_0, f_1, h_0, h_1)(k_1, \ldots, k_n, k+1) =$
  $f_1(k_1, \ldots, k_n, k,$
  $\quad \mathsf{jrec}_1(f_0, f_1, h_0, h_1)(k_1, \ldots, k_n, k),$
  $\quad \mathsf{jrec}_2(f_0, f_1, h_0, h_1)(k_1, \ldots, k_n, k))$;

- $\mathsf{jrec}_2(f_0, f_1, h_0, h_1)(k_1, \ldots, k_n, k+1) =$
  $h_1(k_1, \ldots, k_n, k,$
  $\quad \mathsf{jrec}_1(f_0, f_1, h_0, h_1)(k_1, \ldots, k_n, k),$
  $\quad \mathsf{jrec}_2(f_0, f_1, h_0, h_1)(k_1, \ldots, k_n, k))$.

Exercise 12 (Enumeration of $\mathbb{N}^n$) Let $n \in \mathbb{N}^+$. Show that the maps $J : \mathbb{N}^2 \to \mathbb{N}$, $\mathsf{zigzag} : \mathbb{N} \to \mathbb{N}^2$, $K : \mathbb{N} \to \mathbb{N}$, $L : \mathbb{N} \to \mathbb{N}$ and $\lambda k \,.\, k_{\mathbb{N}^n}$ defined in Chapter 3 are primitive recursive. Show that the inverse $J_{\mathbb{N}^n}$ of $\lambda k \,.\, k_{\mathbb{N}^n}$ is also primitive recursive.

For the purpose of showing that course-of-values recursion preserves primitive recursiveness and also for proving the existence of a primitive recursive predicate, one needs to be able to provide a single map capable of encoding all sequences of natural numbers. Such sequences can be confused with tuples of natural numbers. Therefore, let

$$\mathbb{N}^* = \bigcup_{n \in \mathbb{N}} \mathbb{N}^n.$$

Keeping in mind that $J_{\mathbb{N}^2} = J$, consider the *encoding map*

$$J_{\mathbb{N}^*} : \mathbb{N}^* \to \mathbb{N}$$

illustrated in Figure 12.1 such that:

- $J_{\mathbb{N}^*}(\varepsilon) = 0$;

- $J_{\mathbb{N}^*}(k_1, \ldots, k_n) = J(n - 1, J_{\mathbb{N}^n}(k_n, \ldots, k_1)) + 1$ for each $n \in \mathbb{N}^+$ and $k_1, \ldots, k_n \in \mathbb{N}$.

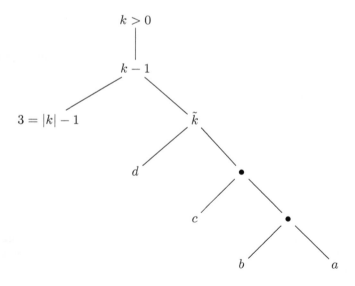

Figure 12.1: $k = J_{\mathbb{N}^*}(a, b, c, d)$.

Observe the reverse ordering in the tree of the elements of the sequence. This reverse ordering facilitates the definition in Exercise 13 of some useful operations on sequences when brought to the realm of their encodings. Note also that $J_{\mathbb{N}^*}$ is injective. Its inverse

$$\lambda\, k \,.\, k_{\mathbb{N}^*}$$

is an injective enumeration of $\mathbb{N}^*$.

Exercise 13 Verify that the following maps are primitive recursive:

1. for each $n \in \mathbb{N}$:

   $(J_{\mathbb{N}^*})|_{\mathbb{N}^n} : \mathbb{N}^n \to \mathbb{N}$.

2. $|\cdot| = \lambda\, k \,.\, \begin{cases} 0 & \text{if } k = 0 \\ (k-1)_{\mathbb{N}^2}[[1]] + 1 & \text{otherwise} \end{cases} : \mathbb{N} \to \mathbb{N}$

   ($|k|$ is the length of the sequence with encoding k).

3. $\tilde{\cdot} = \lambda\, k \,.\, \begin{cases} (k-1)_{\mathbb{N}^2}[[2]] & \text{if } k > 0 \\ 0 & \text{otherwise} \end{cases} : \mathbb{N} \to \mathbb{N}$

   ($\tilde{k}$ is the encoding by $J_{\mathbb{N}^{|k|}}$ of the reverse of the sequence with encoding k when this sequence is not empty, and it is zero otherwise).

4. $\mathsf{app} = \lambda\, k_1, k_2 \,.\, \begin{cases} J(0, k_2) + 1 & \text{if } k_1 = 0 \\ J(|k_1|, J(k_2, \tilde{k}_1)) + 1 & \text{otherwise} \end{cases} : \mathbb{N}^2 \to \mathbb{N}$

   ($\mathsf{app}(k_1, k_2)$ is the encoding of the sequence resulting from appending k_2 to the sequence with encoding k_1).

5. $\mathsf{last} = \lambda\, k \,.\, \begin{cases} \tilde{k}_{\mathbb{N}^2}[[1]] & \text{if } |k| > 1 \\ \tilde{k} & \text{if } |k| = 1 \\ 0 & \text{otherwise} \end{cases} : \mathbb{N} \to \mathbb{N}$

   ($\mathsf{last}(k)$ is the last element of the sequence with encoding k when this sequence is not empty, and it is zero otherwise).

6. $\mathsf{most} = \lambda\, k \,.\, \begin{cases} J(|k| - 2, \tilde{k}_{\mathbb{N}^2}[[2]]) + 1 & \text{if } |k| > 1 \\ 0 & \text{otherwise} \end{cases} : \mathbb{N} \to \mathbb{N}$

   ($\mathsf{most}(k)$ is the encoding of the sequence obtained by dropping the last element from the sequence with encoding k when this sequence is not empty, and it is zero otherwise).

7. $\mathsf{dlast} = \lambda\, k \,.\, \begin{cases} J(|k| - 1, J(\mathsf{pred}(\tilde{k}_{\mathbb{N}^2}[[1]]), \tilde{k}_{\mathbb{N}^2}[[2]])) + 1 & \text{if } |k| > 1 \\ J(0, \mathsf{pred}(\tilde{k})) + 1 & \text{if } |k| = 1 \\ 0 & \text{otherwise} \end{cases} : \mathbb{N} \to \mathbb{N}$

(dlast(k) is the encoding of the sequence obtained by decrementing the last element of the sequence with encoding k when this sequence is not empty, and it is zero otherwise).

8. $\langle\cdot\rangle^{\bullet}_{\cdot} = \lambda\, k, i\, . \begin{cases} \mathsf{last}(k) & \text{if } k > 0, i = 1 \\ \langle\mathsf{most}(k)\rangle^{\bullet}_{i-1} & \text{if } 1 < i \le |k| \\ 0 & \text{otherwise} \end{cases} \quad : \mathbb{N}^2 \to \mathbb{N}$

($\langle k\rangle^{\bullet}_i$ is the i-th element counting from the end of the sequence with encoding k when $1 \le i \le |k|$, and it is zero otherwise).

9. $\langle\cdot\rangle_{\cdot} = \lambda\, k, i\, . \begin{cases} 0 & \text{if } i > |k| \\ \langle k\rangle^{\bullet}_{|k|+1-i} & \text{otherwise} \end{cases} \quad : \mathbb{N}^2 \to \mathbb{N}$

($\langle k\rangle_i$ is the i-th element of the sequence with encoding k when $1 \le i \le |k|$, and it is zero otherwise).

10. for each $m \in \mathbb{N}$:

$$\lambda\, k\, . \, k^{m+1}_{\mathbb{N}^*} = \lambda\, k\, . \begin{cases} (\langle k\rangle_1, \ldots, \langle k\rangle_{m+1}) & \text{if } |k| = m+1 \\ (0, \ldots, 0) & \text{otherwise} \end{cases} \quad : \mathbb{N} \to \mathbb{N}^{m+1}$$

($k^{m+1}_{\mathbb{N}^*}$ is the sequence $k_{\mathbb{N}^*}$ when $|k| = m+1$, and it is the sequence of length $m+1$ composed of zeros otherwise).

Exercise 14 (Course-of-values recursion)
Assume that $f_0 : \mathbb{N}^n \to \mathbb{N}$ and $f_1 : \mathbb{N}^{n+3} \to \mathbb{N}$ are in $p\mathcal{R}$. Show that

$$\mathsf{cvrec}(f_0, f_1),\ \mathsf{cvrec}^*(f_0, f_1) : \mathbb{N}^{n+1} \to \mathbb{N}$$

are both in $p\mathcal{R}$ where

- $\mathsf{cvrec}(f_0, f_1)(k_1, \ldots, k_n, 0) = f_0(k_1, \ldots, k_n)$;
- $\mathsf{cvrec}^*(f_0, f_1)(k_1, \ldots, k_n, 0) = 0$;
- $\mathsf{cvrec}(f_0, f_1)(k_1, \ldots, k_n, k+1) =$
  $f_1(k_1, \ldots, k_n, k,$
  $\quad\quad \mathsf{cvrec}^*(f_0, f_1)(k_1, \ldots, k_n, k), \mathsf{cvrec}(f_0, f_1)(k_1, \ldots, k_n, k))$;
- $\mathsf{cvrec}^*(f_0, f_1)(k_1, \ldots, k_n, k+1) =$
  $\mathsf{app}(\mathsf{cvrec}^*(f_0, f_1)(k_1, \ldots, k_n, k), \mathsf{cvrec}(f_0, f_1)(k_1, \ldots, k_n, k))$.

Map $\mathsf{cvrec}(f_0, f_1)$ is said to be obtained by *course-of-values recursion* from f_0 and f_1. The auxiliary map $\mathsf{cvrec}^*(f_0, f_1)$ provides the encoded list of the previous values of the recursion. Each of such previous values can be obtained with the map $\lambda\, k, i\, . \, \langle k\rangle_i$ and, so, used in the definition of f_1.

Consider now the problem of defining an enumeration $\lambda k \,.\, k_{\mathcal{R}}$ of the set of recursive functions. For readability, let

$$k_{\mathbb{N}^4} = (n, m, c, p)$$

where the components of the tuple are to be used below as follows:

- n for the input arity of the function;

- $m' = m + 1$ for the output arity of the function;

- c for the construction used to build the function (e.g. 7 for minimization);

- p for the parameter of the construction (e.g. the encoding of the arguments of the construction).

Observe that if $c > 0$, then $p < k$. With this notation in mind, the envisaged enumeration

$$\lambda k \,.\, k_{\mathcal{R}}$$

is inductively defined as follows:

p: If $c = 0$, $n = 0$ and $m' = 1$, then

$$k_{\mathcal{R}} = \mathsf{S}^p(0).$$

Z: If $c = 1$, $n = 1$ and $m' = 1$, then

$$k_{\mathcal{R}} = \mathsf{Z}.$$

S: If $c = 2$, $n = 1$ and $m' = 1$, then

$$k_{\mathcal{R}} = \mathsf{S}.$$

P: If $c = 3$, $n \geq 1$, $m' = 1$ and $1 \leq p \leq n$, then

$$k_{\mathcal{R}} = \mathsf{P}^n_p.$$

$\langle \cdot \rangle$: If $c = 4$ and, for each $i = 1, \ldots, m$, $(p_{\mathbb{N}^m}[[i]])_{\mathbb{N}^4}[[1]] = n$ (that is, the input arity of the ith-argument is n) and $(p_{\mathbb{N}^m}[[i]])_{\mathbb{N}^4}[[2]] = 0$ (that is, the output arity of the ith-argument is 1) then

$$k_{\mathcal{R}} = \langle (p_{\mathbb{N}^m}[[1]])_{\mathcal{R}}, \ldots, (p_{\mathbb{N}^m}[[m]])_{\mathcal{R}} \rangle.$$

∘: If $c = 5$, $(p_{\mathbb{N}^2}[[1]])_{\mathbb{N}^4}[[1]] = n$ (that is, the input arity of the first argument is n), $(p_{\mathbb{N}^2}[[2]])_{\mathbb{N}^4}[[2]] = m$ (that is, the output arity of the second argument is m) and $(p_{\mathbb{N}^2}[[1]])_{\mathbb{N}^4}[[2]] = (p_{\mathbb{N}^2}[[2]])_{\mathbb{N}^4}[[1]]$ (that is, the output arity of the first argument is equal to the input arity of the second argument), then

$$k_{\mathcal{R}} = (p_{\mathbb{N}^2}[[2]])_{\mathcal{R}} \circ (p_{\mathbb{N}^2}[[1]])_{\mathcal{R}}.$$

rec: If $c = 6$, $n \geq 1$, $m' = 1$, $(p_{\mathbb{N}^2}[[1]])_{\mathbb{N}^4}[[1]] = n - 1$ (that is, the input arity of the first argument is $n - 1$) $(p_{\mathbb{N}^2}[[1]])_{\mathbb{N}^4}[[2]] = 0$ (that is, the output arity of the first argument is 1), $(p_{\mathbb{N}^2}[[2]])_{\mathbb{N}^4}[[1]] = n+1$ (that is, the input arity of the second argument is $n + 1$) and $(p_{\mathbb{N}^2}[[2]])_{\mathbb{N}^4}[[2]] = 0$ (that is, the output arity of the second argument is 1), then

$$k_{\mathcal{R}} = \mathsf{rec}((p_{\mathbb{N}^2}[[1]])_{\mathcal{R}}, (p_{\mathbb{N}^2}[[2]])_{\mathcal{R}}).$$

min: If $c = 7$, $m' = 1$, $p_{\mathbb{N}^4}[[1]] = n+1$ (that is, the input arity of the argument is $n + 1$) and $p_{\mathbb{N}^4}[[2]] = 0$ (that is, the output arity of the argument is 1), then

$$k_{\mathcal{R}} = \mathsf{min}(p_{\mathcal{R}}).$$

↑: Otherwise,

$$k_{\mathcal{R}} = \lambda x \,.\, \mathrm{undefined} : \mathbb{N}^n \rightharpoonup \mathbb{N}^{m'}.$$

Proposition 15 The map $\lambda k \,.\, k_{\mathcal{R}}$ is an enumeration of $\mathcal{R}$.

Proof:

(a) By looking at the inductive definition of the map, it is easily verified that, for each $k \in \mathbb{N}$, if, for each $k' \in \mathbb{N}$ less than k, the function $k'_{\mathcal{R}}$ is in $\mathcal{R}$, then $k_{\mathcal{R}}$ is also a recursive function. So, by complete induction, for every $k \in \mathbb{N}$, the function $k_{\mathcal{R}}$ is in $\mathcal{R}$.

(b) It remains to prove that, for every recursive function f, there is $k \in \mathbb{N}$ such that $f = k_{\mathcal{R}}$. The proof is easily carried out by complete induction on the structure of any Kleene program provided for computing f.   QED

Exercise 16 Verify that, for each $k \in \mathbb{N}$, the input arity of $k_{\mathcal{R}}$ is $k_{\mathbb{N}^4}[[1]]$ and the output arity of $k_{\mathcal{R}}$ is $k_{\mathbb{N}^4}[[2]] + 1$.

Proposition 17 (Lemma of universal predicate)
There is a primitive recursive map

$$u : \mathbb{N}^4 \to \{0, 1\}$$

such that, for every $k \in \mathbb{N}$, letting $n = k_{\mathbb{N}^4}[[1]]$ and $m = k_{\mathbb{N}^4}[[2]]$, the following conditions hold:

(i)   For every $x_1, \ldots, x_n, y_1, \ldots, y_{m+1} \in \mathbb{N}$,
      $k_{\mathcal{R}}(x_1, \ldots, x_n) = (y_1, \ldots, y_{m+1})$ iff there is $s \in \mathbb{N}$ such that
      $u(k, J_{\mathbb{N}^*}(x_1, \ldots, x_n), J_{\mathbb{N}^*}(y_1, \ldots, y_{m+1}), s) = 1$.

(ii) For every $x, y, s_1, s_2 \in \mathbb{N}$,
      If $u(k, x, y, s_1) = 1$ and $s_1 < s_2$ then $u(k, x, y, s_2) = 1$.

Proof: Again, for readability sake, let $k_{\mathbb{N}^4} = (n, m, c, p)$ and $m' = m + 1$. Define $u : \mathbb{N}^4 \to \{0, 1\}$ as follows:

- $u(k, x, y, s) = 1$ if $|x| = n$, $|y| = m'$ and one of the following conditions holds:

p: $\begin{cases} n = 0, m' = 1, c = 0 \\ y = J_{\mathbb{N}^*}(p); \end{cases}$

Z: $\begin{cases} n = 1, m' = 1, c = 1 \\ y = J_{\mathbb{N}^*}(0); \end{cases}$

S: $\begin{cases} n = 1, m' = 1, c = 2 \\ y = J_{\mathbb{N}^*}(\langle x \rangle_1 + 1); \end{cases}$

P: $\begin{cases} n \geq 1, m' = 1, c = 3 \\ 1 \leq p \leq n \\ y = J_{\mathbb{N}^*}(\langle x \rangle_p); \end{cases}$

$\langle \cdot \rangle$: $\begin{cases} c = 4 \\ (p_{\mathbb{N}^{m'}}[[i]])_{\mathbb{N}^4}[[1]] = n \ \& \ (p_{\mathbb{N}^{m'}}[[i]])_{\mathbb{N}^4}[[2]] = 0 \text{ for } i = 1, \ldots, m' \\ u(p_{\mathbb{N}^{m'}}[[i]], x, J_{\mathbb{N}^*}(\langle y \rangle_i), s) = 1 \text{ for } i = 1, \ldots, m'; \end{cases}$

$\circ$: $\begin{cases} c = 5 \\ (p_{\mathbb{N}^2}[[1]])_{\mathbb{N}^4}[[1]] = n \\ (p_{\mathbb{N}^2}[[2]])_{\mathbb{N}^4}[[2]] = m \\ (p_{\mathbb{N}^2}[[1]])_{\mathbb{N}^4}[[2]] = (p_{\mathbb{N}^2}[[2]])_{\mathbb{N}^4}[[1]] \\ \exists z < s \text{ such that:} \\ \quad \begin{cases} u(p_{\mathbb{N}^2}[[1]], x, z, s) = 1 \\ u(p_{\mathbb{N}^2}[[2]], z, y, s) = 1; \end{cases} \end{cases}$

$$\text{rec:} \begin{cases} n \geq 1, m' = 1, c = 6 \\ (p_{\mathbb{N}^2}[[1]])_{\mathbb{N}^4}[[1]] = n - 1 \\ (p_{\mathbb{N}^2}[[1]])_{\mathbb{N}^4}[[2]] = m \\ (p_{\mathbb{N}^2}[[2]])_{\mathbb{N}^4}[[1]] = n + 1 \\ (p_{\mathbb{N}^2}[[2]])_{\mathbb{N}^4}[[2]] = m \\ \text{if } \mathsf{last}(x) = 0, \\ \qquad \text{then } u(p_{\mathbb{N}^2}[[1]], \mathsf{most}(x), y, s) = 1, \\ \qquad \text{else } \exists\, z < s \text{ such that:} \\ \qquad\qquad \begin{cases} u(k, \mathsf{dlast}(x), J_{\mathbb{N}^*}(z), s) = 1 \\ u(p_{\mathbb{N}^2}[[2]], \mathsf{app}(\mathsf{dlast}(x), z), y, s) = 1; \end{cases} \end{cases}$$

$$\text{min:} \begin{cases} m' = 1, c = 7 \\ p_{\mathbb{N}^4}[[1]] = n + 1 \\ p_{\mathbb{N}^4}[[2]] = m \\ u(p, \mathsf{app}(x, \langle y \rangle_1), 0, s) = 1 \\ \forall\, j < \langle y \rangle_1 \; \exists\, z < s \text{ such that:} \\ \qquad \begin{cases} z > 0 \\ u(p, \mathsf{app}(x, j), J_{\mathbb{N}^*}(z), s) = 1. \end{cases} \end{cases}$$

- $u(k, x, y, s) = 0$ otherwise.

It is necessary to check that u is indeed primitive recursive and that it fulfills the two envisaged properties. The proof of the primitive recursiveness of u is straightforward using Exercises 4–14 above. The proof that u fulfills property (i) is carried out, for arbitrary y, by structural induction on the well founded partial order $\preccurlyeq$ over $\mathbb{N}^2$ defined as follows:

$$(k_1, x_1) \preccurlyeq (k_2, x_2) \text{ if and only if } k_1 < k_2 \text{ or } \begin{cases} k_1 = k_2 \\ x_1 \leq x_2. \end{cases}$$

Observe that induction on k is not enough since in the case of recursion the argument can be k. The proof that u fulfills property (ii) is carried out, for arbitrary y, s_1 and s_2, again by structural induction on $\preccurlyeq$. The details of these proofs are left to the reader.   QED

Intuitively,

$$u(k, x, y, s) = 1$$

means that the k-th recursive function can be checked to produce (the decoding of) y when applied to (the decoding of) x, while bounding searches to s (in existential quantifications).

Note that the restriction of the search space stands in relation to $\mathcal{R}$ as the restriction of execution time stands in relation to $\mathcal{C}$. The following result is an immediate corollary of the universal predicate lemma.[2]

Proposition 18 (Kleene's normal form theorem)
Let $m, n \in \mathbb{N}$ and $f : \mathbb{N}^n \rightharpoonup \mathbb{N}^{m+1}$ be a function in $\mathcal{R}$. Then, there is k such that:

$$f(x_1, \ldots, x_n) = ((\mu\, s \,.\, u(k, J_{\mathbb{N}^*}(x_1, \ldots, x_n), s_{\mathbb{N}^2}[[1]], s))_{\mathbb{N}^2}[[1]])_{\mathbb{N}^*}^{m+1}$$

for every $x_1, \ldots, x_n \in \mathbb{N}$.

Proof: Let $f : \mathbb{N}^n \rightharpoonup \mathbb{N}^{m+1}$ and $k \in \mathbb{N}$ such that $f = k_{\mathcal{R}}$. Given $(x_1, \ldots, x_n) \in \mathbb{N}^n$, we have two cases to consider:

(1) $(x_1, \ldots, x_n) \in \text{dom}\, f$:
Let $f(x_1, \ldots, x_n) = (y_1, \ldots, y_{m+1})$. Then, by (i) of Proposition 17, there is r such that

$$u(k, J_{\mathbb{N}^*}(x_1, \ldots, x_n), J_{\mathbb{N}^*}(y_1, \ldots, y_{m+1}), r) = 1$$

and, so, taking into account (ii) of Proposition 17,

$$u(k, J_{\mathbb{N}^*}(x_1, \ldots, x_n), J_{\mathbb{N}^*}(y_1, \ldots, y_{m+1}), J(J_{\mathbb{N}^*}(y_1, \ldots, y_{m+1}), r)) = 1$$

Thus, there is s such that

$$u(k, J_{\mathbb{N}^*}(x_1, \ldots, x_n), s_{\mathbb{N}^2}[[1]], s) = 1.$$

Furthermore, for every such s, $J_{\mathbb{N}^*}(y_1, \ldots, y_{m+1}) = s_{\mathbb{N}^2}[[1]]$. Therefore,

$$(y_1, \ldots, y_{m+1}) = ((\mu\, s \,.\, u(k, J_{\mathbb{N}^*}(x_1, \ldots, x_n), s_{\mathbb{N}^2}[[1]], s))_{\mathbb{N}^2}[[1]])_{\mathbb{N}^*}^{m+1}.$$

(2) $(x_1, \ldots, x_n) \notin \text{dom}\, f$:
By (i) of Proposition 17, there are no $y, r \in \mathbb{N}$ such that

$$u(k, J_{\mathbb{N}^*}(x_1, \ldots, x_n), y, r) = 1$$

and, so, there is no s such that

$$u(k, J_{\mathbb{N}^*}(x_1, \ldots, x_n), s_{\mathbb{N}^2}[[1]], s) = 1.$$

Therefore,

$$((\mu\, s \,.\, u(k, J_{\mathbb{N}^*}(x_1, \ldots, x_n), s_{\mathbb{N}^2}[[1]], s))_{\mathbb{N}^2}[[1]])_{\mathbb{N}^*}^{m+1}$$

is undefined, as required.   QED

---

[2]Using the μ notation. If p is a predicate, that is, $p : D^{n+1} \to \{0, 1\}$, with totally ordered domain D then $\mu\, a \,.\, p(a, \vec{b})$ denotes the least a such that $p(a, \vec{b}) = 1$ when $\{a : p(a, \vec{b}) = 1\} \neq \emptyset$; otherwise, $\mu\, a \,.\, p(a, \vec{b})$ is undefined. This informal notation is widely used by mathematicians. The explicit relation between the μ notation and Kleene's minimization is left as an exercise.

From Kleene's normal form theorem, the following result concerning the use of minimization is easily established.

Proposition 19 (Normal form corollary)
Every function in $\mathcal{R}$ can be computed by a *Kleene* program containing at most one use of minimization. If needed, the minimization is guarded when the function is total.

Proof: Let $f = k_{\mathcal{R}} : \mathbb{N}^n \to \mathbb{N}^{m+1}$. Consider the map

$$
\begin{aligned}
g &= \min(\lambda\, x_1, \ldots, x_n, s \,.\, \mathsf{neg}(u)(k, J_{\mathbb{N}^*}(x_1, \ldots, x_n), s_{\mathbb{N}^2}[[1]], s)) \\
&= \min(\lambda\, x_1, \ldots, x_n, s \,.\, \mathsf{neg}(u)(k, (J_{\mathbb{N}^*})|_{\mathbb{N}^n}(x_1, \ldots, x_n), s_{\mathbb{N}^2}[[1]], s))
\end{aligned}
$$

Then, since $\mathsf{neg}(u)$, $(J_{\mathbb{N}^*})|_{\mathbb{N}^n}$, $\lambda\, k \,.\, k_{\mathbb{N}^2}[[1]]$ and $\lambda\, k \,.\, k_{\mathbb{N}^*}^{m+1}$ are primitive recursive,

$$
f = \lambda\, x_1, \ldots, x_n \,.\, ((g(x_1, \ldots, x_n))_{\mathbb{N}^2}[[1]])_{\mathbb{N}^*}^{m+1}
$$

is computed by a *Kleene* program containing at most a minimization. Furthermore, if f is total then the minimization used for computing g is guarded, because, thanks to (i) and (ii) of Proposition 17, the search for s is always successful.   QED

Finally, as an immediate consequence of the normal form corollary, we obtain the envisaged result stating that for generating all recursive maps we never need to use a non-guarded minimization and, so, we never need to use a non-total function.

Proposition 20 $\mathcal{R}{\downarrow} \subseteq gm\mathcal{R}$.

In short, every map $h : \mathbb{N}^n \to \mathbb{N}^m$ is computable if and only if it belongs to $gm\mathcal{R}$. As shown in the next section, one can be even more frugal: recursion can be avoided.

12.3 Computability à la Gödel

Nowadays, it is well-known in computer programming that recursion can be replaced by iteration, using a stack to keep track of the intermediate values of the recursion. Kurt Gödel anticipated this programming technique when he understood that $gm\mathcal{R}$ can be obtained without using recursion. To this end, he introduced, without using recursion, a mechanism of encoding sequences of natural numbers (used as stacks) as natural numbers.

More concretely, assume the availability of some map $\alpha : \mathbb{N}^* \to \mathbb{N}$ for encoding sequences of natural numbers as natural numbers, together with a decoding map $\beta : \mathbb{N}^2 \to \mathbb{N}$ such that

$$\beta(\alpha(k_0 \ldots k_j), i) = k_i \quad \text{for } i = 0, \ldots, j.$$

It is then easy to define by minimization any map h introduced by primitive recursion, using, for each $\vec{x}$ and $y > 0$, the sequence of values

$$h(\vec{x}, 0) \ldots h(\vec{x}, y - 1)$$

for computing $h(\vec{x}, y)$.

In fact, the coding of sequences need not be bijective. As the next result shows, it is enough to find a binary map β such that, for every sequence $k_0 \ldots k_j$ of natural numbers, there is a natural number w for which $\beta(w, i) = k_i$ for each $i = 0, \ldots, j$.

Proposition 21 (Lemma of recursion elimination)
Let $\beta : \mathbb{N}^2 \to \mathbb{N}$ be a computable map such that, for every sequence $k_0 \ldots k_j$ of natural numbers, there is a natural number w for which $\beta(w, i) = k_i$ for each $i = 0, \ldots, j$. Let $f_0 : \mathbb{N}^n \to \mathbb{N}$ and $f_1 : \mathbb{N}^{n+2} \to \mathbb{N}$ be computable maps. Then:

1. The set $C_{f_0 f_1}$ such that $(\vec{x}, y, w) \in C_{f_0 f_1}$ if and only if

$$\begin{cases} \beta(w, 0) = f_0(\vec{x}) \\ \beta(w, i+1) = f_1(\vec{x}, i, \beta(w, i)) \text{ for each } i < y \end{cases}$$

   is computable.

2. The map $\mathrm{rec}(f_0, f_1) : \mathbb{N}^{n+1} \to \mathbb{N}$ can be obtained by guarded minimization as follows:

$$\beta \circ \langle \\ \min(\mathrm{neq} \circ \langle \chi_{C_{f_0 f_1}}, \mathsf{Z} \circ \mathsf{P}_1^{n+2} \rangle) \\ , \\ \mathsf{P}_{n+1}^{n+1} \\ \rangle$$

Proof:
1. Since it is assumed that β, f_0 and f_1 are computable maps, it is straightforward to present an algorithm to compute $\chi_{C_{f_0 f_1}}$.
2. Observe that, by construction, letting $f = \mathrm{rec}(f_0, f_1)$,

$(\vec{x}, y, w) \in C_{f_0 f_1}$ if and only if w encodes the sequence $f(\vec{x}, 0) f(\vec{x}, 1) \ldots f(\vec{x}, y)$.

Hence,

$$\mathsf{rec}(f_0, f_1)(\vec{x}, y) = f(\vec{x}, y) = \beta(\mu\, w\,.\,(\vec{x}, y, w) \in C_{f_0 f_1}, y)$$

and the latter can be computed by the proposed *Kleene* program.   QED

Therefore, in order to achieve the desideratum of removing recursion, it is necessary to start by finding some β with the envisaged decoding property and computed without recursion. It is also necessary to verify that $\chi_{C_{f_0 f_1}}$ is computable without recursion when $f_0 : \mathbb{N}^n \to \mathbb{N}$ and $f_1 : \mathbb{N}^{n+2} \to \mathbb{N}$ are computable without recursion. With these goals in mind, Kurt Gödel introduced the following class of maps.

The class $\mathcal{G}$ is the least set of maps with arguments and outputs in $\mathbb{N}$ such that:

- it contains constant k (map of arity zero) for each $\mathsf{k} \in \mathbb{N}$;

- it contains Z (map of arity one);

- it contains the projections (of all different positive arities);

- it contains addition, multiplication and characteristic of relation $<$ (maps of arity two);

- it is closed under aggregation, composition and guarded minimization.

As a tribute to Gödel, the maps in $\mathcal{G}$ will be referred to as *Gödel maps*. Moreover, for each $n \in \mathbb{N}$, a subset of $\mathbb{N}^n$ will be said to be in $\mathcal{G}$ if the corresponding characteristic map is in $\mathcal{G}$. Such sets will be referred to as *Gödel sets*.

Proposition 22 $\mathcal{G} \subseteq gm\mathcal{R}$.

Proof: The proof by induction on the structure of $\mathcal{G}$ is left as an exercise to the reader.   QED

In order to prove the converse, it is first necessary to build a small library of maps and sets in $\mathcal{G}$, a task left to the reader in the following exercise.

Exercise 23 Show that $\mathcal{G}$ is closed under the usual finitary operations on sets (including bounded quantification). Show that $\mathcal{G}$ contains the following maps and sets:

- Map $\lambda\, k\,.\, n$ for every $n \in \mathbb{N}$.

- Binary relations $\leq, >, \geq, =$ and $\neq$.

- Maps J and zigzag.

- Map $\lambda\, x, y \,.\, x \stackrel{.}{-} y$ that returns $x - y$ if $x > y$ and returns 0 otherwise (natural subtraction).

- Map $\lambda\, x, y \,.\, \mathsf{rem}(x, y)$ that returns the remainder of the division of y by x.

- Set $\{(x, y) : x|y\}$ where $x|y$ stands for x divides y, that is, there is $z \in \mathbb{N}$ such that $y = zx$.

- Set $\{x : \mathsf{prime}(x)\}$ of prime numbers.

- Set $\{(x, y) : \mathsf{power}(x, y)\}$ where power(x,y) stands for x is a power of y, that is, there is $z \in \mathbb{N}$ such that $x = y^z$.

The following result provides the first ingredient for capitalizing on Proposition 21 to prove that $\mathcal{G}$ is closed for primitive recursion.

Proposition 24 (Gödel's β map)
There is a map $\beta : \mathbb{N}^2 \to \mathbb{N} \in \mathcal{G}$ such that for each sequence $k_0 \ldots k_j$ of natural numbers there is a natural number w for which $\beta(w, i) = k_i$ for every $i = 0, \ldots, j$.

Proof: Since

$$\beta^* = \lambda\, x, y, i \,.\, \mathsf{rem}(1 + (i + 1)y, x) : \mathbb{N}^3 \to \mathbb{N}$$

is in $\mathcal{G}$,

$$\beta = \lambda\, w, i \,.\, \beta^*(\mathsf{zigzag}(w)[[1]], \mathsf{zigzag}(w)[[2]], i) : \mathbb{N}^2 \to \mathbb{N}$$

is also in $\mathcal{G}$. All that is left is checking that the latter map has the required decoding property: for every sequence $k_0 \ldots k_j$ of natural numbers, there is a natural number w for which $\beta(w, i) = k_i$ for each $i = 0, \ldots, j$.

To this end, a modicum of modular arithmetic is needed (for more details see [33]). Two natural numbers are said to be *coprime* or *relatively prime* if they have no common divisors other than 1. A set of natural numbers is said to be *pairwise coprime* or *pairwise relatively prime* if every two distinct numbers in the set are coprime. The Chinese remainder theorem (originally presented in a third-century AD book by Chinese mathematician Sun Tzu) is at the core of the envisaged proof. Assume that $\{u_0, \ldots, u_j\}$ is a pairwise coprime set. Then, for every given sequence $k_0, \ldots, k_j$ of natural numbers there is a natural number x that solves the following system of congruences:

$$\begin{cases} x \equiv k_0 \bmod u_0 \\ \ldots \\ x \equiv k_j \bmod u_j \end{cases}$$

Furthermore, all solutions of this system are congruent modulo $u_0 \times \cdots \times u_j$.

Let

$$y = \ell! \quad \text{with} \quad \ell = \max\{k_0, \ldots, k_j, j\}.$$

and take

$$u_i = 1 + (i+1)y \quad \text{for } i = 0, \ldots, j.$$

The set $\{u_0, \ldots, u_j\}$ is pairwise coprime as it is now shown. Assume, by contradiction, that it is not the case. That is, assume that there are $a, b \leq j$ and a prime p that divides both $u_a = 1 + (a+1)y$ and $u_b = 1 + (b+1)y$. With no loss of generality, take $a < b$. Then, p divides $u_b - u_a = (b-a)y$ (*). But, p does not divide y (†). Indeed, if so then p would divide $(a+1)y$ and thus it would divide 1, since it divides $1 + (a+1)y$. Moreover, p does not divide $b - a$ (‡). Indeed, otherwise, p would divide $y = \ell!$ since $b - a$ divides y because $(b - a) \leq j \leq \ell$. Therefore, in consequence of (†) and (‡), p does not divide $(b-a)y$, in contradiction with (*).

By the Chinese remainder theorem, there is $x \in \mathbb{N}$ such that

$$\mathsf{rem}(u_i, x) = \mathsf{rem}(u_i, k_i) \quad \text{for each } i = 0, \ldots, j.$$

Therefore, for each $i = 0, \ldots, j$, since $k_i < u_i$,

$$\beta^*(x, y, i) = \mathsf{rem}(1 + (i+1)y, x) = \mathsf{rem}(u_i, x) = \mathsf{rem}(u_i, k_i) = k_i$$

and, so, choosing $w = J(x, y)$,

$$\beta(w, i) = k_i$$

as envisaged. QED

The existence of such $\beta : \mathbb{N}^2 \to \mathbb{N}$ in $\mathcal{G}$ was one of the key contributions made by Gödel towards the incompleteness theorems. Observe that other common ways of encoding sequences of natural numbers (like the one in Exercise 13 or those based on prime factorization) are not directly suitable for the task because they are defined using primitive recursion. With Gödel's β map (usually known as *Gödel's β function*) at hand, it is straightforward to establish:

Proposition 25 If $f_0 : \mathbb{N}^n \to \mathbb{N}$ and $f_1 : \mathbb{N}^{n+2} \to \mathbb{N}$ are maps in $\mathcal{G}$, then $\chi_{C_{f_0 f_1}}$ and $\mathsf{rec}(f_0, f_1)$ are also in $\mathcal{G}$.

Finally, it is possible to prove that $gm\mathcal{R} \subseteq \mathcal{G}$. Thus, denoting the set of computable maps by $\mathcal{C}{\downarrow}$, the following holds:

Proposition 26 (Fundamental lemma of computable maps)

$$\mathcal{G} = gm\mathcal{R} = \mathcal{R}{\downarrow} = \mathcal{C}{\downarrow}.$$

12.4 Representability in Th($\mathbb{N}$)

The goal of this section is to prove that all computable maps and sets are representable in theory Th($\mathbb{N}$). Capitalizing on the results of the previous section, it is sufficient to prove that all Gödel maps and sets are representable in Th($\mathbb{N}$). To this end, a useful technical lemma is proposed in the following exercise.

Exercise 27 Let $\mathrm{fv}_{\Sigma^{\mathbb{N}}}(\varphi) = \{x_1\}$. Show that:

1. $([\varphi]_{\mathbf{0}}^{x_1} \Rightarrow (\ldots \Rightarrow ([\varphi]_{\mathbf{k-1}}^{x_1} \Rightarrow ((x_1 < \mathbf{k}) \Rightarrow \varphi)))) \in \mathrm{Th}(\mathbb{N})$.

2. if $(\neg[\varphi]_{\mathbf{j}}^{x_1}) \in \mathrm{Th}(\mathbb{N})$ for $j = 0, \ldots, k-1$ and $[\varphi]_{\mathbf{k}}^{x_1} \in \mathrm{Th}(\mathbb{N})$,

   then $((\varphi \wedge (\forall x_0 ((x_0 < x_1) \Rightarrow (\neg[\varphi]_{x_0}^{x_1})))) \Leftrightarrow (x_1 \cong \mathbf{k})) \in \mathrm{Th}(\mathbb{N})$.

Proposition 28 (Representability theorem)
Every computable map is representable in Th($\mathbb{N}$).

Proof: The proof is by induction on the structure of the algorithm certifying that the map is in $\mathcal{G}$:

1. Constant $\mathbf{k}$ is represented by
$$(x_1 \cong \mathbf{k}).$$

2. Map Z is represented by
$$((x_2 \cong \mathbf{0}) \wedge (x_1 \cong x_1)).$$

3. Each projection P_i^n is represented by
$$((x_{n+1} \cong x_i) \wedge (x_1 \cong x_1) \wedge \ldots \wedge (x_n \cong x_n)).$$

4. Addition is represented by
$$(x_3 \cong (x_1 + x_2)).$$

5. Multiplication is represented by
$$(x_3 \cong (x_1 \times x_2)).$$

6. Map $\chi_<$ is represented by
$$(((x_1 < x_2) \Rightarrow (x_3 \cong \mathbf{1})) \wedge ((\neg(x_1 < x_2)) \Rightarrow (x_3 \cong \mathbf{0}))).$$

7. Given $h_i : \mathbb{N}^n \to \mathbb{N}$ in $\mathcal{G}$ for $i = 1, \ldots, m$ and so, by the induction hypothesis, with each represented by some formula φ_{h_i}, the aggregation

$$\langle h_1, \ldots, h_m \rangle : \mathbb{N}^n \to \mathbb{N}^m$$

is represented by

$$\varphi_{h_1} \wedge [\varphi_{h_2}]_{x_{n+2}}^{x_{n+1}} \wedge \ldots \wedge [\varphi_{h_m}]_{x_{n+m}}^{x_{n+1}} .$$

8. Given $f : \mathbb{N}^n \to \mathbb{N}^m$ and $g : \mathbb{N}^m \to \mathbb{N}^r$ in $\mathcal{G}$ and so, by the induction hypothesis, represented by some formulas φ_f and φ_g, respectively, the composition

$$g \circ f : \mathbb{N}^n \to \mathbb{N}^r$$

is represented by

$$(\exists x_{j+1} \ldots \exists x_{j+m} ([[\varphi_g]_{x_{j+1}\ldots x_{j+m}}^{x_1 \ldots x_m}]_{x_{n+1},\ldots,x_{n+r}}^{x_{m+1},\ldots,x_{m+r}} \wedge [\varphi_f]_{x_{j+1},\ldots,x_{j+m}}^{x_{n+1},\ldots,x_{n+m}})) ,$$

where $j = \max\{n + m, m + r, n + r\}$.

9. Given $h : \mathbb{N}^{n+1} \to \mathbb{N}$ in $\mathcal{G}$, thus, by the induction hypothesis, represented by some φ_h, such that

for every $(a_1, \ldots, a_n) \in \mathbb{N}^n$ there exists a for which $h(a_1, \ldots, a_n, a) = 0$,

the (hence guarded) minimization

$$\min(h) : \mathbb{N}^n \to \mathbb{N}$$

is represented by

$$([\varphi_h]_{\mathbf{0}}^{x_{n+2}} \wedge (\forall x_{n+3} ((x_{n+3} < x_{n+1}) \Rightarrow (\neg[\varphi_h]_{x_{n+3}\mathbf{0}}^{x_{n+1}x_{n+2}})))) .$$

Showing that these representations are legitimate is left as an exercise. The task is easy in most cases. The case of minimization is helped by fact 2 established in Exercise 27.   QED

Thanks to Proposition 3, representability of all computable maps implies representability of all computable sets. Thus:

Proposition 29 Every computable set is representable in Th(ℕ).

12.5   Representability in **N**

The proof of the following proposition, which will be used in Chapter 13 only for establishing the undecidability of first-order logic as a corollary of the first incompleteness theorem, is left as an exercise. Capitalizing on the results of Section 12.3, it is sufficient to prove that all Gödel maps and sets are representable in **N**. The proof is carried out by induction on the structure of the algorithm certifying that the map is in $\mathcal{G}$, using in each case the same formula as in the proof of Proposition 28. The verification of the legitimacy of the representations is more technically challenging herein than in the proof of Proposition 28. The reader should consult Section 6.7 of Shoenfields's textbook [42] for guidance.

Proposition 30 Every computable map is representable in **N**.

In consequence, again by Proposition 3, computable sets are also representable in this theory:

Proposition 31 Every computable set is representable in **N**.

Note that, as a corollary, one may conclude that computable maps and computable sets are representable in every extension of **N**. Therefore, the representability results for Th(ℕ) could have been obtained by first establishing representability in **N**.

Chapter 13

First incompleteness theorem

Capitalizing on the results of the previous chapter, a very direct route is taken to the first incompleteness theorem (in the strong form known as the Gödel–Rosser version) via Church's theorem.

Afterward, several negative results about theory $\text{Th}(\mathbb{N})$ follow easily. All these results are independent of the fact that computable maps are representable in theory $\mathbf{N}$, but the results in the last section of this chapter (undecidability of FOL) do depend on it.

Section 13.1 starts with Cantor's lemma on diagonalization since this technique plays a central role in the proof of Church's theorem. After some notation and a few technical results on the idea of representing a theory of arithmetic as a subset of $\mathbb{N}^2$ and on the notion of arithmetical extent of a formula in such a theory, a result is proved relating a set with the extent of a formula representing it in a theory assumed to be consistent. It is then possible to show that every computable set is the extent of some formula, provided that in addition the computable maps are representable in the theory. Finally, the main results are established: Church's theorem is proved using the aforementioned diagonalization technique, the Gödel–Rosser strong version of the first-incompleteness theorem appears as an immediate corollary of Church's theorem, and a slight modification of the proof of Church's theorem yields the Gödel-Tarski undefinability theorem.

In Section 13.2, the main negative results about theory $\text{Th}(\mathbb{N})$ are established, including undecidability, non semi-decidability, impossibility of axiomatization, and undefinability.

The undecidability of first-order logic is established in Section 13.3 by capitalizing on the fact that the computable maps are representable in the finitely axiomatized theory $\mathbf{N}$.

13.1   Non computability of arithmetic theories

In order to obtain incompleteness of arithmetic, Kurt Gödel resorted to the *diagonalization technique* due to Georg Cantor.[1] It is worth mentioning here the lemma underlying its variant over the realm of natural numbers:

Proposition 1 (Cantor's lemma)
Let $C \subseteq \mathbb{N}^2$. Then, for every $D \subseteq \mathbb{N}$,

$$\lambda k \,.\, (1 - \chi_C(k,k)) \notin \{\lambda k \,.\, \chi_C(d,k) : d \in D\}.$$

Proof: Left as an exercise to the interested reader.   QED

In other words, for every $D \subseteq \mathbb{N}$,

$$U \notin \{C_d : d \in D\}$$

where:

- $U = \{k : (k,k) \notin C\}$;

- $C_d = \{k : (d,k) \in C\}$ for each $d \in D$.

Exercise 2 Recall the proof of the undecidability of the halting problem given in Chapter 3. Verify that a diagonal argument was used. Reformulate the proof in order to take advantage of Cantor's lemma.

The representation of arithmetical reasoning within arithmetic itself was Kurt Gödel's starting point in his way toward the incompleteness theorems. To this end, one begins by choosing a Gödelization of type $A_{\Sigma_{\mathbb{N}}}$.

Exercise 3 Define a Gödelization g of type $A_{\Sigma_{\mathbb{N}}}$. Hint: Adapt the map given in Exercise 2 of Chapter 4.

Chosen once and for all a Gödelization $g_{\mathbb{N}}$ of type $A_{\Sigma_{\mathbb{N}}}$, the following map is needed in what follows:

$$\mathrm{rsb} = \lambda d, u \,.\, \begin{cases} g_{\mathbb{N}}([g_{\mathbb{N}}^{-1}(d)]_{g_{\mathbb{N}}^{-1}(u)}^{x_1}) & \text{if } \begin{cases} d \in g_{\mathbb{N}}(L_{\Sigma_{\mathbb{N}}}) \\ u \in g_{\mathbb{N}}(T_{\Sigma_{\mathbb{N}}}) \end{cases} : \mathbb{N} \times \mathbb{N} \to \mathbb{N}. \\ d & \text{otherwise} \end{cases}$$

Clearly, the map rsb brings to the realm of natural numbers the binary operation

$$\lambda \varphi, t \,.\, [\varphi]_t^{x_1} : L_{\Sigma_{\mathbb{N}}} \times T_{\Sigma_{\mathbb{N}}} \to L_{\Sigma_{\mathbb{N}}}$$

that given a formula and a term returns the formula with the free occurrences of x_1 replaced by the term.

---

[1]Georg Cantor used this technique for the first time to show that the cardinality of the power set of any given set is strictly greater than the cardinality of the original set.

Exercise 4 Show that the map rsb is computable.

Given a theory of arithmetic Θ, it is also necessary to work with the binary relation

$$\text{ths}_\Theta = \{(g_\mathbb{N}(\delta), k) : [\delta]_\mathbf{k}^{x_1} \in \Theta\} \subseteq \mathbb{N} \times \mathbb{N}$$

said to be the (relational) *extent* of Θ.

Proposition 5 Let Θ be a theory of arithmetic. Then:

$$\text{ths}_\Theta = \{(d, k) : \text{rsb}(d, g_\mathbb{N}(\mathbf{k})) \in g_\mathbb{N}(\Theta)\}.$$

Proof: Let $S_\Theta = \{(d, k) : \text{rsb}(d, g_\mathbb{N}(\mathbf{k})) \in g_\mathbb{N}(\Theta)\}$.

(i) $\text{ths}_\Theta \subseteq S_\Theta$:

if $(g_\mathbb{N}(\delta), k) \in \text{ths}_\Theta$

then $[\delta]_\mathbf{k}^{x_1} \in \Theta$ (definition of ths_Θ)

            $g_\mathbb{N}([\delta]_\mathbf{k}^{x_1}) \in g_\mathbb{N}(\Theta)$

            $g_\mathbb{N}([g_\mathbb{N}^{-1}(g_\mathbb{N}(\delta))]_{g_\mathbb{N}^{-1}(g_\mathbb{N}(\mathbf{k}))}^{x_1}) \in g_\mathbb{N}(\Theta)$

            $\text{rsb}(g_\mathbb{N}(\delta), g_\mathbb{N}(\mathbf{k})) \in g_\mathbb{N}(\Theta)$ (definition of rsb)

            $(g_\mathbb{N}(\delta), k) \in S_\Theta$

(ii) $S_\Theta \subseteq \text{ths}_\Theta$:

First note that if $(d, k) \in S_\Theta$ then $d \in g_\mathbb{N}(L_{\Sigma_\mathbb{N}})$ since otherwise $\text{rsb}(d, g_\mathbb{N}(\mathbf{k})) = d$ and, so, $\text{rsb}(d, g_\mathbb{N}(\mathbf{k})) \notin g_\mathbb{N}(\Theta)$. Thus, it is enough to prove that if $(g_\mathbb{N}(\delta), k) \in S_\Theta$ then $(g_\mathbb{N}(\delta), k) \in \text{ths}_\Theta$. Indeed:

if $(g_\mathbb{N}(\delta), k) \in S_\Theta$

then $\text{rsb}(g_\mathbb{N}(\delta), g_\mathbb{N}(\mathbf{k})) \in g_\mathbb{N}(\Theta)$

            $g_\mathbb{N}([g_\mathbb{N}^{-1}(g_\mathbb{N}(\delta))]_{g_\mathbb{N}^{-1}(g_\mathbb{N}(\mathbf{k}))}^{x_1}) \in g_\mathbb{N}(\Theta)$ (definition of rsb)

            $g_\mathbb{N}([\delta]_\mathbf{k}^{x_1}) \in g_\mathbb{N}(\Theta)$

            $[\delta]_\mathbf{k}^{x_1} \in \Theta$

            $(g_\mathbb{N}(\delta), k) \in \text{ths}_\Theta$ (definition of ths_Θ)

QED

The binary relation ths_Θ captures enough information about the theory Θ for the purpose of establishing Church's theorem by a diagonal argument. To this end, the related concept of extent of a formula in the theory of arithmetic at hand is also needed. Given $\delta \in L_{\Sigma_\mathbb{N}}$, the set

$$\text{ext}_\Theta^\delta = \{k : [\delta]_\mathbf{k}^{x_1} \in \Theta\} \subseteq \mathbb{N}.$$

is said to be the *extent* of δ in Θ. Clearly:

Proposition 6 Let Θ be a theory of arithmetic. Then:

$$\mathrm{ext}_\Theta^\delta = \{k : (g_\mathbb{N}(\delta), k) \in \mathrm{ths}_\Theta\}.$$

Exercise 7 Show that for every $\delta \in L_{\Sigma_\mathbb{N}}$ such that $\mathrm{fv}_{\Sigma_\mathbb{N}}(\delta) = \{x_1\}$:

$$\{\rho(x_1) : \rho \in [\![\delta]\!]_{\Sigma_\mathbb{N}}^\mathbb{N}\} = \mathrm{ext}_{\mathrm{Th}(\mathbb{N})}^\delta.$$

Proposition 8 Let Θ be a consistent theory of arithmetic and suppose $C \subseteq \mathbb{N}$ is represented in Θ by $\varphi \in L_{\Sigma_\mathbb{N}}$. Then,

$$C = \mathrm{ext}_\Theta^\varphi.$$

Proof: The proof is by case analysis:

if	$k \in C$	then
	$[\varphi]_\mathbf{k}^{x_1} \in \Theta$	(representability)
	$k \in \mathrm{ext}_\Theta^\varphi$	(definition of $\mathrm{ext}_\Theta^\varphi$);
if	$k \notin C$	then
	$(\neg[\varphi]_\mathbf{k}^{x_1}) \in \Theta$	(representability)
	$[\varphi]_\mathbf{k}^{x_1} \notin \Theta$	(consistency of Θ)
	$k \notin \mathrm{ext}_\Theta^\varphi$	(definition of $\mathrm{ext}_\Theta^\varphi$).

QED

Proposition 9 Let Θ be a consistent theory of arithmetic where computable maps are representable and suppose $C \subseteq \mathbb{N}$ is computable. Then,

$$C \in \{\mathrm{ext}_\Theta^\delta : \delta \in L_{\Sigma_\mathbb{N}}\}.$$

Proof: By Proposition 3 in Chapter 12, the computable sets are also representable in Θ. Let φ be a formula representing C in Θ. Then, using the previous result, $C = \mathrm{ext}_\Theta^\varphi$. QED

With these results in hand, it is possible to prove the following theorem, due to Alonzo Church, that yields, in a very expedite way, Gödel's first incompleteness theorem, in its stronger version (toward which John Rosser also contributed).

Proposition 10 (Church's theorem)
Any consistent theory of arithmetic where computable maps are representable cannot be computable.

Proof: Let Θ be a consistent theory of arithmetic where computable maps are representable. The proof uses the diagonalization technique applied to the set ths_Θ. Consider the set

$$U = \{k : (k, k) \notin \text{ths}_\Theta\} \subseteq \mathbb{N}$$

with characteristic map

$$\chi_U = \lambda\, k\,.\,(1 - \chi_{\text{ths}_\Theta}(k, k))\,.$$

Using Cantor's lemma, one concludes that

$$\chi_U \notin \{\lambda\, k\,.\,\chi_{\text{ths}_\Theta}(d, k) : d \in D\}$$

for any $D \subseteq \mathbb{N}$. Hence, in particular,

$$\chi_U \notin \{\lambda\, k\,.\,\chi_{\text{ths}_\Theta}(d, k) : d \in g_{\mathbb{N}}(L_{\Sigma_{\mathbb{N}}})\}$$

and, so,

$$
\begin{aligned}
&\chi_U \notin \{\lambda\, k\,.\,\chi_{\text{ths}_\Theta}(g_{\mathbb{N}}(\delta), k) : \delta \in L_{\Sigma_{\mathbb{N}}}\} && \text{that is}\\
&U \notin \{\{k : (g_{\mathbb{N}}(\delta), k) \in \text{ths}_\Theta\} : \delta \in L_{\Sigma_{\mathbb{N}}}\} && \text{that is (Proposition 6)}\\
&U \notin \{\{k : k \in \text{ext}_\Theta^\delta\} : \delta \in L_{\Sigma_{\mathbb{N}}}\} && \text{that is}\\
&U \notin \{\text{ext}_\Theta^\delta : \delta \in L_{\Sigma_{\mathbb{N}}}\}
\end{aligned}
$$

Hence, by Proposition 9, U is not computable and, so, χ_U is not computable. Therefore, since, thanks to Proposition 5,

$$\chi_U = \lambda\, k\,.\,(1 - \chi_{g_{\mathbb{N}}(\Theta)}(\text{rsb}(k, g_{\mathbb{N}}(\mathbf{k}))))$$

and $\lambda\, k\,.\,1 - k : \{0, 1\} \to \{0, 1\}$, $\text{rsb} : \mathbb{N} \times \mathbb{N} \to \mathbb{N}$, $g_{\mathbb{N}} : A_{\Sigma_{\mathbb{N}}}^* \to \mathbb{N}$ and $\lambda\, k\,.\,\mathbf{k} : \mathbb{N} \to T_{\Sigma_{\mathbb{N}}}$ are computable maps, $\chi_{g_{\mathbb{N}}(\Theta)}$ cannot be computable. Thus, $g_{\mathbb{N}}(\Theta)$ is not computable and, so, by Proposition 24 in Chapter 3, Θ is not computable. QED

Proposition 11 (First incompleteness theorem – Gödel–Rosser)
Any axiomatizable theory of arithmetic where computable maps are representable cannot be consistent and exhaustive.

Proof: The result is reached by contradiction. Assume that there is a theory of arithmetic Θ that is exhaustive (1), consistent (2), axiomatizable (3), and where computable maps are representable (4).

Then, applying Proposition 8 in Chapter 6 to (3), it follows that Θ is computably enumerable (5). From (5) and (1), using Proposition 7 in Chapter 6, it follows that Θ is computable (6).

On the other hand, Church's theorem, when applied to (2) and (4), yields that Θ is not computable, in contradiction with (6). QED

The following result, usually attributed to Alfred Tarski, although independently proved by Kurt Gödel, yields another perspective upon the impossibility of formalizing arithmetic.

Proposition 12 (Gödel–Tarski undefinability theorem)
Let Θ be a consistent theory of arithmetic where computable maps are representable. Then, the set $g_{\mathbb{N}}(\Theta)$ is not representable in Θ.

Proof: The proof mimics the proof of Church's theorem, but reasoning about representability instead of computability. Let $U = \{k : (k, k) \notin \text{th}_\Theta\}$. Then, by Cantor's lemma and Proposition 6, it follows that:

$$U \notin \{\text{ext}_\Theta^\delta : \delta \in L_{\Sigma_\mathbb{N}}\}.$$

Thus, by Proposition 8, U is not representable in Θ. Hence, by Proposition 3 in Chapter 12, χ_U is not representable in Θ. On the other hand, by Proposition 5,

$$\begin{aligned} \chi_U &= \lambda k \,.\, (1 \dot{-} \chi_{g_{\mathbb{N}}(\Theta)}(\text{rsb}(k, g_{\mathbb{N}}(\mathbf{k})))) \\ &= \lambda k \,.\, (1 \dot{-} \chi_{g_{\mathbb{N}}(\Theta)}(\text{rsb}(k, g_{\mathbb{N}}(\mathbf{k})))). \end{aligned}$$

Therefore, since the composition of representable maps in Θ is still representable in that theory (see the proof of Proposition 28 of Chapter 12) and the maps $\lambda k \,.\, 1 \dot{-} k : \mathbb{N} \to \mathbb{N}$ and $\lambda k \,.\, \text{rsb}(k, g_{\mathbb{N}}(\mathbf{k})) : \mathbb{N} \to \mathbb{N}$ are representable in Θ because they are computable, $\chi_{g_{\mathbb{N}}(\Theta)}$ cannot be representable in Θ and, so, by Proposition 3 in Chapter 12, $g_{\mathbb{N}}(\Theta)$ is not representable in Θ. QED

Observe that Church's theorem, the Gödel–Rosser theorem and the Gödel–Tarski theorem do not rely upon any property of $\text{Th}(\mathbb{N})$. In particular, they do not rely on the fact that computable maps are representable in $\text{Th}(\mathbb{N})$. However, using this fact, one can establish a number of interesting results about the standard theory of arithmetic, which are the subject of the next section.

On the other hand, if computable maps were not representable, at least in $\text{Th}(\mathbb{N})$, these theorems would be vacuous.

13.2 Impossibility of axiomatizing arithmetic

Proposition 13 (Undecidability of arithmetic)
Theory $\text{Th}(\mathbb{N})$ is not computable.

Proof: The thesis is obtained as an immediate corollary of Church's theorem, since $\text{Th}(\mathbb{N})$ is consistent (recall Chapter 7, Exercise 37) and computable maps are representable in it (see Chapter 12, Proposition 28).

In alternative, the result can be established as a corollary of the Gödel–Rosser theorem, by contradiction. Indeed, assume that $\text{Th}(\mathbb{N})$ is computable. Then, it is axiomatizable (why?). Therefore, since $\text{Th}(\mathbb{N})$ is also consistent and computable maps are representable in it, the Gödel–Rosser theorem yields that it is not exhaustive, contradicting Exercise 37 in Chapter 7.   QED

Proposition 14 (Non semi-decidability of arithmetic)
Theory $\text{Th}(\mathbb{N})$ is not computably enumerable.

Proof: The result is obtained by contradiction. Assume that $\text{Th}(\mathbb{N})$ is computably enumerable. Then, since $\text{Th}(\mathbb{N})$ is exhaustive (Chapter 7, Exercise 37), by Proposition 7 in Chapter 6 it follows that $\text{Th}(\mathbb{N})$ is computable, contradicting the previous proposition.   QED

Proposition 15 (Impossibility of axiomatizing arithmetic)
Theory $\text{Th}(\mathbb{N})$ is not axiomatizable.

Proof: The result is obtained by contradiction. Assume that $\text{Th}(\mathbb{N})$ is axiomatizable. Then, by Proposition 8 in Chapter 6, it follows that $\text{Th}(\mathbb{N})$ is computably enumerable, contradicting the previous proposition.   QED

Proposition 16 (Undefinability of arithmetic)
The set $g_{\mathbb{N}}(\text{Th}(\mathbb{N}))$ is not representable in $\text{Th}(\mathbb{N})$.

Proof: Since $\text{Th}(\mathbb{N})$ is a consistent theory of arithmetic allowing the representation of the computable maps, the Gödel–Tarski theorem (Proposition 12) holds for this theory.   QED

13.3 Undecidability of first-order logic

This section establishes some results, independent of the properties of $\text{Th}(\mathbb{N})$, about consistent extensions of theory **N**. Profiting from the finite axiomatization of **N**, it is also shown that the set of theorems of first-order logic is not always computable.

Proposition 17 No consistent extension of theory **N** is computable.

Proof: This is a particular case of Church's theorem, since computable maps are representable in **N** (Chapter 12, Proposition 30), and, so, in any of its extensions.   QED

Proposition 18 (Undecidability of first-order logic)
The set of theorems of first-order logic over signature Σ is not computable in general.

Proof: The proof is by contradiction. Suppose that the set $\emptyset^{\vdash_\Sigma}$ is computable for every signature Σ. Then, in particular, $\emptyset^{\vdash_{\Sigma_N}}$ would be computable.

However, on the other hand, since the set Ax_N of specific axioms of theory **N** is finite, the following holds:

$$\varphi \in \mathbf{N} \qquad\qquad \text{iff}$$

$$Ax_\mathbf{N} \vdash_{\Sigma_N} \varphi \qquad\qquad \text{iff}$$

$$\left(\bigwedge\nolimits_{\delta \in Ax_\mathbf{N}} \delta\right) \vdash_{\Sigma_N} \varphi \qquad\qquad \text{iff}$$

$$\vdash_{\Sigma_N} \left(\left(\bigwedge\nolimits_{\delta \in Ax_\mathbf{N}} \delta\right) \Rightarrow \varphi\right) \qquad\qquad \text{iff}$$

$$\left(\left(\bigwedge\nolimits_{\delta \in Ax_\mathbf{N}} \delta\right) \Rightarrow \varphi\right) \in \emptyset^{\vdash_{\Sigma_N}}$$

Thus, if $\emptyset^{\vdash_{\Sigma_N}}$ were computable, then the set **N** would also be computable, contradicting the previous proposition. QED

It must be emphasized that one can find useful signatures for which the set of first-order theorems is computable. For example, if signature Σ contains only unary predicates, then $\emptyset^{\vdash_\Sigma}$ is computable.

Proposition 19 No consistent and axiomatizable extension of theory **N** is exhaustive.

Proof: This is a particular case of the Gödel–Rosser theorem, since computable maps are representable in **N** (Chapter 12, Proposition 30), and, hence, in any of its extensions. QED

Proposition 20 No consistent extension Θ of theory **N** allows the representation of $g_\mathbb{N}(\Theta)$.

Proof: This is a particular case of the Gödel–Tarski theorem, since computable maps are representable in **N** (Chapter 12, Proposition 30), and, hence, in any of its extensions. QED

Chapter 14

Second incompleteness theorem

The proof of the second incompleteness theorem is more challenging than the proof of the first incompleteness theorem because it requires in its strongest form the encoding of derivation sequences within arithmetic. Herein, this encoding is avoided by invoking the projection theorem (Proposition 11 of Chapter 3). This approach leads to a pseudo representation of arithmetic derivability,[1] but only allows the proof of the first of the Hilbert–Bernays–Löb conditions.

Section 14.1 motivates, introduces and proves some basic results about pseudo representability of derivability in a suitable theory of arithmetic. Section 14.2 starts with the Hilbert–Bernays–Löb conditions and the fixed point theorem. Afterward, Löb's theorem is proved for any suitable theory of arithmetic fulfilling those conditions. Finally, an abstract[2] version of the second incompleteness theorem is established as an immediate corollary.

14.1 Pseudo representability of derivability

Regardless of the importance of the first incompleteness theorem, one of the questions raised by David Hilbert is still unanswered: is it possible to prove consistency of an axiomatized theory from its axioms?

The negative answer to this question was also provided by Kurt Gödel (second incompleteness theorem): for every sufficiently strong and consistent

---

[1]Given that representability in the sense of Chapter 12 is out of question in consequence of the Gödel–Tarski theorem.

[2]That is, not involving a concrete encoding of derivation sequences.

axiomatization of (a fragment of) arithmetic, there exists a formula stating the consistency of the axioms, but such a formula cannot be derived from them.

In order to establish this result, it is necessary to portray the notion of derivability within the language of arithmetic. More precisely, given a theory Θ of arithmetic, for each formula $\alpha \in L_{\Sigma_\mathbb{N}}$ one must find a formula

$$(\Box_\Theta \alpha)$$

in $cL_{\Sigma_\mathbb{N}}$ stating that α is theorem of Θ. As expected, this desideratum will only be possible via a Gödelization of type $A_{\Sigma_\mathbb{N}}$, say $g_\mathbb{N}$, and only if Θ enjoys some properties.

A theory Θ of arithmetic is said to be *suitable* if it is computably enumerable and if computable maps are representable within it. Thus, every axiomatizable theory of arithmetic extending theory **N** is suitable. In particular, theory **P** is suitable.

Observe that, in consequence of the Gödel–Tarski theorem (Proposition 12 of Chapter 13), there is no hope of finding a formula representing $g_\mathbb{N}(\Theta)$ in a suitable and consistent theory Θ of arithmetic. Actually, this negative result can be proved directly from Church's theorem by taking advantage of the computable enumerability of Θ, as the reader is invited to show in the next exercise.

Exercise 1 Let Θ be a suitable and consistent theory of arithmetic. Show that $g_\mathbb{N}(\Theta)$ is not representable in Θ without using the Gödel–Tarski theorem.

Notwithstanding the negative result above, by making use of the projection theorem, it is possible to find a formula that weakly represents the Gödelization of a given suitable theory in itself. To this end, the following notion is also needed.

A theory Θ of arithmetic is said to be *ω-consistent* if, for each $\alpha \in L_{\Sigma_\mathbb{N}}$ such that $\mathrm{fv}_{\Sigma_\mathbb{N}}(\alpha) = \{y\}$, the following holds:

$$\text{If } \{(\neg[\alpha]_\mathbf{k}^y) : k \in \mathbb{N}\} \subseteq \Theta \text{ then } (\exists y\, \alpha) \notin \Theta.$$

The notion of ω-consistency was proposed by Kurt Gödel in order to establish the first incompleteness theorem. Only later did John Rosser provide a proof of the first incompleteness theorem that did not require that notion.

Proposition 2 (Pseudo representability of derivability)
Let Θ be a suitable theory of arithmetic. Then, there exists a formula $\delta_\Theta \in L_{\Sigma_\mathbb{N}}$ such that:

(1) $\mathrm{fv}_{\Sigma_\mathbb{N}}(\delta_\Theta) = \{x_1\}$;

(2) if $a \in g_{\mathbb{N}}(\Theta)$ then $[\delta_\Theta]_{\mathbf{a}}^{x_1} \in \Theta$;

(3) if $a \notin g_{\mathbb{N}}(\Theta)$ then $[\delta_\Theta]_{\mathbf{a}}^{x_1} \notin \Theta$, provided that, in addition, Θ is ω-consistent.

Proof: First observe that there are many formulas that trivially fulfill requirements (1) and (2), e.g. $(x_1 \cong x_1)$. But it is possible to do much better and provide a formula δ_Θ that does depend on the theory Θ at hand and that fulfills requirement (3).

Indeed, from the computable enumerability of Θ it follows, by Proposition 24 in Chapter 3, that $g_{\mathbb{N}}(\Theta)$ is also computably enumerable. Hence, by the projection theorem (Proposition 11 of Chapter 3), there exists a computable subset R_Θ of $\mathbb{N}^2$ such that:

$a \in g_{\mathbb{N}}(\Theta)$ if and only if there exists $s \in \mathbb{N}$ such that the pair $(a, s) \in R_\Theta$.

Therefore, using the hypothesis of representability of computable maps within Θ, and thus, due to Proposition 3 in Chapter 12, also of computable sets, there exists $\varphi_\Theta \in L_{\Sigma_\mathbb{N}}$ representing R_Θ in theory Θ, that is, such that:

- $\mathrm{fv}_{\Sigma_\mathbb{N}}(\varphi_\Theta) = \{x_1, x_2\}$;

- if $(a, s) \in R_\Theta$, then $[\varphi_\Theta]_{\mathbf{a,s}}^{x_1, x_2} \in \Theta$;

- if $(a, s) \notin R_\Theta$, then $(\neg[\varphi_\Theta]_{\mathbf{a,s}}^{x_1, x_2}) \in \Theta$.

Take δ_Θ to be

$$(\exists x_2 \, \varphi_\Theta).$$

Clearly, $\delta_\Theta \in L_{\Sigma_\mathbb{N}}$, $\mathrm{fv}_{\Sigma_\mathbb{N}}(\delta_\Theta) = \{x_1\}$. It remains to check that δ_Θ enjoys requirements (2) and (3).
Requirement (2):

Let $a \in g_{\mathbb{N}}(\Theta)$. Then, there exists $s \in \mathbb{N}$ such that $(a, s) \in R_\Theta$. Hence, since φ_Θ represents R_Θ in Θ,

$$\text{there exists } s \in \mathbb{N} \text{ such that } [\varphi_\Theta]_{\mathbf{a,s}}^{x_1, x_2} \in \Theta.$$

That is,

$$\text{there exists } s \in \mathbb{N} \text{ such that } [[\varphi_\Theta]_{\mathbf{a}}^{x_1}]_{\mathbf{s}}^{x_2} \in \Theta.$$

Therefore, by Proposition 29 in Section 5.4,

$$(\exists x_2 \, [\varphi_\Theta]_{\mathbf{a}}^{x_1}) \in \Theta.$$

That is,

$$[(\exists x_2 \, \varphi_\Theta)]_{\mathbf{a}}^{x_1} \in \Theta,$$

or yet
$$[\delta_\Theta]_{\mathbf{a}}^{x_1} \in \Theta,$$

as required.

Requirement (3):

Suppose that Θ is ω-consistent and $a \notin g_{\mathbb{N}}(\Theta)$. Then,

there does not exist $s \in \mathbb{N}$ for which $(a, s) \in R_\Theta$.

Equivalently,
$$\text{for any } s \in \mathbb{N},\ (a, s) \notin R_\Theta.$$

Thus, since φ_Θ represents R_Θ in Θ,
$$\text{for any } s \in \mathbb{N},\ (\neg[\varphi_\Theta]_{\mathbf{a},\mathbf{s}}^{x_1,x_2}) \in \Theta.$$

That is,
$$\text{for any } s \in \mathbb{N},\ [(\neg[\varphi_\Theta]_{\mathbf{a}}^{x_1})]_{\mathbf{s}}^{x_2} \in \Theta.$$

Therefore,
$$\{[(\neg[\varphi_\Theta]_{\mathbf{a}}^{x_1})]_{\mathbf{s}}^{x_2} : s \in \mathbb{N}\} \subseteq \Theta,$$

whence, by ω-consistency of Θ, it follows that
$$(\exists x_2\,[\varphi_\Theta]_{\mathbf{a}}^{x_1}) \notin \Theta.$$

That is,
$$[(\exists x_2\,\varphi_\Theta)]_{\mathbf{a}}^{x_1} \notin \Theta,$$

or yet
$$[\delta_\Theta]_{\mathbf{a}}^{x_1} \notin \Theta,$$

as required.

QED

At this point, the reader should wonder how pervasive is ω-consistency among the theories of arithmetic. The next exercise shows that satisfiability by $\mathbb{N}$ is a sufficient condition for ω-consistency. It is usual to say that a theory Θ of arithmetic is *true* if $\mathbb{N} \Vdash_{\Sigma_{\mathbb{N}}} \Theta$. Thus, theories $\mathbf{N}$ and $\mathbf{P}$ are true, and so is, obviously, $\mathrm{Th}(\mathbb{N})$.

Exercise 3 Show that every true theory of arithmetic is ω-consistent. In particular, theories $\mathbf{N}$, $\mathbf{P}$ and $\mathrm{Th}(\mathbb{N})$ are ω-consistent.

As one might expect, ω-consistency is a stronger assumption than consistency. In fact, the former implies the latter:

Proposition 4 Let Θ be a theory of the arithmetic. Then, Θ is consistent whenever Θ is ω-consistent.

Proof: Assume that Θ is a theory of the arithmetic. Then, Θ is a theory with equality and so $(\forall x_1(x_1 \cong x_1)) \in \Theta$. Therefore, by Ax4 and MP,

$$\{(k \cong k) : k \in \mathbb{N}\} \subseteq \Theta,$$

that is,

$$\{(\neg(\neg(k \cong k))) : k \in \mathbb{N}\} \subseteq \Theta.$$

Thus, by ω-consistency,

$$(\exists x_1(\neg(x_1 \cong x_1))) \notin \Theta$$

and so Θ is consistent. QED

Let Θ be a suitable theory of arithmetic. Using the formula δ_Θ defined in the proof of Proposition 2 (pseudo representability of derivability), for each $\alpha \in L_{\Sigma_\mathbb{N}}$, let

$$(\Box_\Theta \alpha) \text{ be an abbreviation of } [\delta_\Theta]^{x_1}_{\lceil\alpha\rceil}$$

where $\lceil\alpha\rceil$ stands for the numeral

$$\overbrace{\prime \cdots \prime}^{g_\mathbb{N}(\alpha)\ \text{times}}$$
$$\mathbf{0}$$

corresponding to the natural number $g_\mathbb{N}(\alpha)$. That is, $\lceil\alpha\rceil$ is $\overline{g_\mathbb{N}(\alpha)}$.

This notation is justified by the fact, not explored in this book, that, for each such Θ, the operator $\Box_\Theta$ behaves like a box modality. The interested reader is referred to [4] for material on this subject.

Proposition 5 Let Θ be a suitable theory of arithmetic. Then:

$$\text{If } \alpha \in \Theta \text{ then } (\Box_\Theta \alpha) \in \Theta.$$

Proof: Let $\alpha \in \Theta$. Then, $g_\mathbb{N}(\alpha) \in g_\mathbb{N}(\Theta)$. Thus, by Proposition 2,

$$[\delta_\Theta]^{x_1}_{\overline{g_\mathbb{N}(\alpha)}} \in \Theta.$$

That is, $[\delta_\Theta]^{x_1}_{\lceil\alpha\rceil} \in \Theta$ and, so, $(\Box_\Theta \alpha) \in \Theta$. QED

Albeit not needed for the proof of the second incompleteness theorem, the following interesting result is obtained by strengthening the assumptions on Θ with ω-consistency.

Proposition 6 Let Θ be a suitable and ω-consistent theory of arithmetic. Then:

$$\alpha \in \Theta \text{ if and only if } (\Box_\Theta \, \alpha) \in \Theta.$$

It is worthwhile to analyze formula $(\Box_\Theta \, \alpha)$ semantically. More precisely, it is worthwhile to establish the conditions under which $(\Box_\Theta \, \alpha)$ is satisfied by the standard structure of arithmetic.

Proposition 7 Let Θ be a suitable and true theory of arithmetic. Then:

$$\mathbb{N} \Vdash_{\Sigma_\mathbb{N}} (\Box_\Theta \, \alpha) \text{ if and only if } \alpha \in \Theta.$$

Proof:

($\rightarrow$) By contraposition:

$\quad$ if $\quad$ $\alpha \notin \Theta$,

$\quad$ then $\quad$ $g_\mathbb{N}(\alpha) \notin g_\mathbb{N}(\Theta)$

$\qquad\qquad$ for any $s \in \mathbb{N}, (g_\mathbb{N}(\alpha), s) \notin R_\Theta$

$\qquad\qquad$ for any $s \in \mathbb{N}, (\neg[\varphi_\Theta]^{x_1,x_2}_{\lceil\alpha\rceil,\mathbf{s}}) \in \Theta$ $\qquad$ (representability)

$\qquad\qquad$ for any $s \in \mathbb{N}, (\neg[\varphi_\Theta]^{x_1,x_2}_{\lceil\alpha\rceil,\mathbf{s}}) \in \text{Th}(\mathbb{N})$ $\qquad$ (trueness of Θ)

$\qquad\qquad$ for any $s \in \mathbb{N}, (\neg[[\varphi_\Theta]^{x_1}_{\lceil\alpha\rceil}]^{x_2}_{\mathbf{s}}) \in \text{Th}(\mathbb{N})$

$\qquad\qquad$ $\{(\neg[[\varphi_\Theta]^{x_1}_{\lceil\alpha\rceil}]^{x_2}_{\mathbf{s}}) : s \in \mathbb{N}\} \subseteq \text{Th}(\mathbb{N})$

$\qquad\qquad$ $(\exists x_2 \, [\varphi_\Theta]^{x_1}_{\lceil\alpha\rceil}) \notin \text{Th}(\mathbb{N})$ $\qquad$ (ω-consistency of $\text{Th}(\mathbb{N})$)

$\qquad\qquad$ $[(\exists x_2 \, \varphi_\Theta)]^{x_1}_{\lceil\alpha\rceil} \notin \text{Th}(\mathbb{N})$

$\qquad\qquad$ $[\delta_\Theta]^{x_1}_{\lceil\alpha\rceil} \notin \text{Th}(\mathbb{N})$

$\qquad\qquad$ $(\Box_\Theta \, \alpha) \notin \text{Th}(\mathbb{N})$

$\qquad\qquad$ $\mathbb{N} \nVdash_{\Sigma_\mathbb{N}} (\Box_\Theta \, \alpha)$

($\leftarrow$) Directly:

$\quad$ if $\quad$ $\alpha \in \Theta$,

$\quad$ then $\quad$ $g_\mathbb{N}(\alpha) \in g_\mathbb{N}(\Theta)$

$\qquad\qquad$ there is $s \in \mathbb{N}$ such that $(g_\mathbb{N}(\alpha), s) \in R_\Theta$

$\qquad\qquad$ there is $s \in \mathbb{N}$ such that $[\varphi_\Theta]^{x_1,x_2}_{\lceil\alpha\rceil,\mathbf{s}} \in \Theta$ $\qquad$ (representability)

$\qquad\qquad$ $(\exists x_2 \, [\varphi_\Theta]^{x_1}_{\lceil\alpha\rceil}) \in \Theta$ $\qquad$ (Section 5.4, 29)

$\qquad\qquad$ $[(\exists x_2 \, \varphi_\Theta)]^{x_1}_{\lceil\alpha\rceil} \in \Theta$

$\qquad\qquad$ $[(\exists x_2 \, \varphi_\Theta)]^{x_1}_{\lceil\alpha\rceil} \in \text{Th}(\mathbb{N})$ $\qquad$ (trueness of Θ)

$\qquad\qquad$ $[\delta_\Theta]^{x_1}_{\lceil\alpha\rceil} \in \text{Th}(\mathbb{N})$

$\qquad\qquad$ $(\Box_\Theta \, \alpha) \in \text{Th}(\mathbb{N})$

$\qquad\qquad$ $\mathbb{N} \Vdash_{\Sigma_\mathbb{N}} (\Box_\Theta \, \alpha)$

QED

Clearly, Proposition 7 amounts to saying that, as envisaged, if Θ is a suitable and true theory of arithmetic then formula $(\Box_\Theta \, \alpha)$ means that α is a theorem of Θ. Observe also that Proposition 7 can be formulated as follows:

$$(\Box_\Theta \, \alpha) \in \mathrm{Th}(\mathbb{N}) \text{ if and only if } \alpha \in \Theta.$$

14.2 Properties of derivability

As they searched for additional conditions strong enough to show that consistency is not a theorem of a consistent, axiomatized theory of arithmetic where computable maps are representable, David Hilbert and Paul Bernays arrived to the following properties, herein presented in a formulation due to the later contribution by Martin Hugo Löb, leading to their being known as *Hilbert–Bernays–Löb conditions* or, in short, *HBL conditions*:

HBL1 For every $\alpha \in cL_{\Sigma_\mathbb{N}}$,

$$\text{if } \alpha \in \Theta \text{ then } (\Box_\Theta \, \alpha) \in \Theta.$$

HBL2 For every $\alpha_1, \alpha_2 \in cL_{\Sigma_\mathbb{N}}$,

$$((\Box_\Theta(\alpha_1 \Rightarrow \alpha_2)) \Rightarrow ((\Box_\Theta \, \alpha_1) \Rightarrow (\Box_\Theta \, \alpha_2))) \in \Theta.$$

HBL3 For every $\alpha \in cL_{\Sigma_\mathbb{N}}$,

$$((\Box_\Theta \, \alpha) \Rightarrow (\Box_\Theta(\Box_\Theta \, \alpha))) \in \Theta.$$

Note that, thanks to Proposition 5, every suitable theory of arithmetic fulfills HBL1. On the other hand, even with the extra assumption of ω-consistency, it is not possible, at least at first sight, to guarantee HBL2 or HBL3, but only weak variants of these conditions as established in the following exercises.

Exercise 8 Show that, if Θ is a suitable and ω-consistent theory of arithmetic, then, for every $\alpha_1, \alpha_2 \in cL_{\Sigma_\mathbb{N}}$,

$$\text{if } (\Box_\Theta(\alpha_1 \Rightarrow \alpha_2)) \in \Theta \text{ and } (\Box_\Theta \, \alpha_1) \in \Theta \text{ then } (\Box_\Theta \, \alpha_2) \in \Theta.$$

Exercise 9 Show that, if Θ is a suitable theory of arithmetic, then, for every $\alpha \in cL_{\Sigma_\mathbb{N}}$,

$$\text{if } (\Box_\Theta \, \alpha) \in \Theta \text{ then } (\Box_\Theta(\Box_\Theta \, \alpha)) \in \Theta.$$

Observe that theory **P** fulfills the HBL2 and HBL3 conditions, but this result requires a lengthy proof, built around the concretization of formula φ_Θ obtained from the Gödelization of derivations, without invoking the projection theorem. The interested reader should refer to George Boolos's presentation of this material in [4].

Proposition 10 Let Θ be a suitable and true theory of arithmetic. Then:

1. For every $\alpha \in L_{\Sigma_\mathbb{N}}$,

$$\text{if } \alpha \in \Theta \text{ then } \mathbb{N} \Vdash_{\Sigma_\mathbb{N}} (\Box_\Theta\, \alpha).$$

2. For every $\alpha_1, \alpha_2 \in L_{\Sigma_\mathbb{N}}$,

$$\mathbb{N} \Vdash_{\Sigma_\mathbb{N}} ((\Box_\Theta(\alpha_1 \Rightarrow \alpha_2)) \Rightarrow ((\Box_\Theta\, \alpha_1) \Rightarrow (\Box_\Theta\, \alpha_2))).$$

3. For every $\alpha \in L_{\Sigma_\mathbb{N}}$,

$$\mathbb{N} \Vdash_{\Sigma_\mathbb{N}} ((\Box_\Theta\, \alpha) \Rightarrow (\Box_\Theta(\Box_\Theta\, \alpha))).$$

Proof:

1. Result already established: Proposition 7.

2. Taking into account that Θ is suitable and true and that each formula $(\Box_\Theta \gamma)$ is closed, Proposition 7 and the lemma of closed formula are used to establish the thesis by contradiction as follows:

if	$\mathbb{N} \nVdash_{\Sigma_\mathbb{N}} ((\Box_\Theta(\alpha_1 \Rightarrow \alpha_2)) \Rightarrow ((\Box_\Theta\, \alpha_1) \Rightarrow (\Box_\Theta\, \alpha_2)))$	
then	$\mathbb{N} \Vdash_{\Sigma_\mathbb{N}} (\Box_\Theta(\alpha_1 \Rightarrow \alpha_2))$	
	and $\mathbb{N} \nVdash_{\Sigma_\mathbb{N}} ((\Box_\Theta\, \alpha_1) \Rightarrow (\Box_\Theta\, \alpha_2))$	(LCF 5)
	$\mathbb{N} \Vdash_{\Sigma_\mathbb{N}} (\Box_\Theta(\alpha_1 \Rightarrow \alpha_2))$	
	and $\mathbb{N} \Vdash_{\Sigma_\mathbb{N}} (\Box_\Theta\, \alpha_1)$	
	and $\mathbb{N} \nVdash_{\Sigma_\mathbb{N}} (\Box_\Theta\, \alpha_2)$	(LCF 5)
	$(\alpha_1 \Rightarrow \alpha_2) \in \Theta$	
	and $\alpha_1 \in \Theta$	
	and $\alpha_2 \notin \Theta$	(Proposition 7)
	$\alpha_2 \in \Theta$ and $\alpha_2 \notin \Theta$	(MP)

3. Left as exercise.   QED

It is worthwhile commenting on the importance of Proposition 10 concerning HBL2 and HBL3 (recall that HBL1 holds for every suitable theory of arithmetic). Observe that $\text{Th}(\mathbb{N})$ contains the HBL2 and HBL3 formulas for every suitable and true theory of arithmetic Θ. Hence, if a suitable and true Θ is strong enough (read close enough to $\mathrm{Th}(\mathbb{N})$), then it will contain the HBL2 and HBL3 formulas. Gödel's second incompleteness theorem is established at the end of this chapter for such strong theories of arithmetic.

Exercise 11 Show that adding the HBL2 and HBL3 formulas to a suitable and true theory of arithmetic preserves suitability as well as trueness, and hence ω-consistency.

The availability of formulas like $(\Box_\Theta \alpha)$ in suitable and true theories of arithmetic immediately raises the following question: is there a theorem η of Θ stating its own theoremhood in Θ? More concretely,

$$\text{is there a formula } \eta \text{ such that } (\eta \Leftrightarrow (\Box_\Theta \eta)) \in \Theta\,?$$

The answer is positive. In order to show this, it is worth establishing the more general result, which will be used later on, of existence of a fixed point for each formula $\beta \in L_{\Sigma_\mathbb{N}}$ such that $\mathrm{fv}_{\Sigma_\mathbb{N}}(\beta) = \{x_1\}$, that is, the existence of a formula $\alpha \in cL_{\Sigma_\mathbb{N}}$ provably equivalent within theory Θ to formula $[\beta]_{\lceil \alpha \rceil}^{x_1}$.

With this in mind, consider the following map:

$$\mathrm{dgn} = \lambda k\,. \begin{cases} g_\mathbb{N}([g_\mathbb{N}^{-1}(k)]_{\mathbf{k}}^{x_1}) & \text{if } k \in g_\mathbb{N}(L_{\Sigma_\mathbb{N}}) \\ k & \text{otherwise} \end{cases} : \mathbb{N} \to \mathbb{N}\,.$$

Map dgn brings to the realm of natural numbers the operation of substituting in a given formula φ the variable x_1 by the numeral corresponding to the Gödelization of φ, that is, the operation

$$\lambda \varphi\,. [\varphi]_{\lceil \varphi \rceil}^{x_1} : L_{\Sigma_\mathbb{N}} \to L_{\Sigma_\mathbb{N}}.$$

Exercise 12 Show that map dgn is computable.

Proposition 13 (Fixed point theorem)
Let Θ be a theory of arithmetic where computable maps are representable and $\beta \in L_{\Sigma_\mathbb{N}}$ be such that $\mathrm{fv}_{\Sigma_\mathbb{N}}(\beta) = \{x_1\}$. Then, there exists $\alpha \in cL_{\Sigma_\mathbb{N}}$ such that:

$$(\alpha \Leftrightarrow [\beta]_{\lceil \alpha \rceil}^{x_1}) \in \Theta.$$

Proof: Suppose that map dgn is represented in Θ by formula ψ. In other words:

- $\mathrm{fv}_{\Sigma_\mathbb{N}}(\psi) = \{x_1, x_2\}$;

- if $\mathrm{dgn}(i) = j$, then $([\psi]_{\mathbf{i}}^{x_1} \Leftrightarrow (x_2 \cong \mathbf{j})) \in \Theta$.

Let γ be the formula $(\forall x_2(\psi \Rightarrow [\beta]_{x_2}^{x_1}))$. Take α to be the formula $[\gamma]_{\lceil\gamma\rceil}^{x_1}$, that is,

$$[(\forall x_2(\psi \Rightarrow [\beta]_{x_2}^{x_1}))]_{\lceil\gamma\rceil}^{x_1},$$

which is clearly closed.

Before proceeding with the verification that α is as envisaged, it is worthwhile to give a rough intuitive account of what is going on, deliberately confusing a formula φ with $g_{\mathbb{N}}(\varphi)$ and its numeral $\lceil\varphi\rceil$, for the sake of simplicity of the explanation.

One should look at formula β as stating that x_1 fulfills some property. Then, the envisaged α is required to state that the property holds for α. Furthermore, this fact should be a theorem of Θ. Observe that the chosen γ states that $\mathrm{dgn}(x_1)$ fulfills the property. So, the proposed α states that $\mathrm{dgn}(\gamma)$ fulfills the property. Furthermore, as defined, α is $\mathrm{dgn}(\gamma)$. Hence, as required, α states that α fulfills the property.

It remains to verify that

$$(\alpha \Leftrightarrow [\beta]_{\lceil\alpha\rceil}^{x_1}) \in \Theta.$$

First, note that

$$\mathrm{dgn}(g_{\mathbb{N}}(\gamma)) = g_{\mathbb{N}}(\alpha)$$

because α was chosen to be $[\gamma]_{\lceil\gamma\rceil}^{x_1}$. Therefore, since ψ represents dgn in Θ,

$$([\psi]_{\lceil\gamma\rceil}^{x_1} \Leftrightarrow (x_2 \cong \lceil\alpha\rceil)) \in \Theta,$$

from which it is possible to establish:

(1) $(\alpha \Rightarrow [\beta]_{\lceil\alpha\rceil}^{x_1}) \in \Theta$;

(2) $([\beta]_{\lceil\alpha\rceil}^{x_1} \Rightarrow \alpha) \in \Theta$.

Indeed:

(a) Consider the following derivation sequence for

$$\Theta, \alpha \vdash_{\Sigma_{\mathbb{N}}} [\beta]_{\lceil\alpha\rceil}^{x_1} \; :$$

1   $[(\forall x_2(\psi \Rightarrow [\beta]_{x_2}^{x_1}))]_{\lceil\gamma\rceil}^{x_1}$ Hyp
     $(\forall x_2 [(\psi \Rightarrow [\beta]_{x_2}^{x_1})]_{\lceil\gamma\rceil}^{x_1})$

2   $[[(\psi \Rightarrow [\beta]_{x_2}^{x_1})]_{\lceil\gamma\rceil}^{x_1}]_{\lceil\alpha\rceil}^{x_2}$ IR 1
     $([\psi]_{\lceil\gamma\rceil,\lceil\alpha\rceil}^{x_1,x_2} \Rightarrow [\beta]_{\lceil\alpha\rceil}^{x_1})$

3   $([\psi]_{\lceil\gamma\rceil}^{x_1} \Leftrightarrow (x_2 \cong \lceil\alpha\rceil))$ $\in \Theta$

4	$(\forall x_2([\psi]_{\lceil\gamma\rceil}^{x_1} \Leftrightarrow (x_2 \cong \lceil\alpha\rceil)))$	Gen 3
5	$([\psi]_{\lceil\gamma\rceil,\lceil\alpha\rceil}^{x_1,x_2} \Leftrightarrow (\lceil\alpha\rceil \cong \lceil\alpha\rceil))$	IR 4
6	$(\lceil\alpha\rceil \cong \lceil\alpha\rceil)$	RE1
7	$[\psi]_{\lceil\gamma\rceil,\lceil\alpha\rceil}^{x_1,x_2}$	Taut 5,6
8	$[\beta]_{\lceil\alpha\rceil}^{x_1}$	MP 2,7

Then, since α is closed, applying the MTD yields requirement (1).

(b) Consider the following derivation sequence for

$$\Theta, [\beta]_{\lceil\alpha\rceil}^{x_1}, [\psi]_{\lceil\gamma\rceil}^{x_1} \vdash_{\Sigma_{\mathbb{N}}} [\beta]_{x_2}^{x_1}$$

assuming, without loss of generality, that x_2, x_3 and x_4 do not occur in β:[3]

1	$[\beta]_{\lceil\alpha\rceil}^{x_1}$	Hyp
2	$[\psi]_{\lceil\gamma\rceil}^{x_1}$	Hyp
3	$([\psi]_{\lceil\gamma\rceil}^{x_1} \Leftrightarrow (x_2 \cong \lceil\alpha\rceil))$	$\in \Theta$
4	$(x_2 \cong \lceil\alpha\rceil)$	Taut 2,3
5	$(x_3 \cong x_4) \Rightarrow ([\beta]_{x_3}^{x_1} \Rightarrow [\beta]_{x_4}^{x_1})$	PSE
6	$(\forall x_4((x_3 \cong x_4) \Rightarrow ([\beta]_{x_3}^{x_1} \Rightarrow [\beta]_{x_4}^{x_1})))$	Gen 5
7	$((x_3 \cong x_2) \Rightarrow ([\beta]_{x_3}^{x_1} \Rightarrow [\beta]_{x_2}^{x_1}))$	IR 6
8	$(\forall x_3((x_3 \cong x_2) \Rightarrow ([\beta]_{x_3}^{x_1} \Rightarrow [\beta]_{x_2}^{x_1})))$	Gen 7
9	$((\lceil\alpha\rceil \cong x_2) \Rightarrow ([\beta]_{\lceil\alpha\rceil}^{x_1} \Rightarrow [\beta]_{x_2}^{x_1}))$	IR 8
10	$((x_2 \cong \lceil\alpha\rceil) \Rightarrow (\lceil\alpha\rceil \cong x_2))$	RE2
11	$(\lceil\alpha\rceil \cong x_2)$	MP 4,10
12	$([\beta]_{\lceil\alpha\rceil}^{x_1} \Rightarrow [\beta]_{x_2}^{x_1})$	MP 9,11
13	$[\beta]_{x_2}^{x_1}$	MP 1,12

Then, since generalizations in this sequence are over variables $x_3, x_4 \notin \{x_2\} = \mathrm{fv}_{\Sigma_{\mathbb{N}}}([\psi]_{\lceil\gamma\rceil}^{x_1})$, applying the MTD yields

$$\Theta, [\beta]_{\lceil\alpha\rceil}^{x_1} \vdash_{\Sigma_{\mathbb{N}}} ([\psi]_{\lceil\gamma\rceil}^{x_1} \Rightarrow [\beta]_{x_2}^{x_1}),$$

whence, by generalization,

$$\Theta, [\beta]_{\lceil\alpha\rceil}^{x_1} \vdash_{\Sigma_{\mathbb{N}}} (\forall x_2([\psi]_{\lceil\gamma\rceil}^{x_1} \Rightarrow [\beta]_{x_2}^{x_1})),$$

---

[3]Otherwise, invoking RS and MTSE, one can work with an equivalent formula with the bound variables x_2, x_3 and x_4 in β replaced by fresh ones.

that is,

$$\Theta, [\beta]_{\lceil \alpha \rceil}^{x_1} \vdash_{\Sigma_{\mathbb{N}}} [(\forall x_2(\psi \Rightarrow [\beta]_{x_2}^{x_1}))]_{\lceil \gamma \rceil}^{x_1},$$

or yet

$$\Theta, [\beta]_{\lceil \alpha \rceil}^{x_1} \vdash_{\Sigma_{\mathbb{N}}} \alpha,$$

whence requirement (2) follows by the MTD since $[\beta]_{\lceil \alpha \rceil}^{x_1}$ is closed.   QED

The fixed point theorem is also known as the diagonalization lemma.[4] This result was used implicitly by Kurt Gödel to prove the incompleteness theorems, and was later made explicit by Rudolf Carnap.

The following exercise asks the reader to establish, as a corollary of Proposition 13, the existence of a closed formula that is provably equivalent in Θ to the formula stating its derivability in Θ. Such a formula is known as a *Henkin formula*.

Exercise 14 Let Θ be a suitable theory of arithmetic. Show that there exists a formula $\eta \in cL_{\Sigma_{\mathbb{N}}}$ such that:

$$(\eta \Leftrightarrow (\Box_{\Theta} \eta)) \in \Theta.$$

With the HBL conditions and the fixed point theorem, one can prove the following result, due to Martin Hugo Löb, which allows Kurt Gödel's second incompleteness theorem to be obtained in a more practical way and formulated in a more general version than the original one.

Proposition 15 (Löb's theorem)
Let Θ be a suitable theory of arithmetic fulfilling HBL2 and HBL3. Then, for every $\gamma \in cL_{\Sigma_{\mathbb{N}}}$, the following holds:

$$\text{If } ((\Box_{\Theta} \gamma) \Rightarrow \gamma) \in \Theta \text{ then } \gamma \in \Theta.$$

Proof: The first step is applying the fixed point theorem to the formula

$$(\delta_{\Theta} \Rightarrow \gamma)$$

in order to conclude that there exists $\alpha \in cL_{\Sigma_{\mathbb{N}}}$ such that

$$(\alpha \Leftrightarrow [(\delta_{\Theta} \Rightarrow \gamma)]_{\lceil \alpha \rceil}^{x_1}) \in \Theta,$$

that is

$$(\alpha \Leftrightarrow ([\delta_{\Theta}]_{\lceil \alpha \rceil}^{x_1} \Rightarrow \gamma)) \in \Theta,$$

---

[4]But it should not be confused with Cantor's diagonalization lemma.

or yet
$$(\alpha \Leftrightarrow ((\square_\Theta \alpha) \Rightarrow \gamma)) \in \Theta.$$

Thus, tautologically, there exists $\alpha \in cL_{\Sigma_N}$ such that
$$\begin{cases} (\dagger) & (\alpha \Rightarrow ((\square_\Theta \alpha) \Rightarrow \gamma)) \in \Theta \\ (\dagger\dagger) & (((\square_\Theta \alpha) \Rightarrow \gamma) \Rightarrow \alpha) \in \Theta. \end{cases}$$

From ($\dagger$), by HBL1, one obtains:
$$(\square_\Theta(\alpha \Rightarrow ((\square_\Theta \alpha) \Rightarrow \gamma))) \in \Theta,$$

whence, by HBL2 and MP,
$$(1) \quad ((\square_\Theta \alpha) \Rightarrow (\square_\Theta((\square_\Theta \alpha) \Rightarrow \gamma))) \in \Theta.$$

On the other hand, again by HBL2, one gets:
$$(2) \quad ((\square_\Theta((\square_\Theta \alpha) \Rightarrow \gamma)) \Rightarrow ((\square_\Theta(\square_\Theta \alpha)) \Rightarrow (\square_\Theta \gamma))) \in \Theta.$$

From (1) and (2), by HS, one reaches:
$$(3) \quad ((\square_\Theta \alpha) \Rightarrow ((\square_\Theta(\square_\Theta \alpha)) \Rightarrow (\square_\Theta \gamma))) \in \Theta.$$

Moreover, by HBL3,
$$(4) \quad ((\square_\Theta \alpha) \Rightarrow (\square_\Theta(\square_\Theta \alpha))) \in \Theta.$$

Thus, tautologically from (3) and (4), one obtains:
$$(5) \quad ((\square_\Theta \alpha) \Rightarrow (\square_\Theta \gamma)) \in \Theta.$$

Then, applying HS to (5) and the hypothesis $((\square_\Theta \gamma) \Rightarrow \gamma) \in \Theta$, it can be concluded that:
$$(6) \quad ((\square_\Theta \alpha) \Rightarrow \gamma) \in \Theta.$$

From ($\dagger\dagger$) and (6), by MP, it follows that:
$$\alpha \in \Theta,$$

whence, again by HBL1,
$$(7) \quad (\square_\Theta \alpha) \in \Theta.$$

Finally, from (6) and (7), by MP, it follows that $\gamma \in \Theta$. QED

Let $\perp_{\mathbb{N}}$ abbreviate the formula $(\neg((\mathbf{0} \cong \mathbf{0}) \Rightarrow (\mathbf{0} \cong \mathbf{0})))$. Clearly, for any theory Θ over the signature $\Sigma_{\mathbb{N}}$,

$$\Theta \text{ is consistent if and only if } \perp_{\mathbb{N}} \notin \Theta.$$

Therefore, if Θ is a suitable and true theory of arithmetic, then the formula

$$(\neg(\Box_\Theta \perp_{\mathbb{N}}))$$

states the consistency of Θ, since, by Proposition 7,

$$\mathbb{N} \Vdash_{\Sigma_{\mathbb{N}}} (\Box_\Theta \perp_{\mathbb{N}}) \text{ if and only if } \perp_{\mathbb{N}} \in \Theta$$

and, so,

$$\mathbb{N} \Vdash_{\Sigma_{\mathbb{N}}} (\neg(\Box_\Theta \perp_{\mathbb{N}})) \text{ if and only if } \perp_{\mathbb{N}} \notin \Theta.$$

Exercise 16 Show that $((\neg\gamma) \Leftrightarrow (\gamma \Rightarrow \perp_{\mathbb{N}}))$ is a tautological formula over signature $\Sigma_{\mathbb{N}}$.

Observe that, as long as $((\neg\gamma) \Leftrightarrow (\gamma \Rightarrow \perp_{\mathbb{N}}))$ is a theorem of the theory at hand for each $\gamma \in L_{\Sigma_{\mathbb{N}}}$, what follows does not depend on the formula chosen for playing the *falsum* role taken here by $\perp_{\mathbb{N}}$.

Proposition 17 (Second incompleteness theorem – Gödel)
Let Θ be a consistent and suitable theory of arithmetic fulfilling HBL2 and HBL3. Then:

$$(\neg(\Box_\Theta \perp_{\mathbb{N}})) \notin \Theta.$$

Proof: The proof is by contradiction. Assume that

$$(\neg(\Box_\Theta \perp_{\mathbb{N}})) \in \Theta.$$

Then, tautologically,

$$((\Box_\Theta \perp_{\mathbb{N}}) \Rightarrow \perp_{\mathbb{N}}) \in \Theta,$$

whence Löb's theorem yields

$$\perp_{\mathbb{N}} \in \Theta,$$

in contradiction with the consistency of Θ.   QED

Therefore, if a given theory Θ of arithmetic is true (and, so, consistent), suitable (computably enumerable and capable of representing the computable maps) and sufficiently strong (fulfilling HBL2 and HBL3), then Θ cannot contain the formula stating the consistency of Θ.

Thus, contrarily to what David Hilbert envisaged, it is not possible, in general, to prove the consistency of a consistent axiomatized theory directly from its axioms, that is, without resorting to models built in other theories.

Incidentally, the second incompleteness theorem provides an example of a closed formula η that is true in $\mathbb{N}$, but such that neither η nor $(\neg\,\eta)$ are derivable within an axiomatizable, true and sufficiently strong theory of arithmetic Θ. Indeed:

- $\mathbb{N} \Vdash_{\Sigma_\mathbb{N}} (\neg(\Box_\Theta \perp_\mathbb{N}))$, that is, $\mathbb{N} \not\Vdash_{\Sigma_\mathbb{N}} (\Box_\Theta \perp_\mathbb{N})$, or yet, by Proposition 7, $\perp_\mathbb{N} \notin \Theta$, which holds because Θ is consistent, since it is a true theory of arithmetic;

- $(\neg(\Box_\Theta \perp_\mathbb{N})) \notin \Theta$, by the second incompleteness theorem;

- $(\Box_\Theta \perp_\mathbb{N}) \notin \Theta$, since otherwise, by trueness of Θ, using Exercise 3 and Proposition 6, it would follow that $\perp_\mathbb{N} \in \Theta$, contradicting the consistency of Θ.

Note that the existence of such a formula η was already a consequence of the first incompleteness theorem. However, only now it was possible to provide a concrete example.

Chapter 15

A decidable theory of arithmetic

The negative results about arithmetic theories established in the two previous chapters rely on the fact that computable maps are representable in those theories. But, as mentioned before, by choosing to work with a less rich language of arithmetic, it is possible to set up a semantic theory of arithmetic where computable maps are not representable. For such weak arithmetics there is still hope of proving some positive results. Indeed, it is even possible to find among them decidable theories.

The purpose of this chapter is to present Presburger arithmetic and prove its decidability. Section 15.1 presents the language and the standard interpretation structure of Presburger arithmetic, as well as those of a slight enrichment of that arithmetic. Decidability is proved in Section 15.2 using a quantifier elimination technique on the aforementioned enrichment of Presburger arithmetic. Decidability of Presburger arithmetic itself follows as a corollary.

The reader may think that the poor language of Presburger arithmetic would make it useless. On the contrary, it is rich enough for many economically significant applications. For instance, the Presburger arithmetic decision algorithm is found at the heart of several software verification tools and in many robotic systems.

15.1 Presburger arithmetic

The *signature of Presburger arithmetic* is the first-order signature $\Sigma_{\mathbb{N}}^P$ such that:

- $F_0 = \{\mathbf{0}\}$;

- $F_1 = \{'\}$;

- $F_2 = \{+\}$;

- $F_n = \emptyset$ for $n > 2$;

- $P_2 = \{\cong\}$;

- $P_n = \emptyset$ for $n \neq 2$.

The *standard structure for Presburger arithmetic* is the interpretation structure over $\Sigma_{\mathbb{N}}^P$

$$\mathbb{N}^P = (\mathbb{N}, \_^{\mathsf{F_N}}, \_^{\mathsf{P_N}})$$

such that:

- $\mathbf{0}_0^{\mathsf{F_N}} = 0$;

- $'^{\mathsf{F_N}}_1 = \lambda\, k\,.\, k + 1$;

- $+^{\mathsf{F_N}}_2 = \lambda\, d_1, d_2\,.\, d_1 + d_2$;

- $\cong^{\mathsf{P_N}}_2 = \lambda\, d_1, d_2\,.\, \begin{cases} 1 & \text{if } d_1 = d_2 \\ 0 & \text{otherwise} \end{cases}$.

Clearly, $\mathbb{N}^P$ is the reduct of the standard interpretation structure $\mathbb{N}$ defined in Chapter 11, along the inclusion $\Sigma_{\mathbb{N}}^P \hookrightarrow \Sigma_{\mathbb{N}}$. It is also worth mentioning that $\Sigma_{\mathbb{N}}^P$ is a signature with equality and that the structure $\mathbb{N}^P$ is normal.

The objective is to show that the standard theory of Presburger arithmetic

$$\mathrm{Th}(\mathbb{N}^P) = \mathrm{Th}_{\Sigma_{\mathbb{N}}^P}(\mathbb{N}^P)$$

is decidable, using a quantifier elimination technique.

However, this theory does not have quantifier elimination. For example, the formula $(\exists x\, (\mathsf{m} \cong (x + x))) \in \mathrm{Th}(\mathbb{N}^P)$ does not have an equivalent quantifier free formula. In fact, the quantifier elimination technique is applied to an enrichment of the theory with natural subtraction and divisibility predicates.

The path towards establishing the decidability of $\mathrm{Th}(\mathbb{N}^P)$ is as follows:

1. Define $\Sigma_{\mathbb{N}}^{P+}$ as an enrichment of signature $\Sigma_{\mathbb{N}}^P$ and the interpretation structure $\mathbb{N}^{P+}$ as an enrichment of $\mathbb{N}^P$, with natural subtraction and divisibility predicates.

2. Observe that $\varphi \in \mathrm{Th}(\mathbb{N}^P)$ iff $\varphi \in \mathrm{Th}(\mathbb{N}^{P+})$ for every formula $\varphi \in L_{\Sigma_{\mathbb{N}}^P}$.

3. Prove that $\mathrm{Th}(\mathbb{N}^{P+})$ has quantifier elimination and, in particular, conclude that for every formula $\varphi \in L_{\Sigma_{\mathbb{N}}^{P}}$ there is a quantifier free formula $\varphi^{\bullet} \in L_{\Sigma_{\mathbb{N}}^{P+}}$ such that

$$(\varphi \Leftrightarrow \varphi^{\bullet}) \in \mathrm{Th}(\mathbb{N}^{P+}).$$

4. Prove that there is an algorithm for deciding if $\varphi^{\bullet} \in \mathrm{Th}(\mathbb{N}^{P+})$.

5. Conclude that there is an algorithm for deciding if $\varphi \in \mathrm{Th}(\mathbb{N}^{P})$.

The envisaged enrichment $\Sigma_{\mathbb{N}}^{P+}$ of the signature of Presburger arithmetic symbols for natural subtraction and divisibility predicates is as follows:

- $F_0^+ = F_0$;
- $F_1^+ = \{'\}$;
- $F_2^+ = \{+, \dot{-}\}$;
- $F_n^+ = \emptyset$ for $n > 2$;
- $P_1^+ = \{|_k : k \geq 1\}$;
- $P_2 = \{\cong, <\}$;
- $P_n = \emptyset$ for $n \neq 2$.

The corresponding standard interpretation structure is as expected:

$$\mathbb{N}^{P+} = (\mathbb{N}, \_^{F_{\mathbb{N}}^+}, \_^{P_{\mathbb{N}}^+})$$

where

- $f_n^{F_{\mathbb{N}}^+} = f_n^{F}$ for every $f \in F_n$;
- $p_n^{P_{\mathbb{N}}^+} = p_n^{P}$ for every $p \in P_n$;
- $\dot{-}_2^{F_{\mathbb{N}}^+} = \lambda\, d_1, d_2\,.\, d_1 \dot{-} d_2$;
- $<_2^{P_{\mathbb{N}}^+} = \lambda\, d_1, d_2\,.\, \begin{cases} 1 & \text{if } d_1 < d_2 \\ 0 & \text{otherwise} \end{cases}$.
- $|_{k1}^{P_{\mathbb{N}}^+} = \lambda\, d\,.\, k|d$ (k divides d).

Clearly, $\mathbb{N}^{P}$ is the reduct of $\mathbb{N}^{P+}$ along the inclusion $\Sigma_{\mathbb{N}}^{P} \hookrightarrow \Sigma_{\mathbb{N}}^{P+}$. In the following, let $\mathrm{Th}(\mathbb{N}^{P+})$ denote the theory of $\mathbb{N}^{P+}$. The proof of the following result is left as an exercise.

Proposition 1 For every formula $\varphi \in L_{\Sigma_{\mathbb{N}}^{P}}$,

$$\varphi \in \mathrm{Th}(\mathbb{N}^{P}) \text{ iff } \varphi \in \mathrm{Th}(\mathbb{N}^{P+}).$$

15.2   Decidability of Presburger arithmetic

The proof of the computability of $\mathrm{Th}(\mathbb{N}^{P+})$ is carried out using the semantic quantifier elimination technique introduced in Section 6.3 (adapted from [14]), and, in particular, Proposition 12. Afterward, the decidability of $\mathrm{Th}(\mathbb{N}^{P})$ follows as an immediate corollary. In the sequel, $\mathsf{m}x$ denotes the term $x + \cdots + x$, m times.

Proposition 2 For each formula β in $B_{\Sigma_{\mathbb{N}}^{P}} \cup \overline{B}_{\Sigma_{\mathbb{N}}^{P}}$ with $\mathrm{fv}_{\Sigma_{\mathbb{N}}^{P}}(\beta) = \{x\}$ there is a Boolean combination β^{*} of atomic formulas in $B_{\Sigma_{\mathbb{N}}^{P+}}$ of the following forms:

$$((\mathsf{m}_1 x) < t), \ (((\mathsf{m}_1 x) + \mathsf{m}_2) < t), \ (t < (\mathsf{m}_1 x)), \ (t < ((\mathsf{m}_1 x) + \mathsf{m}_2)),$$

where $\mathsf{m}_1, \mathsf{m}_2$ are numerals and t is a term with $x \notin \mathrm{var}_{\Sigma_{\mathbb{N}}^{+}}(t)$, in such a way that

$$\vDash_{\Sigma_{\mathbb{N}}^{P+}} (\beta \Leftrightarrow \beta^{*}).$$

Proof: Start by noticing that the equalities between terms can be reduced to literals involving $<$. Indeed, let β be $(u \cong v)$. Observe that

$$\vDash_{\Sigma_{\mathbb{N}}^{P+}} (\beta \Leftrightarrow ((\neg(u < v)) \wedge (\neg(v < u)))).$$

On the other hand, it is very easy to show that every atomic formula in $L_{\Sigma_{\mathbb{N}}^{P+}}$ of the form $(u < v)$ where x occurs free can be written in one of the following forms:

$$((\mathsf{m}_1 x) < t), \ (((\mathsf{m}_1 x) + \mathsf{m}_2) < t), \ (t < (\mathsf{m}_1 x)), \ (t < ((\mathsf{m}_1 x) + \mathsf{m}_2)).$$

Assume that

$$\vDash_{\Sigma_{\mathbb{N}}^{P+}} ((u < v) \Leftrightarrow ((\mathsf{m}_1 x) < t)).$$

Then, taking β^{*} as

$$((\neg((\mathsf{m}_1 x) < t)) \wedge ((\neg(t < (\mathsf{m}_1 x))))),$$

we get $\vDash_{\Sigma_{\mathbb{N}}^{P+}} (\beta \Leftrightarrow \beta^{*})$. The other cases are left as exercises.   QED

Proposition 3 For each formula β in $B_{\Sigma_{\mathbb{N}}^{P}} \cup \overline{B}_{\Sigma_{\mathbb{N}}^{P}}$ with $\mathrm{fv}_{\Sigma_{\mathbb{N}}^{P}}(\beta) = \{x\}$ there is a Boolean combination β^{*} of atomic formulas in $B_{\Sigma_{\mathbb{N}}^{P+}}$ of the following forms:

$$((\mathsf{m}x) < t), \ (t < (\mathsf{m}x))$$

where m is the same for every atomic formula in such a way that

$$\vDash_{\Sigma_{\mathbb{N}}^{P+}} (\beta \Leftrightarrow \beta^{*}).$$

Proof: Observe that if $((m_1 x + m_2) < t)$ holds, then, for sufficiently large q (that is, in such a way that $m_1 q \overset{\cdot}{-} m_2 = m_1 q - m_2$),

$$\vDash_{\Sigma_N^{P+}} ((((m_1 x) + m_2) < t) \Leftrightarrow ((m_1 (x+q)) < (t + (m_1 q) \overset{\cdot}{-} m_2))).$$

Hence, it is possible to work with atomic formulas of the form $((m_j x) < t_j)$. The m can be the same by using the least common multiplier m of all m_i such that $((m_i x) < t_i)$. Then, such formulas are replaced by $(mx) < (mt_i)/m_i$. QED

Taking into account the two technical results above, it is possible to conclude that the set of Boolean combinations of quantifier free formulas in $L_{\Sigma_N^{P+}}$ is generated from the set Q composed by formulas of the following forms:

- $((x < t) \land (|_m(x)))$ and $((t < x) \land (|_m(x)))$;

- $((|_k(x+t)) \land (|_m(x)))$.

Let $\psi^{<,k}$ be the formula obtained from ψ by replacing:

- subformula $(x < t)$ by subformula $(1 < 0)$;

- subformula $(t < x)$ by subformula $(0 < 1)$;

- subterm $|_m(x+t)$ by term $|_m(x+k)$.

Proposition 4 Assume that φ is a Boolean combination of the formulas in Q such that $\mathrm{fv}_{\Sigma_N^{P+}} = \{x_1, \ldots, x_n, x\}$. Then:

1. $\vDash_{\Sigma_N^{P+}} ([\varphi]_k^x \Leftrightarrow \varphi^{<,k})$ for sufficiently large k;

2. $\vDash_{\Sigma_N^{P+}} (\varphi^{<,k} \Leftrightarrow \varphi^{<,k+rs})$.

Proof:
The results follow by induction on the structure of φ. In both situations only one of the base cases is considered. All the other base cases and the step are left as exercises.
1. Let φ be $(x < t)$. By Proposition 16 of Chapter 7, the set

$$\{[\![t]\!]_{\Sigma_N^{P+}}^{\mathbb{N}\rho} : \rho \in X^{\mathbb{N}}\}$$

is finite. Take $k = \max\{[\![t]\!]_{\Sigma_N^{P+}}^{\mathbb{N}\rho} : \rho \in X^{\mathbb{N}}\} + 1$. Then, $\mathbb{N} \nvDash_{\Sigma_N^{P+}} (k < t)$, hence $\mathbb{N} \nvDash_{\Sigma_N^{P+}} [\varphi]_k^x$ and so is equivalent to $(1 < 0)$.

2. Let φ be $|_m(x+t)$. Then, $|_m(x+t)^{<,k}$ and $|_m(x+t)^{<,k+rs}$ are $|_m(k+t)$ and $|_m(k+rs+t)$, respectively. It is enough to observe that if m divides $k+t$, then m divides $k+t+rs$. Assume that m divides $k+t$. Then, there is a such that $k+t = am$. Take $b = a + \frac{rs}{m}$. Then, $k+t+rs = bm$ and so m divides $k+t+rs$. QED

Proposition 5 Theory $\mathrm{Th}(\mathbb{N}^{P+})$ has elimination of quantifiers.

Proof: The aim is to show that $\mathrm{Th}(\mathbb{N}^{P+})$ fulfills the conditions of Proposition 12 of Chapter 6. Let φ be a Boolean combination of quantifier free formulas such that $\mathrm{fv}_{\Sigma_{\mathbb{N}}^{P+}}(\varphi) = \{x_1,\ldots,x_n,x\}$. Let E be the set of terms t occurring in subformulas of φ of the form $(x < t)$. Take ψ as the formula

$$\left(\left(\bigvee_{1 \leq k \leq \ell} \varphi^{<,k}\right) \vee \left(\bigvee_{t \in E}\left(\bigvee_{1 \leq k \leq \ell} [\varphi]_{t-k}^x\right)\right)\right)$$

where ℓ is the least common multiplier of all natural numbers whose associated numerals occur in φ.

It is immediate that $\mathrm{fv}_{\Sigma_{\mathbb{N}}^{P+}}(\psi) = \mathrm{fv}_{\Sigma_{\mathbb{N}}^{P+}}(\varphi) \setminus \{x\}$ and that ψ is a Boolean combination. It remains to show that

$$((\exists x \varphi) \Leftrightarrow \psi) \in \mathrm{Th}(\mathbb{N}^{P+}).$$

(1) $\mathbb{N}^{P+} \Vdash_{\Sigma_{\mathbb{N}}^{P+}} (\psi \Rightarrow (\exists x \varphi))$

Assume that $\mathbb{N}^{P+}\rho \Vdash_{\Sigma_{\mathbb{N}}^{P+}} \psi$. There are two cases to consider:

(i) There are $t \in E$ and $1 \leq k \leq \ell$ such that $\mathbb{N}^{P+}\rho \Vdash_{\Sigma_{\mathbb{N}}^{P+}} [\varphi]_{t-k}^x$. Then,

$$\mathbb{N}^{P+}\rho \Vdash_{\Sigma_{\mathbb{N}}^{P+}} (\exists x\, \varphi).$$

(ii) There is $1 \leq k \leq \ell$ such that $\mathbb{N}^{P+}\rho \Vdash_{\Sigma_{\mathbb{N}}^{P+}} \varphi^{<,k}$. By statement 2 of Proposition 4,

$$\mathbb{N}^{P+}\rho \Vdash_{\Sigma_{\mathbb{N}}^{P+}} \varphi^{<,k+rs}$$

for r sufficiently large. Moreover, by statement 1 of the same proposition

$$\mathbb{N}^{P+}\rho \Vdash_{\Sigma_{\mathbb{N}}^{P+}} [\varphi]_{k+rs}^x$$

and so

$$\mathbb{N}^{P+}\rho \Vdash_{\Sigma_{\mathbb{N}}^{P+}} (\exists x\, \varphi).$$

(2) $\mathbb{N}^{P+} \Vdash_{\Sigma_{\mathbb{N}}^{P+}} ((\exists x \varphi) \Rightarrow \psi)$

There are two cases to consider:

(i) Assume that $\mathbb{N}^{P+}\rho \Vdash_{\Sigma_{\mathbb{N}}^{P+}} [\varphi]_{k+rs}^x$ for every $s \in \mathbb{N}$. Then, for sufficiently large s, by statement 1 of Proposition 4,

$$\mathbb{N}^{P+}\rho \Vdash_{\Sigma_{\mathbb{N}}^{P+}} \varphi^{<,k+rs}$$

and, by statement 2 of the same proposition,

$$\mathbb{N}^{P+}\rho \Vdash_{\Sigma_{\mathbb{N}}^{P+}} \varphi^{<,k}.$$

Hence,

$$\mathbb{N}^{P+}\rho \Vdash_{\Sigma_{\mathbb{N}}^{P+}} \psi.$$

(ii) Assume that s is the smallest natural number such that:

$$\begin{cases} \mathbb{N}^{P+}\rho \Vdash_{\Sigma_{\mathbb{N}}^{P+}} [\varphi]_{k+rs}^{x} \\ \mathbb{N}^{P+}\rho \nVdash_{\Sigma_{\mathbb{N}}^{P+}} [\varphi]_{k+rs'}^{x} \end{cases}$$

That is, there is a formula $(x < t)$ such that

$$\begin{cases} \mathbb{N}^{P+}\rho \Vdash_{\Sigma_{\mathbb{N}}^{P+}} ((k + rs) < t) \\ \mathbb{N}^{P+}\rho \nVdash_{\Sigma_{\mathbb{N}}^{P+}} ((k + rs') < t). \end{cases}$$

Then, there is m such that

$$\mathbb{N}^{P+}\rho \Vdash_{\Sigma_{\mathbb{N}}^{P+}} ((k + rs) \cong (t \dot{-} m)),$$

therefore

$$\mathbb{N}^{P+}\rho \Vdash_{\Sigma_{\mathbb{N}}^{P+}} [\varphi]_{(t\dot{-}m)}^{x}$$

and so

$$\mathbb{N}^{P+}\rho \Vdash_{\Sigma_{\mathbb{N}}^{P+}} \psi.$$

QED

Proposition 6 Theory $\mathrm{Th}(\mathbb{N}^{P+})$ is computable.

Proof:
(a) The set Γ of all Boolean combinations of quantifier free closed formulas in $\mathrm{Th}(\mathbb{N}^{P+})$ is computable.
The characteristic map χ_Γ is such that

$$\chi_\Gamma(\varphi) = \begin{cases} k < m & \text{if } \varphi \text{ is } (k < m) \\ m|k & \text{if } \varphi \text{ is } |_m(k) \\ 1 - \chi_\Gamma(\varphi_1) & \text{if } \varphi \text{ is } (\neg\varphi_1) \\ \mathrm{prod}(\chi_\Gamma(\varphi_1), \chi_\Gamma(\varphi_2)) & \text{if } \varphi \text{ is } (\varphi_1 \wedge \varphi_2). \end{cases}$$

Hence χ_Γ is computable since it is defined by cases over computable maps.
(b) $\mathrm{Th}(\mathbb{N}^{P+})$ is computable.
Let $\varphi \in L_{\Sigma_{\mathbb{N}}^{P+}}$ be a closed formula such that φ is $(\exists x\, \gamma)$. Then, by Proposition 5, there is a closed formula ψ that is a Boolean combination of quantifier free formulas such that

$$(\varphi \Leftrightarrow \psi) \in \mathrm{Th}(\mathbb{N}^{P+}).$$

Hence
$$\varphi \in \text{Th}(\mathbb{N}^{P+}) \text{ if and only if } \psi \in \Gamma.$$

By part (a) Γ is computable and so is $\text{Th}(\mathbb{N}^{P+})$.   QED

Proposition 7 (Decidability of Presburger arithmetic)
Theory $\text{Th}(\mathbb{N}^{P})$ is computable.

Proof: For every formula φ in $L_{\Sigma_{\mathbb{N}}^{P}}$, there is a formula $\varphi^{\bullet}$ in $L_{\Sigma_{\mathbb{N}}^{P+}}$ such that $(\varphi \Leftrightarrow \varphi^{\bullet}) \in \text{Th}(\mathbb{N}^{P+})$. Hence

$$\chi_{\text{Th}(\mathbb{N}^{P})}(\varphi) = 1 \text{ if and only if } \chi_{\text{Th}(\mathbb{N}^{P+})}(\varphi^{\bullet}) = 1.$$

Since, by Proposition 6, $\text{Th}(\mathbb{N}^{P+})$ is computable so is $\text{Th}(\mathbb{N}^{P})$.   QED

Part IV

Answers to selected exercises

Chapter 16

Exercises on computability

The exercises solved in this chapter illustrate several issues addressed in Chapter 3.

Exercise 1 Let C_1 and C_2 be computable sets of type $A_1 \ldots A_n$. Show that $C_1 \cap C_2$ is computable.

Solution: If C_1 and C_2 are sets of type $A_1 \ldots A_n$, that is, if C_1 and C_2 are subsets of $A_1^* \times \cdots \times A_n^*$, their intersection is also a subset of $A_1^* \times \cdots \times A_n^*$, in other words, their intersection is of type $A_1 \ldots A_n$.

Therefore, to show that $C_1 \cap C_2$ is computable, it must be shown that its characteristic map

$$\chi_{C_1 \cap C_2} : A_1^* \times \cdots \times A_n^* \to \{0, 1\}$$

is computable, for which all it takes is to present an algorithm computing it.
Consider

> Function[w,
>   If[$\chi_{C_1}[w] == 1$ && $\chi_{C_2}[w] == 1$,
>   1,
>   0
>   ]
>]

This is an algorithm computing $\chi_{C_1 \cap C_2}$. In fact:

- it is a program because χ_{C_1} and χ_{C_2} are computable (this follows from the fact of C_1 and C_2 being computable);

- it is an algorithm because execution of the program terminates for every value of the argument w, since χ_{C_1} and χ_{C_2} are maps;

- it computes $\chi_{C_1 \cap C_2}$ because execution returns 1 in case $w \in C_1 \cap C_2$, that is, $\chi_{C_1}(w) = 1$ and $\chi_{C_2}(w) = 1$, and it returns 0 otherwise.

$$\nabla$$

The next solution gives a proof of Proposition 14 of Chapter 3.

Exercise 2 Assume that C_1 and C_2 are computably enumerable sets of type $A_1 \ldots A_n$. Show that $C_1 \cap C_2$ is computably enumerable.

Solution: In order to establish that $C_1 \cap C_2$ is computably enumerable, one must show that if $C_1 \cap C_2 \neq \emptyset$, then there is a computable enumeration of $C_1 \cap C_2$, that is, there is a computable surjective map

$$e : \mathbb{N} \to C_1 \cap C_2.$$

Assume that $C_1 \cap C_2 \neq \emptyset$. Then:

- There is at least an element in $C_1 \cap C_2$. Pick some c in $C_1 \cap C_2$.

- Sets C_1 and C_2 are both non empty. Therefore, there exist a computable enumeration $e_1 : \mathbb{N} \to C_1$ of C_1 and a computable enumeration $e_2 : \mathbb{N} \to C_2$ of C_2.

Under these hypotheses,

```
Function[k,
    k1 = kN×N[[1]];
    k2 = kN×N[[2]];
    If[e1[k1] == e2[k2],
        e1[k1],
        c
    ]
]
```

is an algorithm computing an enumeration of $C_1 \cap C_2$. In fact:

- it is a program because e_1, e_2 and $\lambda\, k\,.\, k_{\mathbb{N}\times\mathbb{N}}$ are computable;

- it is an algorithm because execution of the program terminates regardless of the value of the argument k, since e_1, e_2 and $\lambda\, k\,.\, k_{\mathbb{N}\times\mathbb{N}}$ are maps;

- it computes an enumeration of $C_1 \cap C_2$ because,

    – letting $(k_{\mathbb{N}\times\mathbb{N}})_1$ stand for k_1,

    – letting $(k_{\mathbb{N}\times\mathbb{N}})_2$ stand for k_2,

the following hold:

    – execution always returns an element of $C_1 \cap C_2$, since, for every value of the argument k:

        * if $e_1(k_1) = e_2(k_2)$, then $e_1(k_1) \in C_1$ and $e_2(k_2) \in C_2$, because e_1 and e_2 are enumerations of C_1 and C_2, respectively, whence the value $e_1(k_1) = e_2(k_2)$ returned is in $C_1 \cap C_2$;

        * if $e_1(k_1) \neq e_2(k_2)$, then the value c returned is also in $C_1 \cap C_2$;

    – by letting k run over all elements of $\mathbb{N}$, the value returned runs over all elements of $C_1 \cap C_2$, since:

        * $\{k_{\mathbb{N}\times\mathbb{N}} : k \in \mathbb{N}\} = \mathbb{N} \times \mathbb{N}$, because $\lambda k . k_{\mathbb{N}\times\mathbb{N}}$ is an enumeration of $\mathbb{N} \times \mathbb{N}$;

        * whence

$$\{(e_1(k_1), e_2(k_2)) : k \in \mathbb{N}\} = C_1 \times C_2,$$

    since e_1 and e_2 are enumerations of C_1 and C_2, respectively;

        * whence its diagonal

$$\{(e_1(k_1), e_2(k_2)) : e_1(k_1) = e_2(k_2)\}$$

    coincides with $C_1 \cap C_2$, therefore covering $C_1 \cap C_2$.

$$\nabla$$

The following solution is a partial answer to Exercise 26 of Chapter 3, addressing only closure for minimization.

Exercise 3 Show that if $f : \mathbb{N}^2 \rightharpoonup \mathbb{N}$ is computable, then $\min(f) : \mathbb{N} \rightharpoonup \mathbb{N}$ is also computable.

Solution: It must be shown that if $f \in \mathcal{C}$, then $\mathcal{C}$ also contains the function

$$\min(f) = \lambda i . \begin{cases} \text{undefined} & \text{if } W_i = \emptyset \\ \text{minimum of } W_i & \text{otherwise,} \end{cases}$$

where

$$W_i = \{k : (i, j) \in \operatorname{dom} f \text{ for each } j < k \ \& \ f(i, k) = 0\}.$$

In other words, it must be shown that there is a *Mathematica* procedure computing $\min(f)$. To this end, consider the program

$$\text{Function}[i,$$
$$k = 0;$$
$$\text{While}[f[i, k] \mathrel{!=} 0,$$
$$k = k + 1$$
$$]$$
$$]$$

Note that if $(i, k) \notin \operatorname{dom} f$, then the evaluation of the guard of the cycle does not terminate, hence the execution of the program does not end. Execution of the program also does not terminate if there is no zero of the function. In other words, since the search begins with k set to 0, execution terminates if and only if there is k such that $f(i, k) = 0$ and $(i, j) \in \operatorname{dom} f$ and $f(i, j) \neq 0$ for every $j < k$. The formalization of this proof of correction of the proposed program using an enriched Hoare calculus is left to the care of the interested reader. ∇

The following exercises and solutions verify that several functions are recursive. Most of them are used in other chapters of the book.

Exercise 4 Show that the map (natural predecessor)

$$\text{pred} = \lambda k . \begin{cases} k - 1 & \text{if } k > 0 \\ 0 & \text{otherwise} \end{cases} : \mathbb{N} \to \mathbb{N}$$

is in $\mathcal{R}$.

Solution: The map pred is easily obtained by recurrence. Indeed:

- $\text{pred}(0) = 0;$

- $\text{pred}(k + 1) = k.$

Recall the construction rec for the case $n = 1$. If $f_0 :\to \mathbb{N}$ and $f_1 : \mathbb{N}^2 \to \mathbb{N}$, then $\text{rec}(f_0, f_1) : \mathbb{N} \to \mathbb{N}$ is such that:

- $\text{rec}(f_0, f_1)(0) = f_0;$

- $\text{rec}(f_0, f_1)(k + 1) = f_1(k, \text{rec}(f_0, f_1)(k)).$

Therefore, to implement the desired recurrence with rec it is enough to take:

- $f_0 = 0;$

- $f_1 = \lambda k_1, k_2 . k_1.$

Hence,

$$\mathsf{pred} = \mathsf{rec}(0, \mathsf{P}_1^2).$$

Correction of the proposed program is established, by induction on the argument, as follows:

(Basis: $k = 0$)

$$\mathsf{pred}(0) = \mathsf{rec}(0, \mathsf{P}_1^2)(0) = 0.$$

(Step: $k = k' + 1$)

$$\mathsf{pred}(k' + 1) = \mathsf{rec}(0, \mathsf{P}_1^2)(k' + 1) = \mathsf{P}_1^2(k', \mathsf{rec}(0, \mathsf{P}_1^2)(k')) = k'.$$

$\triangledown$

Exercise 5 Show that the map (natural subtraction)

$$\dot{-} = \lambda\, k_1, k_2 . \begin{cases} k_1 - k_2 & \text{if } k_1 \geq k_2 \\ 0 & \text{otherwise} \end{cases} : \mathbb{N}^2 \to \mathbb{N}$$

is in $\mathcal{R}$.

Solution: The map $\dot{-}$ is obtained by recurrence on the second argument:

- $k_1 \dot{-} 0 = k_1$;

- $k_1 \dot{-} k_2 + 1 = \mathsf{pred}(k_1 \dot{-} k_2)$.

Recall the construction rec in the case $n = 2$, that is, with $f_0 : \mathbb{N} \to \mathbb{N}$, $f_1 : \mathbb{N}^3 \to \mathbb{N}$ and $\mathsf{rec}(f_0, f_1) : \mathbb{N}^2 \to \mathbb{N}$,

- $\mathsf{rec}(f_0, f_1)(k_1, 0) = f_0(k_1)$;

- $\mathsf{rec}(f_0, f_1)(k_1, k_2 + 1) = f_1(k_1, k_2, \mathsf{rec}(f_0, f_1)(k_1, k_2))$.

Hence,

$$\dot{-} = \mathsf{rec}(\mathsf{P}_1^1, \mathsf{pred} \circ \mathsf{P}_3^3).$$

The proof of correction of the proposed program is left to the care of the interested reader. Hint: Use induction on the second argument. $\triangledown$

Exercise 6 Show that the map (returning the absolute value of the difference between its two arguments):

$$|-| = \lambda\, k_1, k_2 . \begin{cases} k_1 - k_2 & \text{if } k_1 \geq k_2 \\ k_2 - k_1 & \text{otherwise} \end{cases} : \mathbb{N}^2 \to \mathbb{N}$$

is in $\mathcal{R}$.

Solution: Since

$$|k_1 - k_2| = k_1 \mathrel{\dot-} k_2 + k_2 \mathrel{\dot-} k_1,$$

the following holds:

$$|-| = \mathsf{add} \circ \langle \mathrel{\dot-}, \mathrel{\dot-} \circ \langle \mathsf{P}_2^2, \mathsf{P}_1^2 \rangle \rangle.$$

$\triangledown$

Exercise 7 Show that the map (sign)

$$\mathsf{sg} = \lambda\, k \, . \, \begin{cases} 0 & \text{if } k = 0 \\ 1 & \text{otherwise} \end{cases} : \mathbb{N} \to \mathbb{N}$$

is in $\mathcal{R}$.

Solution: Since

$$\mathsf{sg}(k) = k \mathrel{\dot-} \mathsf{pred}(k),$$

the following holds:

$$\mathsf{sg} = \mathrel{\dot-} \circ \langle \mathsf{P}_1^1, \mathsf{pred} \rangle.$$

$\triangledown$

Exercise 8 Show that the map

$$\mathsf{neq} = \lambda\, k_1, k_2 \, . \, \begin{cases} 1 & \text{if } k_1 \neq k_2 \\ 0 & \text{otherwise} \end{cases} : \mathbb{N}^2 \to \mathbb{N}$$

is in $\mathcal{R}$.

Solution: Since

$$\mathsf{neq}(k_1, k_2) = \mathsf{sg}(|k_1 - k_2|),$$

the following holds:

$$\mathsf{neq} = \mathsf{sg} \circ |-|.$$

$\triangledown$

Exercise 9 Show that the map (partial subtraction):

$$\lambda\, k_1, k_2 \, . \, \begin{cases} k_1 - k_2 & \text{if } k_1 \geq k_2 \\ \text{undefined} & \text{otherwise} \end{cases} : \mathbb{N}^2 \rightharpoonup \mathbb{N}$$

is in $\mathcal{R}$ by checking that it is computed by the following procedure:

$$\min(\text{neq} \circ \langle \mathsf{P}_1^3, \text{add} \circ \langle \mathsf{P}_2^3, \mathsf{P}_3^3 \rangle \rangle).$$

Solution: First observe that

$$\mathsf{P}_1^3 = \lambda\, k_1, k_2, k_3 \,.\, k_1 : \mathbb{N}^3 \to \mathbb{N};$$
$$\mathsf{P}_2^3 = \lambda\, k_1, k_2, k_3 \,.\, k_2 : \mathbb{N}^3 \to \mathbb{N};$$
$$\mathsf{P}_3^3 = \lambda\, k_1, k_2, k_3 \,.\, k_3 : \mathbb{N}^3 \to \mathbb{N}.$$

Therefore, by definition of aggregation,

$$\langle \mathsf{P}_2^3, \mathsf{P}_3^3 \rangle = \lambda\, k_1, k_2, k_3 \,.\, (k_2, k_3) : \mathbb{N}^3 \to \mathbb{N}^2.$$

Hence, by definition of composition,

$$\text{add} \circ \langle \mathsf{P}_2^3, \mathsf{P}_3^3 \rangle = \lambda\, k_1, k_2, k_3 \,.\, k_2 + k_3 : \mathbb{N}^3 \to \mathbb{N}.$$

Thus, again by definition of aggregation,

$$\langle \mathsf{P}_1^3, \text{add} \circ \langle \mathsf{P}_2^3, \mathsf{P}_3^3 \rangle \rangle = \lambda\, k_1, k_2, k_3 \,.\, (k_1, k_2 + k_3) : \mathbb{N}^3 \to \mathbb{N}^2,$$

whence, again by definition of composition, it follows that

$$\text{neq} \circ \langle \mathsf{P}_1^3, \text{add} \circ \langle \mathsf{P}_2^3, \mathsf{P}_3^3 \rangle \rangle = \lambda\, k_1, k_2, k_3 \,.\, \begin{cases} 1 & \text{if } k_1 \neq k_2 + k_3 \\ 0 & \text{otherwise} \end{cases} : \mathbb{N}^3 \to \mathbb{N}.$$

Under these conditions, by definition of minimization,

$$\min(\text{neq} \circ \langle \mathsf{P}_1^3, \text{add} \circ \langle \mathsf{P}_2^3, \mathsf{P}_3^3 \rangle \rangle) =$$

$$\lambda\, k_1, k_2 \,.\, \begin{cases} \text{undefined} & \text{if } W_{k_1,k_2} = \emptyset \\ \text{minimum of } W_{k_1,k_2} & \text{otherwise} \end{cases} : \mathbb{N}^2 \rightharpoonup \mathbb{N}$$

where, taking into account that the function being minimized is a map, for each $k_1, k_2 \in \mathbb{N}$ the set

$$W_{k_1,k_2}$$

coincides with

$$\{k_3 \in \mathbb{N} : \text{neq} \circ \langle \mathsf{P}_1^3, \text{add} \circ \langle \mathsf{P}_2^3, \mathsf{P}_3^3 \rangle \rangle (k_1, k_2, k_3) = 0\} =$$

$$\{k_3 \in \mathbb{N} : k_1 = k_2 + k_3\} =$$

$$\begin{cases} \emptyset & \text{if } k_1 < k_2 \\ \{k_1 - k_2\} & \text{otherwise.} \end{cases}$$

Therefore,

$$\min(\mathsf{neq} \circ \langle \mathsf{P}_1^3, \mathsf{add} \circ \langle \mathsf{P}_2^3, \mathsf{P}_3^3 \rangle \rangle) = \lambda\, k_1, k_2 \,.\, \begin{cases} \text{undefined} & \text{if } k_1 < k_2 \\ k_1 - k_2 & \text{otherwise,} \end{cases}$$

that is, as was originally intended,

$$\min(\mathsf{neq} \circ \langle \mathsf{P}_1^3, \mathsf{add} \circ \langle \mathsf{P}_2^3, \mathsf{P}_3^3 \rangle \rangle) = \lambda\, k_1, k_2 \,.\, \begin{cases} k_1 - k_2 & \text{if } k_1 \geq k_2 \\ \text{undefined} & \text{otherwise.} \end{cases}$$

$$\nabla$$

Chapter 17

Exercises on syntax

The exercises solved here refer to Chapter 4.

Exercise 1 Define a first-order signature Σ_G that is adequate to express properties of groups. Formalize some of those properties within L_{Σ_G}.

Solution: Let $\Sigma_G = (F, P, \tau)$ where:

- $F = \{\mathbf{o}, \mathbf{e}, \mathbf{i}\}$;
- $P = \{\cong\}$;
- $\tau = \{(\mathbf{e}, 0), (\mathbf{i}, 1), (\mathbf{o}, 2), (\cong, 2)\}$.

This same signature could have been defined in the following way:

- $F_0 = \{\mathbf{e}\}$;
- $F_1 = \{\mathbf{i}\}$;
- $F_2 = \{\mathbf{o}\}$;
- $F_n = \emptyset$ for $n > 2$;
- $P_2 = \{\cong\}$;
- $P_n = \emptyset$ for $n \neq 2$.

The following are some examples of properties of groups formalized within L_{Σ_G} (using infix notation for the group's binary operation and for equality).

- Associativity:

$$(\forall x_1 (\forall x_2 (\forall x_3 ((x_1 \mathbf{o} (x_2 \mathbf{o} x_3)) \cong ((x_1 \mathbf{o} x_2) \mathbf{o} x_3))))).$$

- Left unit:

$$(\forall x_1((\mathbf{e} \, \mathbf{o} \, x_1) \cong x_1))\,.$$

- Right inverse:

$$(\forall x_1((x_1 \, \mathbf{o} \, \mathbf{i}(x_1)) \cong \mathbf{e}))\,.$$

- Uniqueness of left unit:

$$(\forall x_1((\forall x_2((x_1 \, \mathbf{o} \, x_2) \cong x_2)) \Rightarrow (x_1 \cong \mathbf{e})))\,.$$

There is also the possibility of consider an alternative signature for groups where $F = \{\mathbf{o}\}$. It is left as an exercise to write, in this context, the left unit and the right inverse properties. ∇

The following provides a solution to Exercise 7 of Chapter 4.

Exercise 2 Show that the map $\lambda\, x, t \, . \, \chi_{\mathrm{var}_\Sigma(t)}(x) : X \times T_\Sigma \to \mathbb{N}$ is computable. Conclude that the set $\mathrm{var}_\Sigma(t)$ is computable for every t.

Solution: Let

$$h = \lambda\, x, t \, . \, \chi_{\mathrm{var}_\Sigma(t)}(x) : X \times T_\Sigma \to \mathbb{N}$$

be the map under consideration. Then, h is computed by the following recursive program:

```
Function[{x, t},
    et = ToExpression[t];
    sh = ToString[et[[0]]];
    na = Length[et];
    Which[
        sh == x,
            1,
        sh ∈ F,
            j = 1;
            b = 0;
            While[j ≤ na && b == 0,
                b = h[x, ToString[et[[j]]]];
                j = j + 1
            ];
            b,
        True,
            0
    ]
]
```

Observe that nothing is demanded from this program if $x \notin X$ or $t \notin T_\Sigma$, whence it is not necessary to verify that the arguments are legitimate nor to take care of the result in the case of illegitimate arguments.

Computability of h carries with it computability, for each $t \in T_\Sigma$, of the set $\mathrm{var}_\Sigma(t)$, since, whenever $x \in X$, $\chi_{\mathrm{var}_\Sigma(t)}(x) = h(x,t)$. In fact, for each $t \in T_\Sigma$, the map

$$\chi_{\mathrm{var}_\Sigma(t)} : A_\Sigma^* \to \{0,1\}$$

is computed by the program

$$\text{Function}[x,$$
$$\text{If}[x \in X, h[x,t], 0]$$
$$]$$

thanks to the fact that X is computable. $\qquad\qquad \triangledown$

The following solution helps in understanding Exercise 16 of Chapter 4, in particular the second assertion.

Exercise 3 Show that:

1. Term c is free for variable x_1 in formula $(\exists x_2(x_1 < x_2))$.

2. Term x_2 is not free for variable x_1 in formula $(\exists x_2(x_1 < x_2))$.

Solution:

1. Intuitively, c is free for x_1 in $(\exists x_2(x_1 < x_2))$ because when x_1 is replaced by c in this formula no variable in the latter term is captured by the quantification over x_2. Rigorously, from the inductive definition of term free for a variable in a formula, the following hold:

$c \triangleright_\Sigma x_1 : (\exists x_2(x_1 < x_2))$	iff
$c \triangleright_\Sigma x_1 : (\neg(\forall x_2(\neg(x_1 < x_2))))$	iff (†)
$c \triangleright_\Sigma x_1 : (\forall x_2(\neg(x_1 < x_2)))$	iff (††)

$$x_2 \text{ is } x_1 \text{ or } \begin{cases} \text{if } x_1 \in \mathrm{fv}_\Sigma((\neg(x_1 < x_2))), \\ \qquad\qquad \text{then } x_2 \notin \mathrm{var}_\Sigma(c) \\ c \triangleright_\Sigma x_1 : (\neg(x_1 < x_2)) \end{cases} \quad \text{iff}$$

false or $\begin{cases} \text{if true then true} \\ c \rhd_\Sigma x_1 : (\neg(x_1 < x_2)) \end{cases}$ iff

false or $\begin{cases} \text{true} \\ c \rhd_\Sigma x_1 : (\neg(x_1 < x_2)) \end{cases}$ iff

false or $c \rhd_\Sigma x_1 : (\neg(x_1 < x_2))$ iff

$c \rhd_\Sigma x_1 : (\neg(x_1 < x_2))$ iff (†)

$c \rhd_\Sigma x_1 : (x_1 < x_2)$ iff (†††)

true

(†) Clause for negation.
(††) Clause for quantification.
(†††) Clause for atomic formulas.

2. Intuitively, x_2 is not free for x_1 in $(\exists x_2(x_1 < x_2))$ because when x_1 is replaced by x_2 in this formula this variable is captured by the quantification over x_2. Rigorously, from the inductive definition of term free for a variable in a formula, the following hold:

$x_2 \rhd_\Sigma x_1 : (\exists x_2(x_1 < x_2))$ iff

$x_2 \rhd_\Sigma x_1 : (\neg(\forall x_2(\neg(x_1 < x_2))))$ iff (†)

$x_2 \rhd_\Sigma x_1 : (\forall x_2(\neg(x_1 < x_2)))$ iff (††)

x_2 is x_1 or $\begin{cases} \text{if } x_1 \in \mathrm{fv}_\Sigma((\neg(x_1 < x_2))), \\ \qquad\qquad \text{then } x_2 \notin \mathrm{var}_\Sigma(x_2) \\ x_2 \rhd_\Sigma x_1 : (\neg(x_1 < x_2)) \end{cases}$ iff

false or $\begin{cases} \text{if true then false} \\ x_2 \rhd_\Sigma x_1 : (\neg(x_1 < x_2)) \end{cases}$ iff

false or $\begin{cases} \text{false} \\ x_2 \rhd_\Sigma x_1 : (\neg(x_1 < x_2)) \end{cases}$ iff

false or false   iff

false

(†) Clause for negation.
(††) Clause for quantification. ∇

The following two solutions are the answer to 1 and 2 in Exercise 19 of Chapter 4.

Exercise 4 Show that if $x \notin \mathrm{fv}_\Sigma(\varphi)$, then $t \rhd_\Sigma x : \varphi$, for any term t.

Solution: The proof is done by induction on the structure of formula φ:

(Basis) φ is $p(u_1, \dots, u_n)$: due to the clause for atomic formulas in the definition of term free for a variable in a formula, $t \rhd_\Sigma x : p(u_1, \dots, u_n)$ without having to use the hypothesis.

(Step) There are three cases to consider:

(Negation) φ is $(\neg \psi)$: due to the clause for negation, $t \rhd_\Sigma x : (\neg \psi)$ if and only if $t \rhd_\Sigma x : \psi$, which holds by the induction hypothesis, since if $x \notin \mathrm{fv}_\Sigma((\neg \psi))$, then $x \notin \mathrm{fv}_\Sigma(\psi)$.

(Implication) φ is $(\psi_1 \Rightarrow \psi_2)$: due to the clause for implication, $t \rhd_\Sigma x : (\psi_1 \Rightarrow \psi_2)$ if and only if $t \rhd_\Sigma x : \psi_1$ and $t \rhd_\Sigma x : \psi_2$, which hold thanks to the induction hypothesis, since if $x \notin \mathrm{fv}_\Sigma((\psi_1 \Rightarrow \psi_2))$, then $x \notin \mathrm{fv}_\Sigma(\psi_1)$ and $x \notin \mathrm{fv}_\Sigma(\psi_2)$.

(Quantification) φ is $(\forall y\, \psi)$: due to the clause for quantification, $t \rhd_\Sigma x : (\forall y\, \psi)$ if and only if

$$y \text{ is } x \text{ or } \begin{cases} \text{if } x \in \mathrm{fv}_\Sigma(\psi) \text{ then } y \notin \mathrm{var}_\Sigma(t) \\ t \rhd_\Sigma x : \psi. \end{cases}$$

If y is x, then $t \rhd_\Sigma x : (\forall y\, \psi)$.

Otherwise, the following must be checked:

- $t \rhd_\Sigma x : \psi$, which holds by the induction hypothesis, since if y is not x and $x \notin \mathrm{fv}_\Sigma((\forall y\, \psi))$, then $x \notin \mathrm{fv}_\Sigma(\psi)$;

- if $x \in \mathrm{fv}_\Sigma(\psi)$, then $y \notin \mathrm{var}_\Sigma(t)$, which holds since $x \notin \mathrm{fv}_\Sigma(\psi)$ as was just seen.

∇

Exercise 5 Show that if $\mathrm{var}_\Sigma(t) \cap \mathrm{bv}_\Sigma(\varphi) = \emptyset$, then $t \rhd_\Sigma x : \varphi$, for every variable x.

Solution: The proof is done by induction on the structure of formula φ:

(Basis) φ is $p(u_1, \dots, u_n)$: due to the clause for atomic formulas in the definition of term free for a variable in a formula, $t \rhd_\Sigma x : p(u_1, \dots, u_n)$ without using the hypothesis.

(Step) There are three cases to consider:

(Negation) φ is $(\neg \psi)$: due to the clause for negation, $t \rhd_\Sigma x : (\neg \psi)$ if and only if $t \rhd_\Sigma x : \psi$, which holds thanks to the induction hypothesis, since if $\mathrm{var}_\Sigma(t) \cap \mathrm{bv}_\Sigma((\neg\,\psi)) = \emptyset$, then $\mathrm{var}_\Sigma(t) \cap \mathrm{bv}_\Sigma(\psi) = \emptyset$.

(Implication) φ is $(\psi_1 \Rightarrow \psi_2)$: due to the clause for implication, $t \rhd_\Sigma x : (\psi_1 \Rightarrow \psi_2)$ if and only if $t \rhd_\Sigma x : \psi_1$ e $t \rhd_\Sigma x : \psi_2$, which holds by the induction hypothesis, since if $\mathrm{var}_\Sigma(t) \cap \mathrm{bv}_\Sigma((\psi_1 \Rightarrow \psi_2)) = \emptyset$, then $\mathrm{var}_\Sigma(t) \cap \mathrm{bv}_\Sigma(\psi_1) = \emptyset$ and $\mathrm{var}_\Sigma(t) \cap \mathrm{bv}_\Sigma(\psi_2) = \emptyset$.

(Quantification) φ is $(\forall y\,\psi)$: due to the clause for quantification, $t \rhd_\Sigma x : (\forall y\,\psi)$ if and only if

$$y \text{ is } x \text{ or } \begin{cases} \text{if } x \in \mathrm{fv}_\Sigma(\psi) \text{ then } y \notin \mathrm{var}_\Sigma(t) \\ t \rhd_\Sigma x : \psi. \end{cases}$$

If y is x, then $t \rhd_\Sigma x : (\forall y\,\psi)$.

Otherwise, the following must be checked:

- $t \rhd_\Sigma x : \psi$, which is true thanks to the induction hypothesis, since if $\mathrm{var}_\Sigma(t) \cap \mathrm{bv}_\Sigma((\forall y\,\psi)) = \emptyset$, then $\mathrm{var}_\Sigma(t) \cap \mathrm{bv}_\Sigma(\psi) = \emptyset$;

- if $x \in \mathrm{fv}_\Sigma(\psi)$, then $y \notin \mathrm{var}_\Sigma(t)$, which holds since $y \in \mathrm{bv}_\Sigma((\forall y\,\psi))$ and, by hypothesis, $\mathrm{var}_\Sigma(t) \cap \mathrm{bv}_\Sigma((\forall y\,\psi)) = \emptyset$.

$\triangledown$

The following is the solution to Exercise 20 of Chapter 4.

Exercise 6 Show that if $y \notin \mathrm{fv}_\Sigma(\varphi)$ and $y \rhd_\Sigma x : \varphi$, then $x \rhd_\Sigma y : [\varphi]_y^x$.

Solution: The proof is done by induction on the structure of formula φ:

(Basis) φ is $p(u_1, \ldots, u_n)$: due to the clause for atomic formulas in the definition of term free for a variable in a formula, $x \rhd_\Sigma y : p([u_1]_y^x, \ldots, [u_n]_y^x)$, that is, $x \rhd_\Sigma y : [p(u_1, \ldots, u_n)]_y^x$.

(Step) There are three cases to consider:

(Negation) φ is $(\neg\,\psi)$: then $x \rhd_\Sigma y : [(\neg\,\psi)]_y^x$, that is, $x \rhd_\Sigma y : (\neg[\psi]_y^x)$, follows, by the clause for negation, from $x \rhd_\Sigma y : [\psi]_y^x$, which holds by the induction hypothesis, since if $y \notin \mathrm{fv}_\Sigma((\neg\,\psi))$ and $y \rhd_\Sigma x : (\neg\,\psi)$, then $y \notin \mathrm{fv}_\Sigma(\psi)$ and $y \rhd_\Sigma x : \psi$.

(Implication) φ is $(\psi_1 \Rightarrow \psi_2)$: then $x \rhd_\Sigma y : [(\psi_1 \Rightarrow \psi_2)]_y^x$, that is, $x \rhd_\Sigma y : ([\psi_1]_y^x \Rightarrow [\psi_2]_y^x)$, follows, by the clause for implication, from $x \rhd_\Sigma y : [\psi_1]_y^x$ and $x \rhd_\Sigma y : [\psi_2]_y^x$, which holds by the induction hypothesis, since if $y \notin \mathrm{fv}_\Sigma((\psi_1 \Rightarrow \psi_2))$ and $y \rhd_\Sigma x : (\psi_1 \Rightarrow \psi_2)$, then $y \notin \mathrm{fv}_\Sigma(\psi_1)$, $y \rhd_\Sigma x : \psi_1$,

$y \notin \mathrm{fv}_\Sigma(\psi_2)$ and $y \rhd_\Sigma x : \psi_2$.

(Quantification) φ is $(\forall z\, \psi)$:

(Case 1) z is x: then $x \rhd_\Sigma y : [(\forall x\, \psi)]_y^x$, that is, $x \rhd_\Sigma y : (\forall x\, \psi)$, which holds since, by hypothesis, $y \notin \mathrm{fv}_\Sigma((\forall x\, \psi))$, using Exercise 19.1 in Chapter 4 — Exercise 4 solved in this chapter.

(Case 2) z is not x: then $x \rhd_\Sigma y : [(\forall z\, \psi)]_y^x$, that is, $x \rhd_\Sigma y : (\forall z\, [\psi]_y^x)$, follows, by the clause for quantification, from

$$z \text{ is } y \text{ or } \begin{cases} \text{if } y \in \mathrm{fv}_\Sigma([\psi]_y^x) \text{ then } z \notin \mathrm{var}_\Sigma(x) \\ x \rhd_\Sigma y : [\psi]_y^x. \end{cases}$$

Hence, if z is y, then $x \rhd_\Sigma y : [(\forall z\, \psi)]_y^x$.

Otherwise, it must be checked that:

- $x \rhd_\Sigma y : [\psi]_y^x$, which holds by the induction hypothesis, since if

$$\begin{cases} y \notin \mathrm{fv}_\Sigma((\forall z\, \psi)) \\ y \rhd_\Sigma x : (\forall z\, \psi) \\ z \text{ is not } y \\ z \text{ is not } x, \end{cases}$$

then $y \notin \mathrm{fv}_\Sigma(\psi)$ and $y \rhd_\Sigma x : \psi$;

- if $y \in \mathrm{fv}_\Sigma([\psi]_y^x)$ then $z \notin \mathrm{var}_\Sigma(x)$, which holds since z is not x.

$\triangledown$

Exercise 7 Let $\Psi \subseteq L_\Sigma$ and

$$\Psi^+ = \Psi \cup \{\alpha : (\alpha \wedge \beta) \in \Psi\} \cup \{\beta : (\alpha \wedge \beta) \in \Psi\}.$$

Justify or refute the following assertion:

If Ψ is computably enumerable, then so is Ψ^+.

Solution: True in general. Indeed, if $\Psi = \emptyset$ then Ψ^+ is also empty and, so, computably enumerable. Otherwise, let $h : \mathbb{N} \to \Psi$ be a computable enumeration of Ψ. In this case, $\Psi^+ \neq \emptyset$ and, so, in order to show that Ψ^+ is computably enumerable one may present a computable enumeration $h^+ : \mathbb{N} \to \Psi^+$. To this end, consider the map computed as follows:

```
Function[k,
    If[OddQ[k],
        h[(k − 1)/2]
    ,
        j = k/2;
        If[OddQ[j],
            s = h[(j − 1)/2];
            e = ToExpression[s];
            If[Length[e] =!= 2 || e[[0]] =!= ∧,
                s
            ,
                a = e[[1]];
                ToString[a]
            ]
        ,
            s = h[j/2];
            e = ToExpression[s];
            If[Length[e] =!= 2 || e[[0]] =!= ∧,
                s
            ,
                b = e[[2]];
                ToString[b]
            ]
        ]
    ]
]
```

Observe that the computation only outputs values in Ψ^{+}. Indeed: (i) when the output is $h[(k − 1)/2]$, $s = h[(j − 1)/2]$ or $s = h[j/2]$, the result is in Ψ; (ii) when the output is ToString[a], the result is in $\{\alpha : (\alpha \wedge \beta) \in \Psi\}$; and (iii) when the output is ToString[b], the result is in $\{\beta : (\alpha \wedge \beta) \in \Psi\}$.

Moreover, (i) when the input k runs over the odd natural numbers, the output $h[(k − 1)/2]$ runs over all elements of Ψ; (ii) when the input k runs over the even natural numbers and $j = k/2$ runs over the odd natural numbers, $s = h[(j − 1)/2]$ runs over all formulas in Ψ and, in particular, over all conjunctions in Ψ, and so, ToString[a] runs over all elements of $\{\alpha : (\alpha \wedge \beta) \in \Psi\}$; and (iii) when the input k runs over the even natural numbers and $j = k/2$ runs over the even natural numbers, $s = h[j/2]$ runs over all formulas in Ψ and, in particular, over all conjunctions in Ψ, and so, ToString[b] runs over all elements of $\{\beta : (\alpha \wedge \beta) \in \Psi\}$. ∇

Chapter 18

Exercises on Hilbert calculus

The solutions presented here concern Chapters 5 and 6.

Exercise 1 Prove the principle of quantifier exchange (PQE):

$$(\forall x(\forall y\,\varphi)) \vdash_\Sigma (\forall y(\forall x\,\varphi))\,.$$

Solution:

1	$(\forall x(\forall y\,\varphi))$	Hyp
2	$((\forall x(\forall y\,\varphi)) \Rightarrow (\forall y\,\varphi))$	Ax4
3	$(\forall y\,\varphi)$	MP 1,2
4	$((\forall y\,\varphi) \Rightarrow \varphi)$	Ax4
5	φ	MP 3,4
6	$(\forall x\,\varphi)$	Gen 5
7	$(\forall y(\forall x\,\varphi))$	Gen 6

∇

Exercise 2 Prove the hypothetical sylogism (HS):

$$(\varphi_1 \Rightarrow \varphi_2), (\varphi_2 \Rightarrow \varphi_3) \vdash_\Sigma (\varphi_1 \Rightarrow \varphi_3)\,.$$

Solution:

1	$(\varphi_1 \Rightarrow \varphi_2)$	Hyp
2	$(\varphi_2 \Rightarrow \varphi_3)$	Hyp
3	$((\varphi_2 \Rightarrow \varphi_3) \Rightarrow (\varphi_1 \Rightarrow (\varphi_2 \Rightarrow \varphi_3)))$	Ax1
4	$(\varphi_1 \Rightarrow (\varphi_2 \Rightarrow \varphi_3))$	MP 2,3
5	$((\varphi_1 \Rightarrow (\varphi_2 \Rightarrow \varphi_3)) \Rightarrow ((\varphi_1 \Rightarrow \varphi_2) \Rightarrow (\varphi_1 \Rightarrow \varphi_3)))$	Ax2
6	$((\varphi_1 \Rightarrow \varphi_2) \Rightarrow (\varphi_1 \Rightarrow \varphi_3))$	MP 4,5
7	$(\varphi_1 \Rightarrow \varphi_3)$	MP 1,6

The reader is invited to provide an alternative proof using the MTD.   ▽

The following solution provides the proof of Proposition 12 of Chapter 5 with respect to the axioms in Ax1. The proof of the computability of the other axioms is similar. Note that the computability of the axioms in Ax4 and Ax5 relies on the computability of the relation freeness of a term for a variable in a formula, and on the computability of the set of free variables that occur in a formula, respectively.

Exercise 3 Show that set Ax1 is computable.

Solution: The following algorithm computes the characteristic map of the set $\{\Rightarrow[\varphi, \Rightarrow[\psi, \varphi]] : \varphi, \psi \in L_{\Sigma}\}$:

$$
\begin{aligned}
&\mathsf{Function}[w, \\
&\quad \mathsf{If}[\chi_{L_{\Sigma}}[w] == 0, \\
&\qquad 0 \\
&\quad , \\
&\qquad \alpha = \mathsf{ToExpression}[w]; \\
&\qquad \mathsf{If}[\alpha[[0]] =!= \Rightarrow, \\
&\qquad\quad 0 \\
&\qquad , \\
&\qquad\quad \varphi = \alpha[[1]]; \\
&\qquad\quad \delta = \alpha[[2]]; \\
&\qquad\quad \mathsf{If}[\delta[[0]] =!= \Rightarrow, \\
&\qquad\qquad 0 \\
&\qquad\quad , \\
&\qquad\qquad \mathsf{If}[\delta[[2]] === \varphi, 1, 0] \\
&\qquad\quad] \\
&\qquad] \\
&\quad] \\
&]
\end{aligned}
$$

          ▽

The formula in the next exercise is sometimes called the *normality* property of universal quantification.

Exercise 4 Establish $\vdash_{\Sigma} ((\forall x(\varphi_1 \Rightarrow \varphi_2)) \Rightarrow ((\forall x\, \varphi_1) \Rightarrow (\forall x\, \varphi_2)))$ by applying the MTD twice.

Solution: Consider the following derivation sequence:

$$
\begin{array}{lll}
1 & (\forall x(\varphi_1 \Rightarrow \varphi_2)) & \text{Hyp} \\
2 & ((\forall x(\varphi_1 \Rightarrow \varphi_2)) \Rightarrow (\varphi_1 \Rightarrow \varphi_2)) & \text{Ax4} \\
3 & (\varphi_1 \Rightarrow \varphi_2) & \text{MP 1,2} \\
4 & (\forall x\, \varphi_1) & \text{Hyp} \\
5 & ((\forall x\, \varphi_1) \Rightarrow \varphi_1) & \text{Ax4} \\
6 & \varphi_1 & \text{MP 4,5} \\
7 & \varphi_2 & \text{MP 3,6} \\
8 & (\forall x\, \varphi_2) & \text{Gen 7}
\end{array}
$$

Generalization was used once, over x and $x \notin \mathrm{fv}_\Sigma((\forall x\, \varphi_1))$. So, there are no essential generalizations over dependents on $(\forall x\, \varphi_1)$ in the above derivation. Therefore, applying the MTD yields

$$
(\forall x(\varphi_1 \Rightarrow \varphi_2)) \vdash_\Sigma ((\forall x\, \varphi_1) \Rightarrow (\forall x\, \varphi_2)).
$$

Observe that $x \notin \mathrm{fv}_\Sigma((\forall x(\varphi_1 \Rightarrow \varphi_2)))$. Hence, the latter sequence also does not contain any essential generalizations over dependents on $(\forall x(\varphi_1 \Rightarrow \varphi_2))$. In the derivation returned by the MTD there are also no essential generalizations over dependents of $(\forall x(\varphi_1 \Rightarrow \varphi_2))$. Hence, applying the MTD again yields the desired result. $\qquad\qquad \triangledown$

The following is the solution to Exercise 25 of Chapter 5.

Exercise 5 Adapting the proof of the MTD, expand the derivation sequence for

$$
(\forall x(\varphi_1 \Rightarrow \varphi_2)), (\forall x\, \varphi_1) \vdash_\Sigma (\forall x\, \varphi_2),
$$

in order to build a derivation sequence for

$$
(\forall x(\varphi_1 \Rightarrow \varphi_2)) \vdash_\Sigma ((\forall x\, \varphi_1) \Rightarrow (\forall x\, \varphi_2)).
$$

Solution:

$$
\begin{array}{lll}
1.1 & (\forall x(\varphi_1 \Rightarrow \varphi_2)) & \text{Hyp} \\
1.2 & ((\forall x(\varphi_1 \Rightarrow \varphi_2)) \Rightarrow ((\forall x\, \varphi_1) \Rightarrow (\forall x(\varphi_1 \Rightarrow \varphi_2)))) & \text{Ax1} \\
1.3 & ((\forall x\, \varphi_1) \Rightarrow (\forall(\varphi_1 \Rightarrow \varphi_2))) & \text{MP 1.1,1.2} \\
2.1 & ((\forall x\, \varphi_1) \Rightarrow ((\forall x\, \varphi_1) \Rightarrow (\forall x\, \varphi_1))) & \text{Ax1} \\
2.2 & ((\forall x\, \varphi_1) \Rightarrow (((\forall x\, \varphi_1) \Rightarrow (\forall x\, \varphi_1)) \Rightarrow (\forall x\, \varphi_1))) & \text{Ax1} \\
2.3 & (((\forall x\, \varphi_1) \Rightarrow (((\forall x\, \varphi_1) \Rightarrow (\forall x\, \varphi_1)) \Rightarrow (\forall x\, \varphi_1))) \Rightarrow \\
 & \quad (((\forall x\, \varphi_1) \Rightarrow \\
 & \qquad ((\forall x\, \varphi_1) \Rightarrow (\forall x\, \varphi_1))) \Rightarrow ((\forall x\, \varphi_1) \Rightarrow (\forall x\, \varphi_1)))) & \text{Ax2}
\end{array}
$$

2.4	$(((\forall x\,\varphi_1) \Rightarrow ((\forall x\,\varphi_1) \Rightarrow (\forall x\,\varphi_1))) \Rightarrow$	
	$\qquad\qquad\qquad ((\forall x\,\varphi_1) \Rightarrow (\forall x\,\varphi_1)))$	MP 2.2,2.3
2.5	$((\forall x\,\varphi_1) \Rightarrow (\forall x\,\varphi_1))$	MP 2.1,2.4
3.1	$((\forall x(\varphi_1 \Rightarrow \varphi_2)) \Rightarrow (\varphi_1 \Rightarrow \varphi_2))$	Ax4
3.2	$(((\forall x(\varphi_1 \Rightarrow \varphi_2)) \Rightarrow (\varphi_1 \Rightarrow \varphi_2)) \Rightarrow$	
	$\qquad ((\forall x\,\varphi_1) \Rightarrow ((\forall x(\varphi_1 \Rightarrow \varphi_2)) \Rightarrow (\varphi_1 \Rightarrow \varphi_2))))$	Ax1
3.3	$((\forall x\,\varphi_1) \Rightarrow ((\forall x(\varphi_1 \Rightarrow \varphi_2)) \Rightarrow (\varphi_1 \Rightarrow \varphi_2)))$	MP 3.1,3.2
4.1	$(((\forall x\,\varphi_1) \Rightarrow ((\forall x(\varphi_1 \Rightarrow \varphi_2)) \Rightarrow (\varphi_1 \Rightarrow \varphi_2))) \Rightarrow$	
	$\quad (((\forall x\,\varphi_1) \Rightarrow$	
	$\quad\;(\forall x(\varphi_1 \Rightarrow \varphi_2))) \Rightarrow ((\forall x\,\varphi_1) \Rightarrow (\varphi_1 \Rightarrow \varphi_2))))$	Ax2
4.2	$(((\forall x\,\varphi_1) \Rightarrow (\forall x(\varphi_1 \Rightarrow \varphi_2))) \Rightarrow$	
	$\qquad\qquad\qquad ((\forall x\,\varphi_1) \Rightarrow (\varphi_1 \Rightarrow \varphi_2)))$	MP 3.3,4.1
4.3	$((\forall x\,\varphi_1) \Rightarrow (\varphi_1 \Rightarrow \varphi_2))$	MP 1.3,4.2
5.1	$((\forall x\,\varphi_1) \Rightarrow \varphi_1)$	Ax4
5.2	$(((\forall x\,\varphi_1) \Rightarrow \varphi_1) \Rightarrow ((\forall x\,\varphi_1) \Rightarrow ((\forall x\,\varphi_1) \Rightarrow \varphi_1)))$	Ax1
5.3	$((\forall x\,\varphi_1) \Rightarrow ((\forall x\,\varphi_1) \Rightarrow \varphi_1))$	MP 5.1,5.2
6.1	$(((\forall x\,\varphi_1) \Rightarrow ((\forall x\,\varphi_1) \Rightarrow \varphi_1)) \Rightarrow$	
	$\qquad (((\forall x\,\varphi_1) \Rightarrow (\forall x\,\varphi_1)) \Rightarrow ((\forall x\,\varphi_1) \Rightarrow \varphi_1)))$	Ax2
6.2	$(((\forall x\,\varphi_1) \Rightarrow (\forall x\,\varphi_1)) \Rightarrow ((\forall x\,\varphi_1) \Rightarrow \varphi_1))$	MP 5.3,6.1
6.3	$((\forall x\,\varphi_1) \Rightarrow \varphi_1)$	MP 2,6.2
7.1	$(((\forall x\,\varphi_1) \Rightarrow (\varphi_1 \Rightarrow \varphi_2)) \Rightarrow$	
	$\qquad ((\forall x\,\varphi_1) \Rightarrow \varphi_1) \Rightarrow (((\forall x\,\varphi_1) \Rightarrow \varphi_2)))$	Ax2
7.2	$(((\forall x\,\varphi_1) \Rightarrow \varphi_1) \Rightarrow ((\forall x\,\varphi_1) \Rightarrow \varphi_2))$	MP 4.3,7,1
7.3	$((\forall x\,\varphi_1) \Rightarrow \varphi_2)$	MP 6.3,7,2
8.1	$(\forall x((\forall x\,\varphi_1) \Rightarrow \varphi_2))$	Gen 7.3
8.2	$((\forall x((\forall x\,\varphi_1) \Rightarrow \varphi_2)) \Rightarrow ((\forall x\,\varphi_1) \Rightarrow (\forall x\,\varphi_2)))$	Ax5
8.3	$((\forall x\,\varphi_1) \Rightarrow (\forall x\,\varphi_2))$	MP 8.1,8.2

In this way a constructive proof is obtained for

$$(\forall x(\varphi_1 \Rightarrow \varphi_2)) \vdash_\Sigma ((\forall x\,\varphi_1) \Rightarrow (\forall x\,\varphi_2))$$

even if that there are nicer derivations of this formula. ∇

Exercise 6 Show that $\vdash_\Sigma ((\varphi \Rightarrow \varphi_1) \Rightarrow ((\varphi \Rightarrow \varphi_2) \Rightarrow (\varphi \Rightarrow (\varphi_1 \wedge \varphi_2))))$.

Solution: Consider the following derivation sequence:

1	$(\varphi \Rightarrow \varphi_1)$	Hyp
2	$(\varphi \Rightarrow \varphi_2)$	Hyp
3	φ	Hyp
4	φ_1	MP 3,1
5	φ_2	MP 3,2
6	$(\varphi_1 \Rightarrow (\varphi_2 \Rightarrow (\varphi_1 \wedge \varphi_2)))$	Thm 5.4(i)
7	$(\varphi_2 \Rightarrow (\varphi_1 \wedge \varphi_2))$	MP 4,6
8	$(\varphi_1 \wedge \varphi_2)$	MP 5,7

The result then follows applying the MTD.   ▽

Exercise 7 Show that $\vdash_\Sigma ((\forall x(\varphi_1 \wedge \varphi_2)) \Rightarrow ((\forall x\, \varphi_1) \wedge (\forall x\, \varphi_2)))$.

Solution:

1	$((\forall x(\varphi_1 \wedge \varphi_2)) \Rightarrow (\varphi_1 \wedge \varphi_2))$	Ax4
2	$((\varphi_1 \wedge \varphi_2) \Rightarrow \varphi_1)$	Thm 5.4(h1)
3	$((\varphi_1 \wedge \varphi_2) \Rightarrow \varphi_2)$	Thm 5.4(h2)
4	$((\forall x(\varphi_1 \wedge \varphi_2)) \Rightarrow \varphi_1)$	HS 1,2
5	$((\forall x(\varphi_1 \wedge \varphi_2)) \Rightarrow \varphi_2)$	HS 1,3
6	$(\forall x((\forall x(\varphi_1 \wedge \varphi_2)) \Rightarrow \varphi_1))$	Gen 4
7	$(\forall x((\forall x(\varphi_1 \wedge \varphi_2)) \Rightarrow \varphi_2))$	Gen 5
8	$((\forall x((\forall x(\varphi_1 \wedge \varphi_2)) \Rightarrow \varphi_1)) \Rightarrow$ $((\forall x(\varphi_1 \wedge \varphi_2)) \Rightarrow (\forall x\, \varphi_1)))$	Ax5
9	$((\forall x((\forall x(\varphi_1 \wedge \varphi_2)) \Rightarrow \varphi_1)) \Rightarrow$ $((\forall x(\varphi_1 \wedge \varphi_2)) \Rightarrow (\forall x\, \varphi_2)))$	Ax5
10	$((\forall x(\varphi_1 \wedge \varphi_2)) \Rightarrow (\forall x\, \varphi_1))$	MP 6,8
11	$((\forall x(\varphi_1 \wedge \varphi_2)) \Rightarrow (\forall x\, \varphi_2))$	MP 7,9
12	$(((\forall x(\varphi_1 \wedge \varphi_2)) \Rightarrow (\forall x\, \varphi_1)) \Rightarrow$ $(((\forall x(\varphi_1 \wedge \varphi_2)) \Rightarrow (\forall x\, \varphi_2)) \Rightarrow$ $((\forall x(\varphi_1 \wedge \varphi_2)) \Rightarrow ((\forall x\, \varphi_1) \wedge (\forall x\, \varphi_2)))))$	Thm Exer. 6
13	$(((\forall x(\varphi_1 \wedge \varphi_2)) \Rightarrow (\forall x\, \varphi_2)) \Rightarrow$ $((\forall x(\varphi_1 \wedge \varphi_2)) \Rightarrow ((\forall x\, \varphi_1) \wedge (\forall x\, \varphi_2))))$	MP 10,12
14	$((\forall x(\varphi_1 \wedge \varphi_2)) \Rightarrow ((\forall x\, \varphi_1) \wedge (\forall x\, \varphi_2)))$	MP 11,13

▽

Exercise 8 Show that $\vdash_\Sigma ((\varphi \wedge \psi) \Leftrightarrow (\psi \wedge \varphi))$.

Solution: Consider the following derivation sequence:

1	$(\varphi \wedge \psi)$	Hyp
2	$((\varphi \wedge \psi) \Rightarrow \varphi)$	Thm 5.4(h1)
3	$((\varphi \wedge \psi) \Rightarrow \psi)$	Thm 5.4(h2)
4	φ	MP 1,2
5	ψ	MP 1,3
6	$(\psi \Rightarrow (\varphi \Rightarrow (\psi \wedge \varphi)))$	Thm 5.4(i)
7	$(\varphi \Rightarrow (\psi \wedge \varphi))$	MP 5,6
8	$(\psi \wedge \varphi)$	MP 4,7

The result then follows applying the MTD.   ▽

Exercise 9 Show that $\vdash_\Sigma ((\forall x\, \varphi) \Rightarrow (\exists x\, \varphi))$.

Solution: Consider the following derivation sequence:

1	$(\forall x \, \varphi)$	Hyp
2	$(\forall x \, (\neg \varphi))$	Hyp
3	$((\forall x \, \varphi) \Rightarrow \varphi)$	Ax4
4	$((\forall x \, (\neg \varphi)) \Rightarrow (\neg \varphi))$	Ax4
5	φ	MP 1,3
6	$(\neg \varphi)$	MP 2,4
7	$(\varphi \Rightarrow ((\neg \varphi) \Rightarrow (\varphi \wedge (\neg \varphi))))$	Thm 5.4(i)
8	$((\neg \varphi) \Rightarrow (\varphi \wedge (\neg \varphi)))$	MP 5,7
9	$(\varphi \wedge (\neg \varphi))$	MP 6,8

Applying the MTC yields $(\forall x \, \varphi) \vdash_\Sigma (\neg(\forall x \, (\neg \varphi)))$. Finally, using the MTD,[1] the desired result follows. ∇

The exercise below shows that from a given derivation sequence it is always possible to extract a derivation sequence for the same formula using the same hypotheses but with no irrelevant steps.

Exercise 10 Let $w = (\psi_1, J_1) \ldots (\psi_n, J_n)$ be a derivation sequence. Step j is said to *depend* on step i in w if $i \in A_j^w$, where this set (of ancestors of j in w) is inductively defined as follows:

- $j \in A_j^w$;

- $i \in A_j^w$ if J_j is Gen k and $i \in A_k^w$;

- $i \in A_j^w$ if J_j is MP$k, k'$ and $i \in A_k^w$ or $i \in A_{k'}^w$.

Step i of sequence w is said to be *irrelevant* if the final step n does not depend on step i. Step i of sequence w is said to be *duplicate* if there exists a step $j \neq i$ in w such that $\psi_i = \psi_j$. The derivation sequence is said to be *sober* if it contains neither irrelevant nor duplicate steps.

(a) Analyze the sobriety of a derivation sequence obtained by applying the MTD.

(b) Show that if $\Gamma \vdash_\Sigma \varphi$, then there exists a derivation sequence for φ from Γ that does not contain irrelevant steps.

Solution:

(a) Regardless of the sobriety of the original derivation sequence, if MP is never applied in it, then the sequence resulting from the application of the MTD does not have irrelevant or duplicate steps, hence it is sober. However, if MP is used

---

[1] Recall that the derivation returned by the MTC does not introduce any collateral generalizations.

in the original sequence, even if it is sober, then the sequence resulting from
applying the MTD may have duplicate steps because of the way in which this
sequence is built in case the original one ends with MP—blind concatenation
of the derivations in both antecedents, which results in duplication whenever a
formula occurs in both sequences. Observe, however, that even in the presence
of MP, the sequence resulting from applying the MTD never contains irrelevant
steps.

(b) Let $w = (\psi_1, J_1) \ldots (\psi_n, J_n)$ be a derivation sequence for $\Gamma \vdash_\Sigma \varphi$. It will
be shown, by induction on n, that it is possible to build a derivation sequence
w' for $\Gamma \vdash_\Sigma \varphi$ containing no irrelevant steps:

(Basis) $n = 1$: since w has no irrelevant steps, one may take $w' = w$.

(Step) Suppose $n = m + 1$. There are three cases to consider.
(i) J_n is Hyp or Ax:
take w' to be the unitary sequence (ψ_n, J_n), which obviously does not contain
irrelevant steps.
(ii) J_n is MP k, j:
$w_{1\ldots k} = (\psi_1, J_1) \ldots (\psi_k, J_k)$ and $w_{1\ldots j} = (\psi_1, J_1) \ldots (\psi_j, J_j)$ are derivation sequences for $\Gamma \vdash_\Sigma \psi_k$ and $\Gamma \vdash_\Sigma \psi_j$, respectively. Hence, by the induction hypothesis, there exist derivation sequences w'' and w''' for $\Gamma \vdash_\Sigma \psi_k$ and $\Gamma \vdash_\Sigma \psi_j$,
respectively, that do not contain irrelevant steps. Let r and s be the lengths
of w'' and w''', respectively. Take w' to be the derivation sequence for $\Gamma \vdash_\Sigma \varphi$
obtained by adding the step

$$r + s + 1 \quad \varphi \quad \text{MP } r, r + s$$

to the concatenation of w'' and w'''. Clearly, w' does not have irrelevant steps.
(iii) J_n is Gen k:
$w_{1\ldots k} = (\psi_1, J_1) \ldots (\psi_k, J_k)$ is a derivation sequence for $\Gamma \vdash_\Sigma \psi_k$. Hence, by
the induction hypothesis, there exists a derivation sequence w'' for $\Gamma \vdash_\Sigma \psi_k$
that does not contain irrelevant steps. Let r be the length of w''. Take w' to
be the derivation sequence for $\Gamma \vdash_\Sigma \varphi$ obtained by adding the step

$$r + 1 \quad \varphi \quad \text{Gen } r$$

to sequence w''. Once again, w' does not have irrelevant steps. ∇

Exercise 11 Show that a theory Θ over Σ is exhaustive if and only if, for
every closed formulas φ and ψ, the following holds:

$$\text{if } \Theta \vdash_\Sigma (\varphi \vee \psi), \text{ then } \Theta \vdash_\Sigma \varphi \text{ or } \Theta \vdash_\Sigma \psi.$$

Solution:
($\rightarrow$): Let Θ be an exhaustive theory over Σ. Suppose that

$$\Theta \vdash_\Sigma (\varphi \vee \psi), \ \Theta \nvdash_\Sigma \varphi \text{ and } \Theta \nvdash_\Sigma \psi$$

for some pair of closed formulas φ and ψ. Observe that under these circumstances Θ is consistent. By exhaustiveness it follows that $\Theta \vdash_\Sigma (\neg\varphi)$ and $\Theta \vdash_\Sigma (\neg\psi)$. Then, $\Theta \vdash_\Sigma ((\neg\varphi) \wedge (\neg\psi))$, and hence $\Theta \vdash_\Sigma (\neg(\varphi \vee \psi))$, contradicting the assumption $\Theta \vdash_\Sigma (\varphi \vee \psi)$.

($\leftarrow$): Let Θ be a theory such that, for every closed formulas φ and ψ, the following holds: if $\Theta \vdash_\Sigma (\varphi \vee \psi)$, then $\Theta \vdash_\Sigma \varphi$ or $\Theta \vdash_\Sigma \psi$. Let φ be a closed formula. Recall that $\vdash_\Sigma ((\neg\varphi) \vee \varphi)$, since $\vdash_\Sigma (\varphi \Rightarrow \varphi)$. Therefore,

$$\Theta \vdash_\Sigma ((\neg\varphi) \vee \varphi).$$

Then, using the hypothesis, it follows that $\Theta \vdash_\Sigma (\neg\varphi)$ or $\Theta \vdash_\Sigma \varphi$, which shows that Θ is exhaustive. ∇

Chapter 19

Exercises on semantics

The solutions presented herein are related to the material introduced in Chapter 7.

Exercise 1 Show that if ρ_1 is Y_1-equivalent to ρ_2 and $Y_1 \subseteq Y_2$, then ρ_1 is Y_2-equivalent to ρ_2.

Solution: Since ρ_1 is Y_1-equivalent to ρ_2, one has $\rho_1(z) = \rho_2(z)$ for each $z \in X \setminus Y_1$. On the other hand, from $Y_1 \subseteq Y_2$ it follows that $X \setminus Y_2 \subseteq X \setminus Y_1$. Therefore, $\rho_1(z) = \rho_2(z)$ for every $z \in X \setminus Y_2$, which corresponds to ρ_1 being Y_2-equivalent to ρ_2. $\qquad\qquad\qquad\nabla$

The solution below is the answer to Exercise 7 of Chapter 7.

Exercise 2 Show that, for each $Y \subseteq X$, the binary relation of Y-equivalence, defined on the set of assignments over a given interpretation structure, is an equivalence relation.

Solution:

(1) Reflexivity: ρ is $\emptyset$-equivalent to ρ, therefore ρ is Y-equivalent to ρ.

(2) Symmetry: if ρ_1 is Y-equivalent to ρ_2, then $\rho_1(z) = \rho_2(z)$ for each $z \in X \backslash Y$, whence $\rho_2(z) = \rho_1(z)$ for each $z \in X \setminus Y$, and therefore ρ_2 is Y-equivalent to ρ_1.

(3) Transitivity: if ρ_1 is Y-equivalent to ρ_2 and ρ_2 is Y-equivalent to ρ_3, then $\rho_1(z) = \rho_2(z)$ for each $z \in X \setminus Y$ and $\rho_2(z) = \rho_3(z)$ for each $z \in X \setminus Y$, whence $\rho_1(z) = \rho_3(z)$ for each $z \in X \setminus Y$, and therefore ρ_1 is Y-equivalent to ρ_3. $\qquad\nabla$

Exercise 3 Show that the formula $((\forall x\, p(x)) \Rightarrow p(x))$ is valid.

Solution: It must be shown that, regardless of the interpretation structure $I = (D, \_^\mathsf{F}, \_^\mathsf{P})$ over the relevant signature Σ, it is the case that

$$I \Vdash_\Sigma ((\forall x\, p(x)) \Rightarrow p(x)).$$

In other words, it must be shown that, for every I and assignment ρ into I, necessarily

$$I\rho \Vdash_\Sigma ((\forall x\, p(x)) \Rightarrow p(x)).$$

Indeed,

$\quad I\rho \Vdash_\Sigma ((\forall x\, p(x)) \Rightarrow p(x))$ iff

$\quad I\rho \not\Vdash_\Sigma (\forall x\, p(x))$ or $I\rho \Vdash_\Sigma p(x)$ iff

$\quad$ there is ρ' x-equivalent to ρ such that $I\rho' \not\Vdash_\Sigma p(x)$
$\qquad$ or $I\rho \Vdash_\Sigma p(x)$ iff

$\quad$ there is ρ' x-equivalent to ρ such that $p_1^\mathsf{P}(\rho'(x)) = 0$
$\qquad$ or $p_1^\mathsf{P}(\rho(x)) = 1$

At this stage, one can consider two cases concerning the interpretation given by I to p.

- $p_1^\mathsf{P} = \lambda d\,.\,1$: in this case $p_1^\mathsf{P}(\rho(x)) = 1$.

- $p_1^\mathsf{P} \neq \lambda d\,.\,1$: in this case there is $d \in D$ such that $p_1^\mathsf{P}(d) = 0$; then it suffices to take ρ' as the assignment x-equivalent to ρ such that $\rho'(x) = d$ in order to guarantee $p_1^\mathsf{P}(\rho'(x)) = 0$.

In short, in each case one of the branches of the disjunction above is guaranteed to be true.

Alternatively, but maybe not as useful for illustrating the concepts at stake, observe that the assertion

$$\text{there is } \rho' \text{ } x\text{-equivalent to } \rho \text{ such that } p_1^\mathsf{P}(\rho'(x)) = 0$$
$$\text{or } p_1^\mathsf{P}(\rho(x)) = 1$$

is true if $p_1^\mathsf{P}(\rho(x)) = 1$. Otherwise, it is also true since choosing $\rho' = \rho$ guarantees that $p_1^\mathsf{P}(\rho'(x)) = 0$. ∇

The next exercise provides the soundness of the generalization rule for the particular case of an atomic formula.

Exercise 4 Show that $p(x) \vDash_\Sigma (\forall x\, p(x))$.

Solution: Letting I be a generic interpretation structure, it must be shown that if $I \Vdash_\Sigma p(x)$, then $I \Vdash_\Sigma (\forall x\, p(x))$.

If $I \Vdash_\Sigma p(x)$, then, by definition of global satisfaction,

$$I\rho \Vdash_\Sigma p(x)$$

for every assignment ρ into I. Therefore, for every assignment ρ into I,

$$I\sigma \Vdash_\Sigma p(x)$$

for every assignment σ that is x-equivalent to ρ. Finally, for every assignment ρ into I,

$$I \Vdash_\Sigma (\forall x\, p(x)).$$

$$\nabla$$

The next exercise provides a semantic counter-example that illustrates the need for restrictions in the application of the metatheorem of deduction. Although

$$p(x) \vDash_\Sigma (\forall x\, p(x))$$

it is not always the case that

$$(p(x) \Rightarrow (\forall x\, p(x))).$$

Exercise 5 Show, providing a counter-example, that the formula

$$(p(x) \Rightarrow (\forall x\, p(x)))$$

is not valid.

Solution: An interpretation structure I over the relevant signature Σ must be found, together with an assignment ρ into I, such that

$$I\rho \not\Vdash_\Sigma (p(x) \Rightarrow (\forall x\, p(x))).$$

Let $I = (D, \_^F, \_^P)$ be such that

- $D = \{d_0, d_1\}$;

- $p_1^P = \lambda d. \begin{cases} 0 & \text{if } d = d_0 \\ 1 & \text{otherwise} \end{cases}.$

Consider ρ such that $\rho(x) = d_1$. Furthermore, let $\bar{\rho}$ be the assignment x-equivalent to ρ such that $\bar{\rho}(x) = d_0$. Note that there are only two assignments that are x-equivalent to ρ: precisely ρ and $\bar{\rho}$. Then:

$$I\rho \Vdash_\Sigma (p(x) \Rightarrow (\forall x\, p(x))) \qquad\qquad \text{iff}$$

$$I\rho \nVdash_\Sigma p(x) \text{ or } I\rho \Vdash_\Sigma (\forall x\, p(x)) \qquad\qquad \text{iff}$$

$$p_1^{\mathsf{P}}(\rho(x)) = 0 \text{ or}$$
$$\quad I\rho' \Vdash_\Sigma p(x) \text{ for any } \rho'\ x\text{-equivalent to } \rho \qquad \text{iff}$$

$$p_1^{\mathsf{P}}(\rho(x)) = 0 \text{ or}$$
$$\quad p_1^{\mathsf{P}}(\rho'(x)) = 1 \text{ for any } \rho'\ x\text{-equivalent to } \rho \qquad \text{iff}$$

$$p_1^{\mathsf{P}}(d_1) = 0 \text{ or}$$
$$\quad p_1^{\mathsf{P}}(\rho'(x)) = 1 \text{ for any } \rho'\ x\text{-equivalent to } \rho \qquad \text{iff}$$

$$\text{false or}$$
$$\quad p_1^{\mathsf{P}}(\rho'(x)) = 1 \text{ for any } \rho'\ x\text{-equivalent to } \rho \qquad \text{iff}$$

$$p_1^{\mathsf{P}}(\rho'(x)) = 1 \text{ for any } \rho'\ x\text{-equivalent to } \rho \qquad\qquad \text{iff}$$

$$p_1^{\mathsf{P}}(\rho(x)) = 1 \text{ and } p_1^{\mathsf{P}}(\bar{\rho}(x)) = 1 \qquad\qquad \text{iff}$$

$$p_1^{\mathsf{P}}(d_1) = 1 \text{ and } p_1^{\mathsf{P}}(d_0) = 0 \qquad\qquad \text{iff}$$

$$\text{true and false} \qquad\qquad \text{iff}$$

$$\text{false}$$

$$\nabla$$

Exercise 6 Show that formula

$$((\forall x(\varphi \Rightarrow \psi)) \Rightarrow ((\forall x\, \varphi) \Rightarrow (\forall x\, \psi)))$$

is valid.

Solution: It must be shown that, for every interpretation structure I and assignment ρ into I, the following holds:

$$I\rho \Vdash_\Sigma ((\forall x(\varphi \Rightarrow \psi)) \Rightarrow ((\forall x\, \varphi) \Rightarrow (\forall x\, \psi))).$$

Two alternative proofs are given below of this fact.

Direct proof:

For any interpretation structure I and assignment ρ into I:

$$I\rho \Vdash_\Sigma ((\forall x(\varphi \Rightarrow \psi)) \Rightarrow ((\forall x\, \varphi) \Rightarrow (\forall x\, \psi))) \qquad\qquad \text{iff}$$

$$(\text{not } I\rho \Vdash_\Sigma (\forall x(\varphi \Rightarrow \psi))) \text{ or } I\rho \Vdash_\Sigma ((\forall x\, \varphi) \Rightarrow (\forall x\, \psi)) \qquad \text{iff}$$

(there is σ x-equivalent to ρ, such that $I\sigma \not\Vdash_\Sigma (\varphi \Rightarrow \psi)$)
 or
$(I\rho \not\Vdash_\Sigma (\forall x\,\varphi)$ or $I\rho \Vdash_\Sigma (\forall x\,\psi))$ iff

(a) (there is σ x-equivalent to ρ,
      such that $I\sigma \Vdash_\Sigma \varphi$ and $I\sigma \not\Vdash_\Sigma \psi$)
 or
(b) (there is σ' x-equivalent to ρ, such that $I\sigma' \not\Vdash_\Sigma \varphi$)
 or
(c) (for every σ'' x-equivalent to ρ, $I\sigma'' \Vdash_\Sigma \psi$)   iff ($*$)

 true

The only step to be detailed is $(*)$. There are two cases to consider:
(1) If (a) holds, then either (a) or (b) or (c) holds trivially.
(2) If (a) does not hold, then, for every assignment ρ' that is x-equivalent to ρ, it is the case that:
$$I\rho' \not\Vdash_\Sigma \varphi \text{ or } I\rho' \Vdash_\Sigma \psi,$$
whence:
(2.1) If there is an assignment ρ' that is x-equivalent to ρ and such that
$$I\rho' \not\Vdash_\Sigma \varphi$$
then (b) holds, and therefore (a) or (b) or (c) holds.
(2.2) If there is no assignment ρ' that is x-equivalent to ρ and such that
$$I\rho' \not\Vdash_\Sigma \varphi$$
then (c) holds, and therefore (a) or (b) or (c) holds.

Proof by contradiction:
Assume that there exist an interpretation structure I and an assignment ρ into I such that:
$$I\rho \not\Vdash_\Sigma ((\forall x(\varphi \Rightarrow \psi)) \Rightarrow ((\forall x\,\varphi) \Rightarrow (\forall x\,\psi))).$$
First observe that:

 not $I\rho \Vdash_\Sigma ((\forall x(\varphi \Rightarrow \psi)) \Rightarrow ((\forall x\,\varphi) \Rightarrow (\forall x\,\psi)))$ iff

$I\rho \Vdash_\Sigma (\forall x(\varphi \Rightarrow \psi))$ and (not $I\rho \Vdash_\Sigma ((\forall x\,\varphi) \Rightarrow (\forall x\,\psi)))$ iff

$I\rho \Vdash_\Sigma (\forall x(\varphi \Rightarrow \psi))$
 and
$I\rho \Vdash_\Sigma (\forall x\,\varphi)$
 and
(not $I\rho \Vdash_\Sigma (\forall x\,\psi)$)   iff

(a) $I\rho \Vdash_\Sigma (\forall x(\varphi \Rightarrow \psi))$

and

(b) $I\rho \Vdash_\Sigma (\forall x\, \varphi)$

and

(c) there is an assignment σ that is x-equivalent to ρ
    such that $I\sigma \not\Vdash_\Sigma \psi$

Therefore, there is an assignment σ that is x-equivalent to ρ and such that:

$$\begin{cases} (a') & I\sigma \Vdash_\Sigma (\varphi \Rightarrow \psi) \\ (b') & I\sigma \Vdash_\Sigma \varphi \\ (c') & I\sigma \not\Vdash_\Sigma \psi \end{cases}$$

Hence, there exists an assignment σ that is x-equivalent to ρ such that:

$$\begin{cases} I\sigma \Vdash_\Sigma (\varphi \Rightarrow \psi) \\ I\sigma \not\Vdash_\Sigma (\varphi \Rightarrow \psi) \end{cases}$$

which is a contradiction. ∇

The following is a solution for Exercise 8 of Chapter 7.

Exercise 7 Prove the following assertions over the connectives and quantifier introduced as abbreviations:

1. $I\rho \Vdash_\Sigma (\varphi \vee \psi)$ if and only if $I\rho \Vdash_\Sigma \varphi$ or $I\rho \Vdash_\Sigma \psi$;

2. $I\rho \Vdash_\Sigma (\varphi \wedge \psi)$ if and only if $I\rho \Vdash_\Sigma \varphi$ and $I\rho \Vdash_\Sigma \psi$;

3. $I\rho \Vdash_\Sigma (\varphi \Leftrightarrow \psi)$ if and only if $I\rho \Vdash_\Sigma \varphi$ and $I\rho \Vdash_\Sigma \psi$ or $I\rho \not\Vdash_\Sigma \varphi$ and $I\rho \not\Vdash_\Sigma \psi$;

4. $I\rho \Vdash_\Sigma (\exists x\, \varphi)$ if and only if there exists an assignment ρ' that is x-equivalent to ρ such that $I\rho' \Vdash_\Sigma \varphi$.

Solution:

1.

$$\begin{array}{ll} I\rho \Vdash_\Sigma (\varphi \vee \psi) & \text{iff} \\ I\rho \Vdash_\Sigma ((\neg \varphi) \Rightarrow \psi) & \text{iff} \\ \text{not } I\rho \Vdash_\Sigma (\neg \varphi) \text{ or } I\rho \Vdash_\Sigma \psi & \text{iff} \\ \text{not not } I\rho \Vdash_\Sigma \varphi \text{ or } I\rho \Vdash_\Sigma \psi & \text{iff} \\ I\rho \Vdash_\Sigma \varphi \text{ or } I\rho \Vdash_\Sigma \psi. & \end{array}$$

2.

$$I\rho \Vdash_\Sigma (\varphi \wedge \psi) \qquad\qquad\qquad \text{iff}$$
$$I\rho \Vdash_\Sigma (\neg(\varphi \Rightarrow (\neg\psi))) \qquad\qquad \text{iff}$$
$$\text{not } I\rho \Vdash_\Sigma (\varphi \Rightarrow (\neg\psi)) \qquad\qquad \text{iff}$$
$$\text{not } (\text{not } I\rho \Vdash_\Sigma \varphi \text{ or } I\rho \Vdash_\Sigma (\neg\psi)) \quad \text{iff}$$
$$\text{not } (\text{not } I\rho \Vdash_\Sigma \varphi \text{ or not } I\rho \Vdash_\Sigma \psi) \quad \text{iff}$$
$$I\rho \Vdash_\Sigma \varphi \text{ and } I\rho \Vdash_\Sigma \psi\,.$$

3.

$$I\rho \Vdash_\Sigma (\varphi \Leftrightarrow \psi) \qquad\qquad\qquad\qquad\qquad\qquad\qquad\qquad\qquad \text{iff}$$
$$I\rho \Vdash_\Sigma (\neg((\varphi \Rightarrow \psi) \Rightarrow (\neg(\psi \Rightarrow \varphi)))) \qquad\qquad\qquad \text{iff}$$
$$\text{not } I\rho \Vdash_\Sigma ((\varphi \Rightarrow \psi) \Rightarrow (\neg(\psi \Rightarrow \varphi))) \qquad\qquad\qquad \text{iff}$$
$$\text{not } (\text{not } I\rho \Vdash_\Sigma (\varphi \Rightarrow \psi) \text{ or } I\rho \Vdash_\Sigma (\neg(\psi \Rightarrow \varphi))) \quad \text{iff}$$
$$\text{not } (\text{not } I\rho \Vdash_\Sigma (\varphi \Rightarrow \psi) \text{ or not } I\rho \Vdash_\Sigma (\psi \Rightarrow \varphi)) \quad \text{iff}$$
$$I\rho \Vdash_\Sigma (\varphi \Rightarrow \psi) \text{ and } I\rho \Vdash_\Sigma (\psi \Rightarrow \varphi) \qquad\qquad\qquad \text{iff}$$
$$(\text{not } I\rho \Vdash_\Sigma \varphi \text{ or } I\rho \Vdash_\Sigma \psi) \text{ and } (\text{not } I\rho \Vdash_\Sigma \psi \text{ or } I\rho \Vdash_\Sigma \varphi) \quad \text{iff}$$
$$(\text{not } I\rho \Vdash_\Sigma \varphi \text{ and } (\text{not } I\rho \Vdash_\Sigma \psi \text{ or } I\rho \Vdash_\Sigma \varphi)) \text{ or}$$
$$\qquad\qquad (I\rho \Vdash_\Sigma \psi \text{ and } (\text{not } I\rho \Vdash_\Sigma \psi \text{ or } I\rho \Vdash_\Sigma \varphi)) \quad \text{iff}$$
$$(\text{not } I\rho \Vdash_\Sigma \varphi \text{ and not } I\rho \Vdash_\Sigma \psi) \text{ or}$$
$$\qquad\qquad (\text{not } I\rho \Vdash_\Sigma \varphi \text{ and } I\rho \Vdash_\Sigma \varphi) \text{ or}$$
$$\qquad (I\rho \Vdash_\Sigma \psi \text{ and not } I\rho \Vdash_\Sigma \psi) \text{ or}$$
$$\qquad\qquad (I\rho \Vdash_\Sigma \psi \text{ and } I\rho \Vdash_\Sigma \varphi) \quad \text{iff}$$
$$(\text{not } I\rho \Vdash_\Sigma \varphi \text{ and not } I\rho \Vdash_\Sigma \psi) \text{ or false or}$$
$$\qquad \text{false or } (I\rho \Vdash_\Sigma \psi \text{ and } I\rho \Vdash_\Sigma \varphi) \quad \text{iff}$$
$$(\text{not } I\rho \Vdash_\Sigma \varphi \text{ and not } I\rho \Vdash_\Sigma \psi) \text{ or } (I\rho \Vdash_\Sigma \psi \text{ and } I\rho \Vdash_\Sigma \varphi)$$

4.

$$I\rho \Vdash_\Sigma (\exists x\, \varphi) \qquad\qquad\qquad\qquad\qquad\qquad \text{iff}$$
$$I\rho \Vdash_\Sigma (\neg(\forall x(\neg \varphi))) \qquad\qquad\qquad\qquad \text{iff}$$
$$\text{not } I\rho \Vdash_\Sigma (\forall x(\neg \varphi)) \qquad\qquad\qquad\qquad \text{iff}$$
$$\text{not } (I\rho' \Vdash_\Sigma (\neg \varphi) \text{ for any } \rho' \text{ x-equivalent to } \rho) \quad \text{iff}$$
$$\text{there is } \rho' \text{ x-equivalent to } \rho \text{ such that not } I\rho' \Vdash_\Sigma (\neg \varphi) \quad \text{iff}$$
$$\text{there is } \rho' \text{ x-equivalent to } \rho \text{ such that not not } I\rho' \Vdash_\Sigma \varphi \quad \text{iff}$$
$$\text{there is } \rho' \text{ x-equivalent to } \rho \text{ such that } I\rho' \Vdash_\Sigma \varphi$$

$\triangledown$

Exercise 8 Show that there exist formulas φ_1 and φ_2 such that

$$((\forall x(\varphi_1 \vee \varphi_2)) \Rightarrow ((\forall x\, \varphi_1) \vee (\forall x\, \varphi_2)))$$

is not valid.

Solution: Take φ_1 to be $p(x)$ and φ_2 to be $q(x)$, with $p, q \in P_1$. The goal is to find an interpretation structure I and an assignment ρ into I such that

$$I\rho \not\Vdash_\Sigma ((\forall x((p(x) \vee q(x)))) \Rightarrow ((\forall x\, p(x)) \vee (\forall x\, q(x))))$$

that is, $I\rho \Vdash_\Sigma (\forall x(p(x) \vee q(x)))$ e $I\rho \not\Vdash_\Sigma ((\forall x\, p(x)) \vee (\forall x\, q(x)))$.

Let I be an interpretation structure such that:

- $D = \{d_1, d_2\}$;

- $p_1^\mathsf{P}(d_1) = 1$, $p_1^\mathsf{P}(d_2) = 0$, $q_1^\mathsf{P}(d_1) = 0$ and $q_1^\mathsf{P}(d_2) = 1$.

Take an assignment ρ such that $\rho(x) = d_1$. Note that

$$I\rho \Vdash_\Sigma p(x) \quad \text{and} \quad I\rho \not\Vdash_\Sigma q(x).$$

The first step is checking that $I\rho \Vdash_\Sigma (\forall x(p(x) \vee q(x)))$, that is, that

$$I\sigma \Vdash_\Sigma (p(x) \vee q(x))$$

for every assignment σ that is x-equivalent to ρ. There are two cases to consider: (i) if $\sigma(x) = d_1 = \rho(x)$, then $\sigma = \rho$, whence $I\sigma \Vdash_\Sigma p(x)$, and therefore $I\sigma \Vdash_\Sigma (p(x) \vee q(x))$; (ii) if $\sigma(x) = d_2$, then $I\sigma \Vdash_\Sigma q(x)$, and therefore $I\sigma \Vdash_\Sigma (p(x) \vee q(x))$.

Finally, one must show that $I\rho \not\Vdash_\Sigma (\forall x\, p(x))$ and $I\rho \not\Vdash_\Sigma (\forall x\, q(x))$, that is, one must find assignments σ' and σ'' that are x-equivalent to ρ such that $I\sigma' \not\Vdash_\Sigma p(x)$ and $I\sigma'' \not\Vdash_\Sigma q(x)$. Take σ'' to be ρ and pick $\sigma'(x) = d_2$. ∇

The following is the solution to Exercise 28 of Chapter 7.

Exercise 9 Show that $((\forall x(\exists y\, p(x, y))) \Rightarrow (\exists y\, p(y, y)))$ is not valid taking as counter-example the interpretation structure I such that :

- D is $\mathbb{N}$;

- $p_2^\mathsf{P} : \mathbb{N}^2 \to \{0, 1\}$ is such that $p_2^\mathsf{P}(d_0, d_1) = 1$ if and only if $d_0 < d_1$.

Solution: It is necessary to show that there is ρ such that:

$$I\rho \not\Vdash_\Sigma ((\forall x(\exists y\, p(x, y))) \Rightarrow (\exists y\, p(y, y))).$$

In fact, it is proved that this is so for arbitrary ρ which is not unexpected since the formula is closed.

$I\rho \not\Vdash_\Sigma ((\forall x(\exists y\, p(x, y))) \Rightarrow (\exists y\, p(y, y)))$ iff

not $((\text{not } I\rho \Vdash_\Sigma (\forall x(\exists y\, p(x, y)))) \text{ or } I\rho \Vdash_\Sigma (\exists y\, p(y, y)))$ iff

(a) $I\rho \Vdash_\Sigma (\forall x(\exists y\, p(x, y)))$ and (b) (not $I\rho \Vdash_\Sigma (\exists y\, p(y, y)))$

Assertions (a) and (b) are checked as shown below.

(a) $I\rho \Vdash_\Sigma (\forall x(\exists y\, p(x,y)))$:

$\quad I\rho \Vdash_\Sigma (\forall x(\exists y\, p(x,y)))$ iff

$\quad\quad$ for every assignment σ that is x-equivalent to ρ,
$\quad\quad\quad I\sigma \Vdash_\Sigma (\exists y\, p(x,y)$ iff

$\quad\quad$ for every assignment σ that is x-equivalent to ρ,
$\quad\quad\quad$ there exists σ' that is y-equivalent to σ, $I\sigma' \Vdash_\Sigma p(x,y)$ iff $\;(*)$

$\quad\quad$ true

$(*)$ Once assignment σ is fixed, it suffices to take an assignment σ' that is y-equivalent to σ such that

$$\sigma'(y) = \sigma(x) + 1\,.$$

Then,
$$I\sigma' \Vdash_\Sigma p(x,y)$$

because $p_2^P(\sigma'(x),\sigma'(y))) = 1$, since $\sigma'(x) < \sigma'(y)$ and $\sigma(x) = \sigma'(x)$.

(b) not $I\rho \Vdash_\Sigma (\exists y\, p(y,y))$:

$\quad$ not $I\rho \Vdash_\Sigma (\exists y\, p(y,y))$ iff

$\quad$ $I\rho \Vdash_\Sigma (\forall y\,(\neg p(y,y)))$ iff

$\quad$ for every assignment σ such that y-equivalent to ρ,
$\quad\quad I\sigma \Vdash_\Sigma (\neg p(y,y))$ iff

$\quad$ for every assignment σ such that y-equivalent to ρ,
$\quad\quad I\sigma \not\Vdash_\Sigma p(y,y)$ iff

$\quad$ for every assignment σ such that y-equivalent to ρ,
$\quad\quad p_2^P(\sigma(y),\sigma(y)) = 0$ iff

$\quad\quad$ true

Observe that this example shows that Ax4 is not valid when the term is not free for the variable in the formula. More specifically, the substitution is of x by y, y is not free for x in $(\exists y p(x,y))$, since $x \in \mathrm{fv}_\Sigma((\exists y p(x,y)))$, and y is a variable of term y. ∇

The following is the solution of Exercise 46 of Chapter 7. It shows that, given an interpretation structure I and a set $D' \supseteq D$, it is possible to define a structure I' in such a way that I is an elementary substructure of I'.

Exercise 10 Let $I = (D, \_^\mathsf{F}, \_^\mathsf{P})$ be an interpretation structure over Σ, $D' \supseteq D$ and $a \in D$. Consider the map $(\_)_a : D' \to D$ such that

$$(d')_a = \begin{cases} d' & \text{whenever } d' \in D \\ a & \text{otherwise} \end{cases}$$

1. Consider the triple $I' = (D', \_^{\mathsf{F}'}, \_^{\mathsf{P}'})$ where

   - $f_n^{\mathsf{F}'}(d'_1, \dots, d'_n) = f_n^{\mathsf{F}}((d'_1)_a, \dots, (d'_n)_a)$;
   - $p_n^{\mathsf{P}'}(d'_1, \dots, d'_n) = p_n^{\mathsf{P}}((d'_1)_a, \dots, (d'_n)_a)$.

   Show that I' is an interpretation structure over Σ.

2. Show that:

   - $f_n^{\mathsf{F}'}|_D = f_n^{\mathsf{F}}$;
   - $p_n^{\mathsf{P}'}|_D = p_n^{\mathsf{P}}$.

3. Given $\rho' : X \to D'$, take $\rho'_a : X \to D$ such that $\rho'_a(x) = (\rho'(x))_a$. Show that:

   (a) $[\![t]\!]_\Sigma^{I\rho'_a} = ([\![t]\!]_\Sigma^{I'\rho'})_a$;

   (b) $I\rho'_a \Vdash_\Sigma \varphi$ if and only if $I'\rho' \Vdash_\Sigma \varphi$.

4. Classify the relationship between I and I'.

Solution:
1. I' is an interpretation structure, since $D' \neq \emptyset$ because $D \neq \emptyset$ and $D \subseteq D'$. Furthermore, each $f_n^{\mathsf{F}'}$ is a map from D'^n into D' and each $p_n^{\mathsf{P}'}$ is a map from D'^n into $\{0, 1\}$.

2. $f_n^{\mathsf{F}'}|_D = f_n^{\mathsf{F}}$: $f_n^{\mathsf{F}'}|_D(d_1, \dots d_n) = f_n^{\mathsf{F}}((d_1)_a, \dots, (d_n)_a) = f_n^{\mathsf{F}}(d_1, \dots, d_n)$, since $(d)_a = d$ whenever $d \in D$. A similar reasoning establishes $p_n^{\mathsf{P}'}|_D = p_n^{\mathsf{P}}$.

3.(a) One shows that

$$[\![t]\!]_\Sigma^{I\rho'_a} = ([\![t]\!]_\Sigma^{I'\rho'})_a$$

by induction on the structure of t.
(Basis) There are two cases to consider.
(i) t is x:

$$[\![x]\!]_\Sigma^{I\rho'_a} = \rho'_a(x) = (\rho'(x))_a = ([\![x]\!]_\Sigma^{I'\rho'})_a$$

(ii) t is c with $c \in F_0$:

$$[\![c]\!]_\Sigma^{I\rho'_a} = c_0^{\mathsf{F}} = c_0^{\mathsf{F}'} = [\![c]\!]_\Sigma^{I'\rho'} = ([\![c]\!]_\Sigma^{I'\rho'})_a$$

(Step) t is $f(t_1, \ldots, t_n)$.

$$[\![f(t_1, \ldots, t_n)]\!]_\Sigma^{I\rho'_a} = f_n^{\mathsf{F}}([\![t_1]\!]_\Sigma^{I\rho'_a}, \ldots, [\![t_n]\!]_\Sigma^{I\rho'_a})$$
$$= f_n^{\mathsf{F}}(([\![t_1]\!]_\Sigma^{I'\rho'})_a, \ldots, ([\![t_n]\!]_\Sigma^{I'\rho'})_a) \quad \text{(by IH)}$$
$$= f_n^{\mathsf{F}'}([\![t_1]\!]_\Sigma^{I'\rho'}, \ldots, [\![t_n]\!]_\Sigma^{I'\rho'}) \quad \text{(by definition of } I')$$
$$= [\![f(t_1, \ldots, t_n)]\!]_\Sigma^{I'\rho'}$$

3.(b) It will now be shown that

$$I'\rho' \Vdash_\Sigma \varphi \text{ if and only if } I\rho'_a \Vdash_\Sigma \varphi$$

by induction on the structure of φ:

(Basis) φ is $p(t_1, \ldots, t_n)$:

$$
\begin{array}{ll}
I'\rho' \Vdash_\Sigma p(t_1, \ldots, t_n) & \text{iff} \\
p_n^{\mathsf{P}'}([\![t_1]\!]_\Sigma^{I'\rho'}, \ldots, [\![t_n]\!]_\Sigma^{I'\rho'}) = 1 & \text{iff (by definition of } p_n^{\mathsf{P}'}) \\
p_n^{\mathsf{P}}(([\![t_1]\!]_\Sigma^{I'\rho'})_a, \ldots, ([\![t_n]\!]_\Sigma^{I'\rho'})_a) = 1 & \text{iff (by IH)} \\
p_n^{\mathsf{P}}([\![t_1]\!]_\Sigma^{I\rho'_a}, \ldots, [\![t_n]\!]_\Sigma^{I\rho'_a}) = 1 & \text{iff} \\
I\rho'_a \Vdash_\Sigma p(t_1, \ldots, t_n).
\end{array}
$$

(Step) There are three cases to consider:

(i) φ is $(\neg \psi)$:

$$
\begin{array}{ll}
I'\rho' \Vdash_\Sigma (\neg \psi) & \text{iff} \\
I'\rho' \nVdash_\Sigma \psi & \text{iff (by IH)} \\
I\rho'_a \nVdash_\Sigma \psi & \text{iff} \\
I\rho'_a \Vdash_\Sigma (\neg \psi).
\end{array}
$$

(ii) φ is $(\psi \Rightarrow \delta)$:

$$
\begin{array}{ll}
I'\rho' \Vdash_\Sigma (\psi \Rightarrow \delta) & \text{iff} \\
I'\rho' \nVdash_\Sigma \psi \text{ or } I'\rho' \Vdash_\Sigma \delta & \text{iff (by IH)} \\
I\rho'_a \nVdash_\Sigma \psi \text{ or } I\rho'_a \Vdash_\Sigma \delta & \text{iff} \\
I\rho'_a \Vdash_\Sigma (\psi \Rightarrow \delta).
\end{array}
$$

(iii) φ is $(\forall x\, \psi)$:

($\rightarrow$) By contraposition:

Suppose that $I\rho'_a \not\Vdash_\Sigma (\forall x\,\psi)$. Then, there exists an assignment σ that is x-equivalent to ρ'_a and such that

$$(1)\quad I\sigma \not\Vdash_\Sigma \psi\,.$$

Consider an assignment σ', x-equivalent to ρ', such that $\sigma'(x) = \sigma(x)$. Then, $\sigma = \sigma'_a$: $\sigma(y) = \rho'_a(y) = \sigma'_a(y)$ and $\sigma(x) = \sigma'_a(x)$ (because $(d')_a = d'$ if $d' \in D$).

Then, from (1), it follows that

$$I\sigma'_a \not\Vdash_\Sigma \psi$$

whence the induction hypothesis yields

$$I'\sigma' \not\Vdash_\Sigma \psi\,.$$

Since assignment σ' is x-equivalent to ρ', also

$$I'\rho' \not\Vdash_\Sigma (\forall x\,\psi)\,.$$

$(\leftarrow)$ By contraposition:
Suppose that $I'\rho' \not\Vdash_\Sigma (\forall x\,\psi)$. Then, there exists an assignment σ' that is x-equivalent to ρ' and such that

$$I'\sigma' \not\Vdash_\Sigma \psi$$

whence follows, by the induction hypothesis, that

$$I\sigma'_a \not\Vdash_\Sigma \psi\,.$$

Since assignment σ'_a is x-equivalent to ρ'_a (because assignment σ' is x-equivalent to ρ'),

$$I\rho'_a \not\Vdash_\Sigma (\forall x\,\psi)\,.$$

4. I is a substructure of I'. Furthermore, I is an elementary substructure of I'. In other words,

$$I\rho \Vdash_\Sigma \varphi \text{ if and only if } I'\rho \Vdash_\Sigma \varphi$$

for every assignment ρ into I. It suffices to note that ρ is also an assignment into I' and such that ρ_a is ρ. ∇

Chapter 20

Exercises on completeness

The exercises solved herein complement Chapter 8.

Exercise 1 Show that $\vdash_\Sigma ((\varphi \Rightarrow (\neg \varphi)) \Rightarrow (\neg \varphi))$.

Solution: Consider the following derivation sequence:

1	$(((\neg(\neg\varphi)) \Rightarrow (\neg\varphi)) \Rightarrow (\neg\varphi))$	Thm 5.4(b)
2	$((\neg(\neg\varphi)) \Rightarrow \varphi)$	Thm 5.4(a)
3	$(\varphi \Rightarrow (\neg(\neg\varphi)))$	Thm 5.4(d)
4	$((((\neg(\neg\varphi)) \Rightarrow \varphi) \Rightarrow$ $((\varphi \Rightarrow (\neg(\neg\varphi))) \Rightarrow ((\neg(\neg\varphi)) \Leftrightarrow \varphi))))$	Thm 5.4(i)
5	$((\varphi \Rightarrow (\neg(\neg\varphi))) \Rightarrow ((\neg(\neg\varphi)) \Leftrightarrow \varphi))$	MP 2,4
6	$((\neg(\neg\varphi)) \Leftrightarrow \varphi)$	MP 3,5
7	$((((\neg(\neg\varphi)) \Rightarrow (\neg\varphi)) \Rightarrow (\neg\varphi)) \Leftrightarrow$ $((\varphi \Rightarrow (\neg\varphi)) \Rightarrow (\neg\varphi)))$	MTSE 6
8	$(((((\neg(\neg\varphi)) \Rightarrow (\neg\varphi)) \Rightarrow (\neg\varphi)) \Leftrightarrow$ $((\varphi \Rightarrow (\neg\varphi)) \Rightarrow (\neg\varphi)))$ $\Rightarrow$ $((((\neg(\neg\varphi)) \Rightarrow (\neg\varphi)) \Rightarrow (\neg\varphi)) \Rightarrow$ $((\varphi \Rightarrow (\neg\varphi)) \Rightarrow (\neg\varphi))))$	Thm 5.4(h1)
9	$((((\neg(\neg\varphi)) \Rightarrow (\neg\varphi)) \Rightarrow (\neg\varphi)) \Rightarrow$ $((\varphi \Rightarrow (\neg\varphi)) \Rightarrow (\neg\varphi)))$	MP 7,8
10	$((\varphi \Rightarrow (\neg\varphi)) \Rightarrow (\neg\varphi))$	MP 1,9

$\triangledown$

The following is a solution to Exercise 4 of Chapter 8 on the notion of maximal consistency.

Exercise 2 A set of formulas Γ is said to be *maximal consistent* w.r.t. Σ if it is consistent w.r.t. Σ and, for every formula ψ over Σ, if $\Gamma \nvdash_\Sigma \psi$, then $\Gamma \cup \{\psi\}$ is not consistent w.r.t. Σ. Show that a set is maximal consistent if and only if it is consistent and exhaustive.

Solution:

($\rightarrow$) By contraposition, assume that Γ is not exhaustive. Then, there is a closed formula δ such that $\Gamma \nvdash_\Sigma \delta$ and $\Gamma \nvdash_\Sigma (\neg \delta)$. From the second assertion, applying Proposition 3, it follows that $\Gamma \cup \{\delta\}$ is consistent. Hence Γ is not maximal consistent.

($\leftarrow$) Suppose Γ is consistent and exhaustive. Then, $(\forall \psi) \notin \Gamma^{\vdash_\Sigma}$ for every $\psi \notin \Gamma^{\vdash_\Sigma}$, by Chapter 5, Exercise 2. Hence, by exhaustiveness of Γ, $(\neg(\forall \psi)) \in \Gamma^{\vdash_\Sigma}$, whence, by monotonicity of derivation, $\Gamma \cup \{\psi\} \vdash_\Sigma (\neg(\forall \psi))$.

On the other hand, due to extensivity of derivation and again using Chapter 5, Exercise 2, one concludes that $\Gamma \cup \{\psi\} \vdash_\Sigma (\forall \psi)$.

Therefore, using Proposition 2, it follows that $\Gamma \cup \{\psi\}$ is not consistent for any $\psi \notin \Gamma^{\vdash_\Sigma}$.

Hence, observing that Γ is consistent, Γ is maximal consistent. ∇

The following exercise is related to Proposition 10 of Chapter 8 where a specific interpretation structure is induced from a set Γ of formulas by taking as domain the set of closed terms. Herein, it is shown that the proposed triple is in fact an interpretation structure. That is, the domain is not empty and the denotation maps of both function and predicate symbols are well defined.

Exercise 3 Let $\Sigma = (F, P, \tau)$ be a first-order signature and $\Psi \subseteq L_\Sigma$. Check under which conditions the following triple provides an interpretation structure over Σ:

$$\mathrm{IS}_\Sigma(\Psi) = (D, \_^F, \_^P)$$

where:

- $D = cT_\Sigma$;

- $c_0^F = c$;

- $f_n^F = \lambda d_1, \ldots, d_n. f(d_1, \ldots, d_n)$ for each $n \in \mathbb{N}^+$;

- $p_n^P = \lambda d_1, \ldots, d_n. \begin{cases} 1 & \text{if } \Psi \vdash_\Sigma p(d_1, \ldots, d_n) \\ 0 & \text{otherwise} \end{cases}$ for every $n \in \mathbb{N}^+$.

Solution: According to the definition of interpretation structure introduced at the beginning of Chapter 7, three conditions must be satisfied:

Domain:

The set D must be non empty, that is, $cT_\Sigma \neq \emptyset$, for which a necessary and sufficient condition is the existence of some $c \in F$ such that $\tau(c) = 0$, or, in other words, the existence of a constant symbol in Σ.

Interpretation of function symbols:
For each $n \in \mathbb{N}$ and $f \in F_n$, one must have $f_n^F : D^n \to D$, that is,

$$f_n^F : cT_\Sigma^n \to cT_\Sigma.$$

Indeed: (i) $c_0^F = c$ and c is in cT_Σ; (ii) for every $n \in \mathbb{N}^+$, $f_n^F(d_1, \ldots, d_n) \in cT_\Sigma$ whenever $(d_1, \ldots, d_n) \in cT_\Sigma$.

Interpretation of predicate symbols:
For each $n \in \mathbb{N}^+$ and $p \in P_n$, one must have $p_n^P : D^n \to \{0, 1\}$, that is,

$$p_n^P : cT_\Sigma^n \to \{0, 1\}.$$

Indeed, for every $n \in \mathbb{N}^+$, $p_n^F(d_1, \ldots, d_n) \in \{0, 1\}$ whenever $(d_1, \ldots, d_n) \in cT_\Sigma$.

In short, as long as Σ contains at least one constant, $\mathrm{IS}_\Sigma(\Psi)$ is an interpretation structure over Σ, known as the *canonical interpretation structure* of set Ψ of formulas over Σ. $\qquad\nabla$

The following exercise shows that the denotation of a closed term in $\mathrm{IS}_\Sigma(\Psi)$ is the term itself.

Exercise 4 Show that if $t \in cT_\Sigma$, then

$$[\![t]\!]_\Sigma^{\mathrm{IS}_\Sigma(\Psi)\rho} = t$$

for every assignment ρ over $\mathrm{IS}_\Sigma(\Psi)$.

Solution: If $cT_\Sigma = \emptyset$, then the thesis holds vacuously. Otherwise, it is obtained by induction on the structure of closed term t:

(Basis) t is $c \in F_0$:
$$[\![c]\!]_\Sigma^{\mathrm{IS}_\Sigma(\Psi)\rho} = c_0^F = c.$$

(Step) t is $f(t_1, \ldots, t_n)$ with $t_1, \ldots, t_n \in cT_\Sigma$:

$$
\begin{aligned}
[\![f(t_1, \ldots, t_n)]\!]_\Sigma^{\mathrm{IS}_\Sigma(\Psi)\rho} &= \\
f_n^F([\![t_1]\!]_\Sigma^{\mathrm{IS}_\Sigma(\Psi)\rho}, \ldots, [\![t_n]\!]_\Sigma^{\mathrm{IS}_\Sigma(\Psi)\rho}) &= \\
f([\![t_1]\!]_\Sigma^{\mathrm{IS}_\Sigma(\Psi)\rho}, \ldots, [\![t_n]\!]_\Sigma^{\mathrm{IS}_\Sigma(\Psi)\rho}) &= \text{(by the induction hypothesis)} \\
f(t_1, \ldots, t_n). &
\end{aligned}
$$

$\qquad\nabla$

Exercise 5 Show that $\vdash_\Sigma ((\neg(\varphi \Rightarrow \psi)) \Rightarrow \varphi)$.

Solution: Consider the following derivation sequence:

1	$((\neg \varphi) \Rightarrow (\varphi \Rightarrow \psi))$	Thm 5.4(g)
2	$((\varphi \Rightarrow \psi) \Rightarrow (\neg(\neg(\varphi \Rightarrow \psi))))$	Thm 5.4(d)
3	$((\neg \varphi) \Rightarrow (\neg(\neg(\varphi \Rightarrow \psi))))$	HS 1,2
4	$(((\neg \varphi) \Rightarrow (\neg(\neg(\varphi \Rightarrow \psi)))) \Rightarrow ((\neg(\varphi \Rightarrow \psi)) \Rightarrow \varphi))$	Ax3
5	$((\neg(\varphi \Rightarrow \psi)) \Rightarrow \varphi)$	MP 3,4

$\triangledown$

Exercise 6 Show that $\vdash_\Sigma ((\neg(\varphi \Rightarrow \psi)) \Rightarrow (\neg \psi))$.

Solution: Consider the following derivation sequence:

1	$(\psi \Rightarrow (\varphi \Rightarrow \psi))$	Ax1
2	$((\psi \Rightarrow (\varphi \Rightarrow \psi)) \Rightarrow ((\neg(\varphi \Rightarrow \psi)) \Rightarrow (\neg \psi)))$	Thm 5.4(f)
3	$((\neg(\varphi \Rightarrow \psi)) \Rightarrow (\neg \psi))$	MP 1,2

$\triangledown$

Exercise 7 Show that every set of formulas that has a model is consistent.

Solution: Let I be a model of Γ. It will be shown by contradiction that Γ is consistent.

Suppose that Γ is not consistent. Then, for every closed formula φ, the following hold:

$$\begin{cases} \Gamma \vdash_\Sigma \varphi \\ \Gamma \vdash_\Sigma (\neg \varphi). \end{cases}$$

From the soundness of the Hilbert calculus, it follows that:

$$\begin{cases} \Gamma \vDash_\Sigma \varphi \\ \Gamma \vDash_\Sigma (\neg \varphi). \end{cases}$$

Hence, using the fact that I is a model of Γ:

$$\begin{cases} I \Vdash_\Sigma \varphi \\ I \Vdash_\Sigma (\neg \varphi). \end{cases}$$

Therefore, for every assignment ρ into I, one has:

$$\begin{cases} I\rho \Vdash_\Sigma \varphi \\ I\rho \Vdash_\Sigma (\neg \varphi) \end{cases}$$

whence

$$\begin{cases} I\rho \Vdash_\Sigma \varphi \\ I\rho \not\Vdash_\Sigma \varphi \end{cases}$$

which is absurd. $\triangledown$

Exercise 8 Show that entailment is compact, that is,

$$\Gamma^{\vDash_\Sigma} = \bigcup_{\Phi \in \wp_{\mathrm{fin}}\Gamma} \Phi^{\vDash_\Sigma}.$$

Solution:

$$\Gamma^{\vDash_\Sigma} = \Gamma^{\vdash_\Sigma} \qquad \text{(soundness and completeness)}$$

$$= \bigcup_{\Phi \in \wp_{\mathrm{fin}}\Gamma} \Phi^{\vdash_\Sigma} \quad \text{(compactness } \vdash_\Sigma)$$

$$= \bigcup_{\Phi \in \wp_{\mathrm{fin}}\Gamma} \Phi^{\vDash_\Sigma} \quad \text{(soundness and completeness)}$$

$\triangledown$

Exercise 9 Justify or refute the following assertion:

If $\Gamma, p(x_1) \vdash_\Sigma q(x_1)$ then $\Gamma \vdash_\Sigma (\forall x_1 (p(x_1) \Rightarrow q(x_1)))$.

Solution: Thanks to the soundness and the completeness of FOL, if the assertion holds then so would

If $\Gamma, p(x_1) \vDash_\Sigma q(x_1)$ then $\Gamma \vDash_\Sigma (\forall x_1 (p(x_1) \Rightarrow q(x_1)))$ (†).

Indeed, from $\Gamma, p(x_1) \vDash_\Sigma q(x_1)$ by completeness one would get $\Gamma, p(x_1) \vdash_\Sigma q(x_1)$ and, so, $\Gamma \vdash_\Sigma (\forall x_1 (p(x_1) \Rightarrow q(x_1)))$ from whence by soundness one would conclude $\Gamma \vDash_\Sigma (\forall x_1 (p(x_1) \Rightarrow q(x_1)))$.

However, (†) does not hold in general. If p and q are distinct predicate symbols, one can find Γ such that $\Gamma, p(x_1) \vDash_\Sigma q(x_1)$ holds, but, on the other hand, $\Gamma \vDash_\Sigma (\forall x_1 (p(x_1) \Rightarrow q(x_1)))$ does not hold.

Indeed, take $\Gamma = \{((\forall x_1 p(x_1)) \Rightarrow (\forall x_1 q(x_1)))\}$. Then,

$$((\forall x_1 p(x_1)) \Rightarrow (\forall x_1 q(x_1))), p(x_1) \vDash_\Sigma q(x_1)$$

because (by soundness of MP)

$$((\forall x_1 p(x_1)) \Rightarrow (\forall x_1 q(x_1))), (\forall x_1 p(x_1)) \vDash_\Sigma (\forall x_1 q(x_1))$$

and (by soundness of generalization and instantiation)

$$\begin{cases} p(x_1) \vDash_\Sigma (\forall x_1\, p(x_1)); \\ (\forall x_1\, q(x_1)) \vDash_\Sigma q(x_1). \end{cases}$$

On the other hand,

$$((\forall x_1\, p(x_1)) \Rightarrow (\forall x_1\, q(x_1))) \nvDash_\Sigma (\forall x_1\, (p(x_1) \Rightarrow q(x_1)))$$

since one can find an I satisfying $((\forall x_1\, p(x_1)) \Rightarrow (\forall x_1\, q(x_1)))$ but not satisfying $(\forall x_1\, (p(x_1) \Rightarrow q(x_1)))$. Indeed, take the domain of I to be the set $\{a, b\}$ and let

$$\begin{cases} p_1^P(a) = q_1^P(b) = 1 \\ p_1^P(b) = q_1^P(a) = 0. \end{cases}$$

Clearly, I satisfies $((\forall x_1\, p(x_1)) \Rightarrow (\forall x_1\, q(x_1)))$ because $I \nVdash_\Sigma (\forall x_1\, p(x_1))$ since $I\rho \nVdash_\Sigma p(x_1)$ when $\rho(x_1) = b$. Furthermore, since $I\rho \Vdash_\Sigma p(x_1)$ and $I\rho \nVdash_\Sigma q(x_1)$ when $\rho(x_1) = a$, I does not satisfy $(\forall x_1\, (p(x_1) \Rightarrow q(x_1)))$.   ▽

Exercise 10 Assume that a formula is satisfied by every interpretation structure whose domain has cardinality greater than $\aleph_0 = \#\mathbb{N}$. Show that the formula is valid.

Solution: Assume that a formula is satisfied by every interpretation structure whose domain has cardinality greater than $\aleph_0 = \#\mathbb{N}$. Suppose, by contradiction, that φ is not a valid formula. Then, there is an interpretation structure J with domain E such that

$$\#E \leq \aleph_0 \text{ and } J \nVdash_\Sigma \varphi$$

and so

$$J \nVdash_\Sigma (\forall \varphi).$$

Hence, by the Lemma of the closed formula

$$J \Vdash_\Sigma (\neg(\forall \varphi)).$$

So, by the Skölem-Lowenheim Theorem, there is an interpretation structure I with domain D such that

$$\#D = \aleph_0 \text{ and } I \Vdash_\Sigma (\neg(\forall \varphi)).$$

Then, by the Cardinality Theorem,

$$I' \Vdash_\Sigma (\neg(\forall x\, \varphi)) \text{ for some } I' \text{ such that } D \subseteq D' \text{ and } \#D' > \aleph_0.$$

Thus, for some I' such that $D \subseteq D'$ and $\#D' > \aleph_0$, $I' \nVdash_\Sigma (\forall x\, \varphi)$ and, so, $I' \nVdash_\Sigma \varphi$, in contradiction with the hypothesis.   ▽

Chapter 21

Exercises on equality

This chapter complements Chapter 10.

Exercise 1 Consider the signature $\Sigma_{\cong}$ such that:

- $F_n = \emptyset$ for every $n \in \mathbb{N}$;

- $P_2 = \{\cong\}$;

- $P_n = \emptyset$ for every $n \neq 2$.

Let the *theory of equality* be the theory over $\Sigma_{\cong}$ with the following specific axioms:

Ref $(\forall x_1(x_1 \cong x_1))$;

Sym $(\forall x_1(\forall x_2((x_1 \cong x_2) \Rightarrow (x_2 \cong x_1))))$;

Trans $(\forall x_1(\forall x_2 (\forall x_3((x_1 \cong x_2) \Rightarrow ((x_2 \cong x_3) \Rightarrow (x_1 \cong x_3))))))$.

Show that the theory of equality is a theory with equality.

Solution: It must be shown that the theory of equality contains formula **E1**, the formulas of type **E2** and the formulas of type **E3**.

Note that formula **E1** is precisely formula **Ref**, and, since $F = \emptyset$, there are no formulas of type **E2** over the signature of the theory of equality.

Note also that $P = \{\cong\}$. Hence, it remains only to show that the theory of equality contains the following formula:

$$(\forall x_1(\forall x_2(\forall x_3(\forall x_4((x_1 \cong x_3) \Rightarrow ((x_2 \cong x_4) \Rightarrow ((x_1 \cong x_2) \Rightarrow (x_3 \cong x_4)))))))).$$

The first step is showing that, for any variables y_1 and y_2, not necessarily distinct, formula

$$\text{gSym} \quad ((y_1 \cong y_2) \Rightarrow (y_2 \cong y_1))$$

belongs to the theory of equality. Indeed, choosing an auxiliary variable z different from x_2, it is possible to build the following derivation sequence:

1	$(\forall x_1 (\forall x_2 ((x_1 \cong x_2) \Rightarrow (x_2 \cong x_1))))$	Sym
2	$((\forall x_1 (\forall x_2 ((x_1 \cong x_2) \Rightarrow (x_2 \cong x_1)))) \Rightarrow$	
	$\qquad\qquad (\forall x_2 ((z \cong x_2) \Rightarrow (x_2 \cong z))))$	Ax4
3	$(\forall x_2 ((z \cong x_2) \Rightarrow (x_2 \cong z)))$	MP 1,2
4	$((\forall x_2 ((z \cong x_2) \Rightarrow (x_2 \cong z))) \Rightarrow ((z \cong y_2) \Rightarrow (y_2 \cong z)))$	Ax4
5	$((z \cong y_2) \Rightarrow (y_2 \cong z))$	MP 3,4
6	$(\forall z ((z \cong y_2) \Rightarrow (y_2 \cong z)))$	Gen 5
7	$((\forall z ((z \cong y_2) \Rightarrow (y_2 \cong z))) \Rightarrow ((y_1 \cong y_2) \Rightarrow (y_2 \cong y_1)))$	Ax4
8	$((y_1 \cong y_2) \Rightarrow (y_2 \cong y_1))$	MP 6,7

Showing that formula

$$\text{gTrans} \quad ((y_1 \cong y_2) \Rightarrow ((y_2 \cong y_3) \Rightarrow (y_1 \cong y_3)))$$

belongs to the theory of equality, for any variables y_1, y_2 and y_3, not necessarily distinct, is left to the care of the reader.

Having in mind these results, consider the following derivation sequence:

1	$(x_1 \cong x_3)$	Hyp
2	$(x_2 \cong x_4)$	Hyp
3	$(x_1 \cong x_2)$	Hyp
4	$((x_1 \cong x_3) \Rightarrow (x_3 \cong x_1))$	gSym
5	$(x_3 \cong x_1)$	MP 1,4
6	$((x_3 \cong x_1) \Rightarrow ((x_1 \cong x_2) \Rightarrow (x_3 \cong x_2)))$	gTrans
7	$((x_1 \cong x_2) \Rightarrow (x_3 \cong x_2))$	MP 5,6
8	$(x_3 \cong x_2)$	MP 3,7
9	$((x_3 \cong x_2) \Rightarrow ((x_2 \cong x_4) \Rightarrow (x_3 \cong x_4)))$	gTrans
10	$((x_2 \cong x_4) \Rightarrow (x_3 \cong x_4))$	MP 8,9
11	$(x_3 \cong x_4)$	MP 2,10

Since this derivation does not contain any essential generalizations over dependents of the hypothesis, it follows, by repeated application of the MTD, that formula

$$((x_1 \cong x_3) \Rightarrow ((x_2 \cong x_4) \Rightarrow ((x_1 \cong x_2) \Rightarrow (x_3 \cong x_4))))$$

belongs to the theory of equality. Finally, the desired result is obtained applying generalizations. $\triangledown$

Exercise 2 Show that, given $I \in \text{Mod}_\Theta$, there is $h' : I \to I'$ in Mod_Θ such that:

- I' is an object of nMod_Θ;

- for every $g : I \to I''$ with I'' an object of nMod_Θ, there is a unique $g' : I' \to I''$ such that $g' \circ h' = g$.

Solution: Clearly, the candidates for I' and h' are $I/\cong$ and $\lambda d.\, [d]_{\cong_2^P}$, respectively. Assume that $g : I \to I''$, where I'' is a normal model of Θ, is an homomorphism. It must be shown that there is a unique homomorphism $g' : I/\cong \to I''$ such that $g' \circ h' = g$. Take

$$g' = \lambda [d]_{\cong_2^P} . g(d).$$

Observe that g' is well defined since $g(d_1) = g(d_2)$ whenever $d_1 \cong_2^P d_2$. The uniqueness of g' is obvious. It remains to prove that g is a homomorphism. Indeed, for every function symbol f in Σ:

$$
\begin{aligned}
g'(f_n^{\mathsf{F}/\cong}([d_1]_{\cong_2^P}, \ldots, [d_n]_{\cong_2^P})) &= g'([f_n^{\mathsf{F}}(d_1, \ldots, d_n)]_{\cong_2^P}) \\
&= g(f_n^{\mathsf{F}}(d_1, \ldots, d_n)) \\
&= f_n^{\mathsf{F}'}(g(d_1), \ldots, g(d_n)) \\
&= f_n^{\mathsf{F}'}((g' \circ h')(d_1), \ldots, (g' \circ h')(d_n)) \\
&= f_n^{\mathsf{F}'}(g'([d_1]_{\cong_2^P}), \ldots, g'([d_n]_{\cong_2^P}))
\end{aligned}
$$

The homomorphism condition is verified for each predicate symbol in a similar way. ▽

Exercise 3 Show that there is no theory with equality that admits normal models with domains of every finite cardinality and no normal models with infinite domains.

Solution: Assume that such a theory Θ exists. Let $\varphi^{\leq n}$ be a formula that solves question 1 of Exercise 9 in Chapter 10. Consider the set of formulas

$$\Theta \cup \{\varphi^{>n} : n \in \mathbb{N}\}$$

where $\varphi^{>n}$ is $(\neg \varphi^{\leq n})$. Then, use a compactness argument to reach a contradiction. ▽

Chapter 22

Exercises on arithmetic

The exercises solved herein complement the material in Chapters 11, 12, 13 and 14.

Exercise 1 Show that $[\![k]\!]_{\Sigma_N}^{N\rho} = k$ for every $k \in \mathbb{N}$.

Solution: The result is early proved by induction as follows:
(Basis) $k = 0$:

$$[\![0]\!]_{\Sigma_N}^{N\rho} = 0_0^{F_N} = 0.$$

(Step) $k = j + 1$:

$$[\![k]\!]_{\Sigma_N}^{N\rho} \overset{(a)}{=} [\![j']\!]_{\Sigma_N}^{N\rho} = '_1^{F_N}([\![j]\!]_{\Sigma_N}^{N\rho}) = [\![j]\!]_{\Sigma_N}^{N\rho} + 1 \overset{(b)}{=} j + 1 = k$$

since:

$$\overbrace{\hspace{2cm}}^{k \text{ times}} \qquad \overbrace{\hspace{2cm}}^{j \text{ times}}$$

(a) $\mathbf{k}$ is $\mathbf{j}'$ because $\mathbf{k}$ is $\mathbf{0}' \cdots '$, $\mathbf{j}$ is $\mathbf{0}' \cdots '$ and $k = j + 1$;

(b) $[\![j]\!]_{\Sigma_N}^{N\rho} = j$ by the induction hypothesis. $\qquad\qquad \triangledown$

Exercise 2 Show that $(\neg(x_1 \cong x_1'))$ is not a theorem of $\mathbf{N}$.

Solution: The proof is by contradiction, using soundness of first-order logic. Suppose that $(\neg(x_1 \cong x_1')) \in \mathbf{N}$. That is, assume that

$$\mathrm{Ax_N} \vdash_{\Sigma_N} (\neg(x_1 \cong x_1')).$$

Then, by the soundness of the Hilbert calculus,

$$\mathrm{Ax_N} \vDash_{\Sigma_N} (\neg(x_1 \cong x_1')).$$

and, so, for every interpretation structure I over $\Sigma_{\mathbb{N}}$,

$$(1)\quad \text{if } I \Vdash_{\Sigma_{\mathbb{N}}} \text{Ax}_{\mathbf{N}} \text{ then } I \Vdash_{\Sigma_{\mathbb{N}}} (\neg(x_1 \cong x_1')).$$

On the other hand, consider the interpretation structure

$$\overline{\mathbb{N}} = (\overline{\mathbb{N}}, \_^{F_{\overline{\mathbb{N}}}}, \_^{P_{\overline{\mathbb{N}}}})$$

over $\Sigma_{\mathbb{N}}$ such that:

- $\overline{\mathbb{N}} = \mathbb{N} \cup \{\infty\}$;

- $0_0^{F_{\overline{\mathbb{N}}}} = 0$;

- $'^{F_{\overline{\mathbb{N}}}}_1 = \lambda d \, . \begin{cases} d+1 & \text{if } d \in \mathbb{N} \\ \infty & \text{otherwise} \end{cases}$;

- $+_2^{F_{\overline{\mathbb{N}}}} = \lambda d_1, d_2 \, . \begin{cases} d_1 + d_2 & \text{if } d_1, d_2 \in \mathbb{N} \\ \infty & \text{otherwise} \end{cases}$;

- $\times_2^{F_{\overline{\mathbb{N}}}} = \lambda d_1, d_2 \, . \begin{cases} d_1 \times d_2 & \text{if } d_1, d_2 \in \mathbb{N}^+ \\ 0 & \text{if } d_1 = 0 \text{ or } d_2 = 0; \\ \infty & \text{otherwise} \end{cases}$

- $\cong_2^{P_{\overline{\mathbb{N}}}} = \lambda d_1, d_2 \, . \begin{cases} 1 & \text{if } d_1 = d_2 \\ 0 & \text{otherwise} \end{cases}$;

- $<_2^{P_{\overline{\mathbb{N}}}} = \lambda d_1, d_2 \, . \begin{cases} 1 & \text{if } d_1, d_2 \in \mathbb{N} \ \& \ d_1 < d_2 \\ 1 & \text{if } d_2 = \infty \\ 0 & \text{otherwise} \end{cases}$.

The reader should have no trouble in showing that:

$$\begin{cases} (2) \ \overline{\mathbb{N}} \Vdash_{\Sigma_{\mathbb{N}}} \text{Ax}_{\mathbf{N}} \\ (3) \ \overline{\mathbb{N}} \not\Vdash_{\Sigma_{\mathbb{N}}} (\neg(x_1 \cong x_1')). \end{cases}$$

But from (1) and (2) it follows that $\overline{\mathbb{N}} \Vdash_{\Sigma_{\mathbb{N}}} (\neg(x_1 \cong x_1'))$, contradicting assertion (3). ∇

Exercise 3 Show that $(\neg(x_1 \cong x_1'))$ is a theorem of **P**.

Solution: The thesis follows by Ax4 and MP from $(\forall x_1(\neg(x_1 \cong x_1'))) \in \mathbf{P}$, and this assertion is proved, using the induction principle of $\mathbf{P}$, as follows:

(i) $[(\neg(x_1 \cong x_1'))]_{\mathbf{0}}^{x_1} \in \mathbf{P}$, since

$$[(\neg(x_1 \cong x_1'))]_{\mathbf{0}}^{x_1} \text{ is } (\neg(\mathbf{0} \cong \mathbf{0}'))$$

and this formula is derivable from $\mathrm{Ax_P}$:

1	$(\forall x_1(\neg(x_1' \cong \mathbf{0})))$	N1
2	$((\forall x_1(\neg(x_1' \cong \mathbf{0}))) \Rightarrow (\neg(\mathbf{0}' \cong \mathbf{0})))$	Ax4
3	$(\neg(\mathbf{0}' \cong \mathbf{0}))$	MP 1,2
4	$((\mathbf{0} \cong \mathbf{0}') \Rightarrow (\mathbf{0}' \cong \mathbf{0}))$	RE2
5	$(((\mathbf{0} \cong \mathbf{0}') \Rightarrow (\mathbf{0}' \cong \mathbf{0})) \Rightarrow ((\neg(\mathbf{0}' \cong \mathbf{0})) \Rightarrow (\neg(\mathbf{0} \cong \mathbf{0}'))))$	Thm 5.4(f)
6	$((\neg(\mathbf{0}' \cong \mathbf{0})) \Rightarrow (\neg(\mathbf{0} \cong \mathbf{0}')))$	MP 4,5
7	$(\neg(\mathbf{0} \cong \mathbf{0}'))$	MP 3,6

(ii) $(\forall x_1((\neg(x_1 \cong x_1')) \Rightarrow [(\neg(x_1 \cong x_1'))]_{x_1'}^{x_1})) \in \mathbf{P}$, since

$$(\forall x_1((\neg(x_1 \cong x_1')) \Rightarrow [(\neg(x_1 \cong x_1'))]_{x_1'}^{x_1}))$$

is

$$(\forall x_1((\neg(x_1 \cong x_1')) \Rightarrow (\neg(x_1' \cong x_1''))))$$

and this formula is derivable from $\mathrm{Ax_P}$ as follows:

1	$(\forall x_1(\forall x_2((x_1' \cong x_2') \Rightarrow (x_1 \cong x_2))))$	N2
2	$((\forall x_1(\forall x_2((x_1' \cong x_2') \Rightarrow (x_1 \cong x_2)))) \Rightarrow (\forall x_2((x_1' \cong x_2') \Rightarrow (x_1 \cong x_2))))$	Ax4
3	$(\forall x_2((x_1' \cong x_2') \Rightarrow (x_1 \cong x_2)))$	MP 1,2
4	$((\forall x_2((x_1' \cong x_2') \Rightarrow (x_1 \cong x_2))) \Rightarrow ((x_1' \cong x_1'') \Rightarrow (x_1 \cong x_1')))$	Ax4
5	$((x_1' \cong x_1'') \Rightarrow (x_1 \cong x_1'))$	MP 3,4
6	$(((x_1' \cong x_1'') \Rightarrow (x_1 \cong x_1')) \Rightarrow ((\neg(x_1 \cong x_1')) \Rightarrow (\neg(x_1' \cong x_1''))))$	Thm 5.4(f)
7	$((\neg(x_1 \cong x_1')) \Rightarrow (\neg(x_1' \cong x_1'')))$	MP 5,6
8	$(\forall x_1((\neg(x_1 \cong x_1')) \Rightarrow (\neg(x_1' \cong x_1''))))$	Gen 7

(iii) Finally, the derivation sequence

1	$[(\neg(x_1 \cong x_1'))]_{\mathbf{0}}^{x_1}$	Thm (i)
2	$(\forall x_1((\neg(x_1 \cong x_1')) \Rightarrow [(\neg(x_1 \cong x_1'))]_{x_1'}^{x_1}))$	Thm (ii)
3	$([[(\neg(x_1 \cong x_1'))]_{\mathbf{0}}^{x_1} \Rightarrow ((\forall x_1((\neg(x_1 \cong x_1')) \Rightarrow [(\neg(x_1 \cong x_1'))]_{x_1'}^{x_1})) \Rightarrow (\forall x_1(\neg(x_1 \cong x_1'))))$	Ind
4	$((\forall x_1((\neg(x_1 \cong x_1')) \Rightarrow [(\neg(x_1 \cong x_1'))]_{x_1'}^{x_1})) \Rightarrow (\forall x_1(\neg(x_1 \cong x_1'))))$	MP 1,3
5	$(\forall x_1(\neg(x_1 \cong x_1')))$	MP 2,4

establishes that $(\forall x_1(\neg(x_1 \cong x_1'))) \in \mathbf{P}$. ∇

Exercise 4 Show that $\mathbb{N} \Vdash_{\Sigma_{\mathbb{N}}} ([\varphi]_0^x \Rightarrow ((\forall x(\varphi \Rightarrow [\varphi]_{x'}^x)) \Rightarrow (\forall x\, \varphi)))$.

Solution: The proof is by contradiction as follows. Assume that

$$\mathbb{N} \nVdash_{\Sigma_{\mathbb{N}}} ([\varphi]_0^x \Rightarrow ((\forall x(\varphi \Rightarrow [\varphi]_{x'}^x)) \Rightarrow (\forall x\, \varphi))).$$

Then, there exists ρ over $\mathbb{N}$ such that

$$\mathbb{N}\rho \nVdash_{\Sigma_{\mathbb{N}}} ([\varphi]_0^x \Rightarrow ((\forall x(\varphi \Rightarrow [\varphi]_{x'}^x)) \Rightarrow (\forall x\, \varphi))),$$

whence

(1) $\mathbb{N}\rho \Vdash_{\Sigma_{\mathbb{N}}} [\varphi]_0^x$;

(2) $\mathbb{N}\rho \Vdash_{\Sigma_{\mathbb{N}}} (\forall x(\varphi \Rightarrow [\varphi]_{x'}^x))$;

(3) $\mathbb{N}\rho \nVdash_{\Sigma_{\mathbb{N}}} (\forall x\, \varphi)$.

From (3), it follows that there is an assignment σ that is x-equivalent to ρ such that $\mathbb{N}\sigma \nVdash_{\Sigma_{\mathbb{N}}} \varphi$. Hence, the set

$$C_\rho = \{\sigma : \sigma \text{ is } x\text{-equivalent to } \rho \ \& \ \mathbb{N}\sigma \nVdash_{\Sigma_{\mathbb{N}}} \varphi\}$$

is non empty. Let $\sigma' \in C_\rho$ be the assignment such that

$$\sigma'(x) \le \sigma(x), \text{ for any } \sigma \in C_\rho.$$

In other words, σ' is the assignment in C_ρ that assigns the least value to variable x. Then,

$$(3.1)\ \mathbb{N}\sigma' \nVdash_{\Sigma_{\mathbb{N}}} \varphi$$

Using Exercise 1 in Chapter 11, it follows that

$$(3.2)\ [\![\,\overline{\sigma'(x)}\,]\!]_{\Sigma_{\mathbb{N}}}^{\mathbb{N}\rho} = \sigma'(x).$$

Furthermore, by choice of σ', for every assignment v that is x-equivalent to σ' (and hence to ρ) and such that $v(x) < \sigma'(x)$, the following holds:

$$(3.3)\ \mathbb{N}v \Vdash_{\Sigma_{\mathbb{N}}} \varphi.$$

Taking into account[1] that, for every $j \in \mathbb{N}$, either $j = 0$ or there exists $k \in \mathbb{N}$ such that $j = k + 1$, the proof proceeds by case analysis on the value of $\sigma'(x)$:

(i) $\sigma'(x) = 0$:

---

[1]Which is easily shown by structural induction over the natural numbers.

From (3.1) and (3.2), using the lemma of substitution in formula, it follows that

$$\mathbb{N}\rho \not\Vdash_{\Sigma_\mathbb{N}} [\varphi]_0^x$$

contradicting (1).

(ii) $\sigma'(x) = k + 1$:

Let σ'' be the assignment that is x-equivalent to σ' (and hence to ρ) and such that $\sigma''(x) = k$. Then,

$$\sigma'(x) = k + 1 = [\![x']\!]_{\Sigma_\mathbb{N}}^{\mathbb{N}\sigma''} .$$

Hence, by (3.1), again invoking the lemma of substitution in formula, it follows that

$$(4.1) \quad \mathbb{N}\sigma'' \not\Vdash_{\Sigma_\mathbb{N}} [\varphi]_{x'}^x .$$

On the other hand, since $\sigma''(x) < \sigma'(x)$, (3.3) yields

$$(4.2) \quad \mathbb{N}\sigma'' \Vdash_{\Sigma_\mathbb{N}} \varphi .$$

From (4.1) and (4.2) one obtains

$$(4.1) \quad \mathbb{N}\sigma'' \not\Vdash_{\Sigma_\mathbb{N}} (\varphi \Rightarrow [\varphi]_{x'}^x) ,$$

whence, since σ'' is x-equivalent to ρ,

$$\mathbb{N}\rho \not\Vdash_{\Sigma_\mathbb{N}} (\forall x(\varphi \Rightarrow [\varphi]_{x'}^x)) ,$$

contradicting (2). $\qquad\qquad\qquad\qquad\qquad\qquad\qquad \triangledown$

Exercise 5 Show that every constant $k \in \mathbb{N}$ is representable within $Th(\mathbb{N})$.

Solution: The formula $(x_1 \cong k)$ represents constant k, that is, the map $k :\rightarrow \mathbb{N}$, or yet, the map $k : \mathbb{N}^0 \rightarrow \mathbb{N}^1$, since $fv_{\Sigma_\mathbb{N}}((x_1 \cong k)) = \{x_1\}$ and

$$((x_1 \cong k) \Leftrightarrow (x_1 \cong k)) \in Th(\mathbb{N}),$$

that is,

$$\mathbb{N} \Vdash_{\Sigma_\mathbb{N}} ((x_1 \cong k) \Leftrightarrow (x_1 \cong k)) .$$

Indeed, $((x_1 \cong k) \Leftrightarrow (x_1 \cong k))$ is a tautological formula, and hence, by the lemma of tautological formula in Chapter 7, it is valid. Therefore, it is satisfied by every interpretation structure over $\Sigma_\mathbb{N}$, hence in particular by $\mathbb{N}$. $\qquad \triangledown$

Exercise 6 Show that the map $Z : \mathbb{N} \rightarrow \mathbb{N}$ is representable within $Th(\mathbb{N})$.

Solution: The formula $((x_2 \cong \mathbf{0}) \wedge (x_1 \cong x_1))$ represents the map Z, as shown below.

Since $\mathsf{Z}(a) = 0$ for every $a \in \mathbb{N}$, it must be shown that

$$([((x_2 \cong \mathbf{0}) \wedge (x_1 \cong x_1))]_{\mathbf{a}}^{x_1} \Leftrightarrow (x_2 \cong \mathbf{0})) \in \mathrm{Th}(\mathbb{N}),$$

that is,

$$\mathbb{N} \Vdash_{\Sigma_{\mathbb{N}}} (((x_2 \cong \mathbf{0}) \wedge (\mathbf{a} \cong \mathbf{a})) \Leftrightarrow (x_2 \cong \mathbf{0})).$$

Indeed:

$$\mathbb{N} \Vdash_{\Sigma_{\mathbb{N}}} (((x_2 \cong \mathbf{0}) \wedge (\mathbf{a} \cong \mathbf{a})) \Leftrightarrow (x_2 \cong \mathbf{0})) \qquad\text{iff}$$
$$\text{for every } \rho, \mathbb{N}\rho \Vdash_{\Sigma_{\mathbb{N}}} (((x_2 \cong \mathbf{0}) \wedge (\mathbf{a} \cong \mathbf{a})) \Leftrightarrow (x_2 \cong \mathbf{0})) \qquad\text{iff}$$
$$\text{for every } \rho, (\rho(x_2) = 0 \text{ and } a = a) \text{ if and only if } (\rho(x_2) = 0) \qquad\text{iff}$$
$$\text{true}$$

$$\nabla$$

Exercise 7 Check that the map $\lambda k.\, k + 1$ is represented within $\mathrm{Th}(\mathbb{N})$ by the formula $(x_2 \cong x_1')$.

Solution: Let a, b be natural numbers with $b = a + 1$. It must be shown that

$$([(x_2 \cong x_1')]_{\mathbf{a}}^{x_1} \Leftrightarrow (x_2 \cong \mathbf{b})) \in \mathrm{Th}(\mathbb{N}),$$

that is,

$$((x_2 \cong \mathbf{a}') \Leftrightarrow (x_2 \cong \mathbf{b})) \in \mathrm{Th}(\mathbb{N}),$$

or yet

$$\mathbb{N} \Vdash_{\Sigma_{\mathbb{N}}} ((x_2 \cong \mathbf{a}') \Leftrightarrow (x_2 \cong \mathbf{b})),$$

that is, for every assignment ρ over $\mathbb{N}$,

$$\mathbb{N}\rho \Vdash_{\Sigma_{\mathbb{N}}} ((x_2 \cong \mathbf{a}') \Leftrightarrow (x_2 \cong \mathbf{b})).$$

Indeed:

$$\mathbb{N}\rho \Vdash_{\Sigma_{\mathbb{N}}} ((x_2 \cong \mathbf{a}') \Leftrightarrow (x_2 \cong \mathbf{b})) \qquad\begin{array}{l}\text{iff}\\ \text{(Chap. 19, Exer. 7)}\end{array}$$

$$\begin{cases}\mathbb{N}\rho \Vdash_{\Sigma_{\mathbb{N}}} (x_2 \cong \mathbf{a}') \\ \mathbb{N}\rho \Vdash_{\Sigma_{\mathbb{N}}} (x_2 \cong \mathbf{b})\end{cases} \text{or} \begin{cases}\mathbb{N}\rho \nVdash_{\Sigma_{\mathbb{N}}} (x_2 \cong \mathbf{a}') \\ \mathbb{N}\rho \nVdash_{\Sigma_{\mathbb{N}}} (x_2 \cong \mathbf{b})\end{cases} \qquad\text{iff}$$

$$\begin{cases} \rho(x_2) = \llbracket \mathbf{a'} \rrbracket^{\mathbb{N}\rho}_{\Sigma_{\mathbb{N}}} \\ \rho(x_2) = \llbracket \mathbf{b} \rrbracket^{\mathbb{N}\rho}_{\Sigma_{\mathbb{N}}} \end{cases} \text{or} \quad \begin{cases} \rho(x_2) \neq \llbracket \mathbf{a'} \rrbracket^{\mathbb{N}\rho}_{\Sigma_{\mathbb{N}}} \\ \rho(x_2) \neq \llbracket \mathbf{b} \rrbracket^{\mathbb{N}\rho}_{\Sigma_{\mathbb{N}}} \end{cases} \qquad \text{iff}$$

$$\begin{cases} \rho(x_2) = \llbracket \mathbf{a} \rrbracket^{\mathbb{N}\rho}_{\Sigma_{\mathbb{N}}} + 1 \\ \rho(x_2) = \llbracket \mathbf{b} \rrbracket^{\mathbb{N}\rho}_{\Sigma_{\mathbb{N}}} \end{cases} \text{or} \quad \begin{cases} \rho(x_2) \neq \llbracket \mathbf{a} \rrbracket^{\mathbb{N}\rho}_{\Sigma_{\mathbb{N}}} + 1 \\ \rho(x_2) \neq \llbracket \mathbf{b} \rrbracket^{\mathbb{N}\rho}_{\Sigma_{\mathbb{N}}} \end{cases} \qquad \text{iff}$$

$$\begin{cases} \rho(x_2) = a + 1 \\ \rho(x_2) = b \end{cases} \text{or} \quad \begin{cases} \rho(x_2) \neq a + 1 \\ \rho(x_2) \neq b \end{cases} \qquad \begin{aligned} &\text{iff} \\ &(\text{since } b = a + 1) \end{aligned}$$

$$\begin{cases} \rho(x_2) = a + 1 \\ \rho(x_2) = a + 1 \end{cases} \text{or} \quad \begin{cases} \rho(x_2) \neq a + 1 \\ \rho(x_2) \neq a + 1 \end{cases} \qquad \text{iff}$$

$$\rho(x_2) = a + 1 \ \text{ or } \ \rho(x_2) \neq a + 1 \qquad\qquad \text{iff}$$

true

$$\triangledown$$

Exercise 8 Present a formula γ representing the map

$$g = \lambda\, a, b, c\,.\, \langle a, b + c \rangle : \mathbb{N}^3 \to \mathbb{N}^2$$

in Th($\mathbb{N}$). Justify.

Solution: Consider the formula

$$\gamma = ((x_4 \cong x_1) \wedge (x_5 \cong (x_2 + x_3))).$$

This formula is as required because

$$\mathrm{fv}_{\Sigma_{\mathbb{N}}}(\gamma) = \{x_1, x_2, x_3, x_4, x_5\}$$

(taking into account that g has input arity 3 and output arity 2) and

$$\frac{g(a, b, c) = (r, s)}{([\gamma]^{x_1 x_2 x_3}_{\mathbf{a,b,c}} \Leftrightarrow ((x_4 \cong \mathbf{r}) \wedge (x_5 \cong \mathbf{s}))) \in \mathrm{Th}(\mathbb{N})}.$$

Indeed,

$$\frac{g(a, b, c) = (r, s)}{\mathbb{N} \Vdash_{\Sigma_{\mathbb{N}}} ([\gamma]^{x_1 x_2 x_3}_{\mathbf{a,b,c}} \Leftrightarrow ((x_4 \cong \mathbf{r}) \wedge (x_5 \cong \mathbf{s})))}$$

as established in the following.

Assume that $g(a, b, c) = (r, s)$. Then, $r = a$ and $s = b + c$. Therefore, for every assignment ρ,

$$\mathbb{N}\rho \Vdash_{\Sigma_{\mathbb{N}}} ([\gamma]^{x_1 x_2 x_3}_{\mathbf{a,b,c}} \Leftrightarrow ((x_4 \cong \mathbf{r}) \wedge (x_5 \cong \mathbf{s})))$$

iff
$$\mathbb{N}\rho \Vdash_{\Sigma_{\mathrm{N}}} (((x_4 \cong \mathbf{a}) \wedge (x_5 \cong (\mathbf{b}+\mathbf{c}))) \Leftrightarrow ((x_4 \cong \mathbf{r}) \wedge (x_5 \cong \mathbf{s})))$$

iff
$$\begin{cases} \mathbb{N}\rho \Vdash_{\Sigma_{\mathrm{N}}} ((x_4 \cong \mathbf{a}) \wedge (x_5 \cong (\mathbf{b}+\mathbf{c}))) \\ \mathbb{N}\rho \Vdash_{\Sigma_{\mathrm{N}}} ((x_4 \cong \mathbf{r}) \wedge (x_5 \cong \mathbf{s})) \end{cases}$$
$$\text{or}$$
$$\begin{cases} \mathbb{N}\rho \not\Vdash_{\Sigma_{\mathrm{N}}} ((x_4 \cong \mathbf{a}) \wedge (x_5 \cong (\mathbf{b}+\mathbf{c}))) \\ \mathbb{N}\rho \not\Vdash_{\Sigma_{\mathrm{N}}} ((x_4 \cong \mathbf{r}) \wedge (x_5 \cong \mathbf{s})) \end{cases}$$

iff
$$\begin{cases} [\![x_4]\!]^{\mathbb{N}\rho}_{\Sigma_{\mathrm{N}}} = [\![\mathbf{a}]\!]^{\mathbb{N}\rho}_{\Sigma_{\mathrm{N}}} \\ [\![x_5]\!]^{\mathbb{N}\rho}_{\Sigma_{\mathrm{N}}} = [\![\mathbf{b}+\mathbf{c}]\!]^{\mathbb{N}\rho}_{\Sigma_{\mathrm{N}}} \\ [\![x_4]\!]^{\mathbb{N}\rho}_{\Sigma_{\mathrm{N}}} = [\![\mathbf{r}]\!]^{\mathbb{N}\rho}_{\Sigma_{\mathrm{N}}} \\ [\![x_5]\!]^{\mathbb{N}\rho}_{\Sigma_{\mathrm{N}}} = [\![\mathbf{s}]\!]^{\mathbb{N}\rho}_{\Sigma_{\mathrm{N}}} \end{cases}$$
$$\text{or}$$
$$\begin{cases} [\![x_4]\!]^{\mathbb{N}\rho}_{\Sigma_{\mathrm{N}}} \neq [\![\mathbf{a}]\!]^{\mathbb{N}\rho}_{\Sigma_{\mathrm{N}}} \text{ or } [\![x_5]\!]^{\mathbb{N}\rho}_{\Sigma_{\mathrm{N}}} \neq [\![\mathbf{b}+\mathbf{c}]\!]^{\mathbb{N}\rho}_{\Sigma_{\mathrm{N}}} \\ [\![x_4]\!]^{\mathbb{N}\rho}_{\Sigma_{\mathrm{N}}} \neq [\![\mathbf{r}]\!]^{\mathbb{N}\rho}_{\Sigma_{\mathrm{N}}} \text{ or } [\![x_5]\!]^{\mathbb{N}\rho}_{\Sigma_{\mathrm{N}}} \neq [\![\mathbf{s}]\!]^{\mathbb{N}\rho}_{\Sigma_{\mathrm{N}}} \end{cases}$$

iff (since $[\![\mathbf{k}]\!]^{\mathbb{N}\rho}_{\Sigma_{\mathrm{N}}} = k$ for every $k \in \mathbb{N}$)

$$\rho(x_4) = a \text{ and } \rho(x_5) = b+c \text{ and } \rho(x_4) = r \text{ and } \rho(x_5) = s$$
$$\text{or}$$
$$(\rho(x_4) \neq a \text{ or } \rho(x_5) \neq b+c) \text{ and } (\rho(x_4) \neq r \text{ or } \rho(x_5) \neq s)$$

iff (since $r = a$ and $s = b+c$)

$$\rho(x_4) = a \text{ and } \rho(x_5) = b+c \text{ and } \rho(x_4) = a \text{ and } \rho(x_5) = b+c$$
$$\text{or}$$
$$(\rho(x_4) \neq a \text{ or } \rho(x_5) \neq b+c) \text{ and } (\rho(x_4) \neq a \text{ or } \rho(x_5) \neq b+c)$$

iff

$$\rho(x_4) = a \text{ and } \rho(x_5) = b+c$$
$$\text{or}$$
$$(\rho(x_4) \neq a \text{ or } \rho(x_5) \neq b+c)$$

iff

$$\rho(x_4) = a \text{ and } \rho(x_5) = b+c$$
$$\text{or}$$
$$\text{not } (\rho(x_4) = a \text{ and } \rho(x_5) = b+c)$$

iff

$$\text{true.}$$

$$\nabla$$

Exercise 9 Check that the map $\lambda k.\, k + 1$ is represented within **N** by the formula $(x_2 \cong (x_1 + 1))$.

Solution: Let a, b be natural numbers with $b = a + 1$. It must be shown

$$([(x_2 \cong (x_1 + 1))]_{\mathbf{a}}^{x_1} \Leftrightarrow (x_2 \cong \mathbf{b})) \in \mathbf{N},$$

that is,

$$((x_2 \cong (\mathbf{a} + 1)) \Leftrightarrow (x_2 \cong \mathbf{b})) \in \mathbf{N},$$

First, observe that $\mathbf{a}'$ coincides with $\mathbf{b}$, that is, $\mathbf{a}'$ and $\mathbf{b}$ are abbreviations of the same numeral. Indeed, $\mathbf{a}$ abbreviates

$$\overbrace{\mathbf{0}\ '\cdots\ '}^{a \text{ times}}$$

whence $\mathbf{a}'$ is an abbreviation of

$$\overbrace{\mathbf{0}\ '\cdots\ '}^{a+1 \text{ times}}$$

which is precisely the numeral

$$\overbrace{\mathbf{0}\ '\cdots\ '}^{b \text{ times}}$$

that is abbreviated to $\mathbf{b}$.

Next, it is shown that $((\mathbf{a} + 1) \cong \mathbf{b}) \in \mathbf{N}$, that is, that

$$((\mathbf{a} + \mathbf{0}') \cong \mathbf{a}') \in \mathbf{N},$$

since 1 is an abbreviation of $\mathbf{0}'$ and it has been seen that $\mathbf{b}$ and $\mathbf{a}'$ coincide. To this end, consider the following derivation sequence:

1	$(\forall x_1 (\forall x_2 ((x_1 + x_2') \cong (x_1 + x_2)')))$	N4
2	$((\forall x_1 (\forall x_2 ((x_1 + x_2') \cong (x_1 + x_2)'))) \Rightarrow$	
	$\qquad ((\mathbf{a} + \mathbf{0}') \cong (\mathbf{a} + \mathbf{0})'))$	Ax4
3	$((\mathbf{a} + \mathbf{0}') \cong (\mathbf{a} + \mathbf{0})')$	MP 1,2
4	$(\forall x_1 ((x_1 + \mathbf{0}) \cong x_1))$	N3
5	$((\forall x_1 ((x_1 + \mathbf{0}) \cong x_1)) \Rightarrow ((\mathbf{a} + \mathbf{0}) \cong \mathbf{a}))$	Ax4
6	$((\mathbf{a} + \mathbf{0}) \cong \mathbf{a})$	MP 4,5
7	$(\forall x_1 (\forall x_2 ((x_1 \cong x_2) \Rightarrow (x_1' \cong x_2'))))$	E2a
8	$((\forall x_1 (\forall x_2 ((x_1 \cong x_2) \Rightarrow (x_1' \cong x_2')))) \Rightarrow$	
	$\qquad (\forall x_2 (((\mathbf{a} + \mathbf{0}) \cong x_2) \Rightarrow ((\mathbf{a} + \mathbf{0})' \cong x_2'))))$	Ax4
9	$(\forall x_2 (((\mathbf{a} + \mathbf{0}) \cong x_2) \Rightarrow ((\mathbf{a} + \mathbf{0})' \cong x_2')))$	MP 7,8
10	$((\forall x_2 (((\mathbf{a} + \mathbf{0}) \cong x_2) \Rightarrow ((\mathbf{a} + \mathbf{0})' \cong x_2'))) \Rightarrow$	
	$\qquad (((\mathbf{a} + \mathbf{0}) \cong \mathbf{a}) \Rightarrow ((\mathbf{a} + \mathbf{0})' \cong \mathbf{a}')))$	Ax4
11	$(((\mathbf{a} + \mathbf{0}) \cong \mathbf{a}) \Rightarrow ((\mathbf{a} + \mathbf{0})' \cong \mathbf{a}'))$	MP 9,10
12	$((\mathbf{a} + \mathbf{0})' \cong \mathbf{a}')$	MP 6,11

Finally, the intended result is a consequence of the principle of substitution by equal (Chapter 10, Exercise 5):

1	$((a+1) \cong b)$	$\in \mathbf{N}$
2	$((x_3 \cong x_4) \Rightarrow ((x_2 \cong x_3) \Rightarrow (x_2 \cong x_4)))$	PSE
3	$(\forall x_3((x_3 \cong x_4) \Rightarrow ((x_2 \cong x_3) \Rightarrow (x_2 \cong x_4))))$	Gen 2
4	$((\forall x_3((x_3 \cong x_4) \Rightarrow ((x_2 \cong x_3) \Rightarrow (x_2 \cong x_4)))) \Rightarrow$	
	$\quad (((a+1) \cong x_4) \Rightarrow ((x_2 \cong (a+1)) \Rightarrow (x_2 \cong x_4))))$	Ax4
5	$(((a+1) \cong x_4) \Rightarrow ((x_2 \cong (a+1)) \Rightarrow (x_2 \cong x_4)))$	MP 3,4
6	$(\forall x_4(((a+1) \cong x_4) \Rightarrow ((x_2 \cong (a+1)) \Rightarrow (x_2 \cong x_4))))$	Gen 5
7	$((\forall x_4(((a+1) \cong x_4) \Rightarrow ((x_2 \cong (a+1)) \Rightarrow (x_2 \cong x_4))))$	
	$\quad \Rightarrow$	
	$\quad (((a+1) \cong b) \Rightarrow ((x_2 \cong (a+1)) \Rightarrow (x_2 \cong b))))$	Ax4
8	$(((a+1) \cong b) \Rightarrow ((x_2 \cong (a+1)) \Rightarrow (x_2 \cong b)))$	MP 6,7
9	$((x_2 \cong (a+1)) \Rightarrow (x_2 \cong b))$	MP 1,8

$$\nabla$$

Exercise 10 Assume that $f : \mathbb{N} \to \mathbb{N}$ is representable in $\mathrm{Th}(\mathbb{N})$ by formula φ. Provide a formula γ representing the map

$$g = \lambda k . f(k+1)$$

in $\mathrm{Th}(\mathbb{N})$ and verify that it does so.

Solution: Take γ to be $[\varphi]_{x_1+\bar{1}}^{x_1}$. In order to show that $[\varphi]_{x_1+\bar{1}}^{x_1}$ does represent g in $\mathrm{Th}(\mathbb{N})$ one has to verify that, for arbitrary $a, b \in \mathbb{N}$, if

$$g(a) = b \quad (*)$$

then

$$([[\varphi]_{x_1+\bar{1}}^{x_1}]_{\bar{a}}^{x_1} \Leftrightarrow (x_2 \cong \bar{b})) \in \mathrm{Th}(\mathbb{N}) \quad (**).$$

Indeed, $(*)$ can be restated as

$$f(a+1) = b$$

from whence one gets

$$([\varphi]_{\overline{a+1}}^{x_1} \Leftrightarrow (x_2 \cong \bar{b})) \in \mathrm{Th}(\mathbb{N})$$

using the fact that φ represents f in $\mathrm{Th}(\mathbb{N})$. Therefore, invoking the PSE (principle of substitution by equal) instantiating x to $\overline{a+1}$ and y to $(\bar{a}+\bar{1})$, one obtains

$$([\varphi]_{\bar{a}+\bar{1}}^{x_1} \Leftrightarrow (x_2 \cong \bar{b})) \in \mathrm{Th}(\mathbb{N}) \quad (\dagger)$$

since
$$(\overline{a+1} \cong (\overline{a} + \overline{1})) \in \mathrm{Th}(\mathbb{N}) \quad (\ddagger)$$

as can easily be shown (see below).

Finally, one gets (∗∗) from (†) taking into account that

$$[\varphi]^{x_1}_{\overline{a+1}} \text{ is } [[\varphi]^{x_1}_{x_1+\overline{1}}]^{x_1}_{\overline{a}}.$$

It remains to prove (‡). Recall that

$$[\![\overline{k}]\!]^{N\rho}_{\Sigma_{\mathbb{N}}} = k.$$

Therefore,

$$[\![\overline{a+1}]\!]^{N\rho}_{\Sigma_{\mathbb{N}}} = a + 1 = [\![\overline{a}]\!]^{N\rho}_{\Sigma_{\mathbb{N}}} + [\![\overline{1}]\!]^{N\rho}_{\Sigma_{\mathbb{N}}} = [\![\overline{a} + \overline{1}]\!]^{N\rho}_{\Sigma_{\mathbb{N}}}$$

and, so,

$$\mathbb{N} \Vdash_{\Sigma_{\mathbb{N}}} (\overline{a+1} \cong (\overline{a} + \overline{1}))$$

as envisaged. ▽

Exercise 11 Suppose that the maps $f : \mathbb{N} \to \mathbb{N}$ and $g : \mathbb{N} \to \mathbb{N}$ are representable within $\mathrm{Th}(\mathbb{N})$. Show that their composition $g \circ f$ is also representable within $\mathrm{Th}(\mathbb{N})$.

Solution: Suppose that formulas φ_f and φ_g represent f and g, respectively. Observe that $\mathrm{fv}_{\Sigma_{\mathbb{N}}}(\varphi_f) = \mathrm{fv}_{\Sigma_{\mathbb{N}}}(\varphi_g) = \{x_1, x_2\}$.

Formula

$$(\exists x_3 ([\varphi_g]^{x_1}_{x_3} \wedge [\varphi_f]^{x_2}_{x_3})),$$

abbreviated here to $\varphi_{g \circ f}$, represents $g \circ f$, as shown below.

Suppose that $(g \circ f)(a) = b$. It must be shown that

$$([\varphi_{g \circ f}]^{x_1}_{\mathbf{a}} \Leftrightarrow (x_2 \cong \mathbf{b})) \in \mathrm{Th}(\mathbb{N}),$$

that is,

$$\mathbb{N} \Vdash_{\Sigma_{\mathbb{N}}} ([\varphi_{g \circ f}]^{x_1}_{\mathbf{a}} \Leftrightarrow (x_2 \cong \mathbf{b})).$$

Note that formula $[([\varphi_g]^{x_1}_{x_3} \wedge [\varphi_f]^{x_2}_{x_3})]^{x_1}_{\mathbf{a}}$ coincides with formula

$$([\varphi_g]^{x_1}_{x_3} \wedge [\varphi_f]^{x_1 x_2}_{\mathbf{a} x_3}),$$

since $x_1 \notin \mathrm{fv}_{\Sigma_{\mathbb{N}}}([\varphi_g]^{x_1}_{x_3})$.

Assume that $f(a) = c$. Then, $g(c) = b$. Hence, by hypothesis,

(1) $([\varphi_g]^{x_1}_{\mathbf{c}} \Leftrightarrow (x_2 \cong \mathbf{b})) \in \mathrm{Th}(\mathbb{N})$;

(2) $([\varphi_f]^{x_1}_{\mathbf{a}} \Leftrightarrow (x_2 \cong \mathbf{c})) \in \mathrm{Th}(\mathbb{N})$.

Let ρ be an arbitrary assignment into $\mathbb{N}$. Then, by (1),

$$(3) \ \ \mathbb{N}\rho \Vdash_{\Sigma_{\mathbb{N}}} ([\varphi_g]_{\mathbf{c}}^{x_1} \Leftrightarrow (x_2 \cong \mathbf{b}))$$

and, by (2),

$$(4) \ \ \mathbb{N}\rho \Vdash_{\Sigma_{\mathbb{N}}} ([\varphi_f]_{\mathbf{a}}^{x_1} \Leftrightarrow (x_2 \cong \mathbf{c})).$$

Consider an assignment σ that is x_3-equivalent to ρ and such that $\sigma(x_3) = c$. Then, from (3), it follows that

$$(5) \ \ \mathbb{N}\sigma \Vdash ([\varphi_g]_{x_3}^{x_1} \Leftrightarrow (x_2 \cong \mathbf{b}))$$

using the lemma of substitution in formula from Chapter 7, since $c \rhd_{\Sigma_{\mathbb{N}}} x_3 :$ $[\varphi_g]_{x_3}^{x_1}$. Furthermore, from (4), observing that $c \rhd_{\Sigma_{\mathbb{N}}} x_3 : [\varphi_f]_{\mathbf{a}\,x_3}^{x_1\,x_2}$, the same lemma yields

$$(6) \ \ \mathbb{N}\sigma \Vdash_{\Sigma_{\mathbb{N}}} ([\varphi_f]_{\mathbf{a}\,x_3}^{x_1\,x_2} \Leftrightarrow (\mathbf{c} \cong \mathbf{c})).$$

From (6) it follows that

$$(7) \ \ \mathbb{N}\sigma \Vdash_{\Sigma_{\mathbb{N}}} [\varphi_f]_{\mathbf{a}\,x_3}^{x_1\,x_2}$$

and, from (5) and (7),

$$(8) \ \ \mathbb{N}\sigma \Vdash_{\Sigma_{\mathbb{N}}} (([\varphi_f]_{\mathbf{a}\,x_3}^{x_1\,x_2} \wedge [\varphi_g]_{x_3}^{x_1}) \Leftrightarrow (x_2 \cong \mathbf{b})).$$

From (8), applying 4 of Chapter 19, Exercise 7, it follows that

$$\mathbb{N}\rho \Vdash_{\Sigma_{\mathbb{N}}} ((\exists x_3([\varphi_f]_{\mathbf{a}\,x_3}^{x_1\,x_2} \wedge [\varphi_g]_{x_3}^{x_1})) \Leftrightarrow (x_2 \cong \mathbf{b})),$$

that is,

$$\mathbb{N}\rho \Vdash_{\Sigma_{\mathbb{N}}} ([\varphi_{g \circ f}]_{\mathbf{a}}^{x_1} \Leftrightarrow (x_2 \cong \mathbf{b})).$$

Since this result was established for any ρ, it follows that

$$\mathbb{N} \Vdash_{\Sigma_{\mathbb{N}}} ([\varphi_{g \circ f}]_{\mathbf{a}}^{x_1} \Leftrightarrow (x_2 \cong \mathbf{b})).$$

$$\nabla$$

Exercise 12 Reformulate the proof given in Chapter 3 of the undecidability of the halting problem in order to make use of Cantor's lemma.

Solution: The requested proof is also by contradiction. Let $\lambda k.k_{\mathcal{P}}$ be an enumeration of set $\mathcal{P}$. Recall the injective enumeration $\lambda k.k_{A_M^*}$ of A_M^* and its inverse ord, both computable. Consider the following sets:

- $C = \{(q, v) \in \mathbb{N}^2 : q_{\mathcal{P}}(v_{A_M^*})\!\downarrow\}$;
- $U = \{q \in \mathbb{N} : q_{\mathcal{P}}(q_{A_M^*})\!\uparrow\}$;

- $C_q = \{v \in \mathbb{N} : q_P(v_{A_M^*})\!\downarrow\}$.

Observe that

$$(C_q)_{A_M^*} = \{v_{A_M^*} : v \in C_q\}$$

coincides with the domain of convergence of program q_P. Applying Cantor's lemma, it follows that:

$$(1) \quad U \notin \{C_q : q \in \mathbb{N}\}.$$

However, if the halting problem were decidable, then set C would be computable. Hence set U would also be computable, and hence computably enumerable. Under these conditions, recalling Proposition 10 in Chapter 3, the function

$$h = \lambda q . \begin{cases} 1 & \text{if } q \in U \\ \text{undefined} & \text{otherwise,} \end{cases}$$

would also be computable, and likewise function $h \circ \mathrm{ord}$. Thus, there would exist $u \in \mathbb{N}$ such that $u_P = h \circ \mathrm{ord}$. Thus, since $\mathrm{ord}(v_{A_M^*}) = v$, one would obtain

$$(2) \quad U = C_u,$$

contradicting (1), since:

$$
\begin{aligned}
U \;=\; & \{v \in \mathbb{N} : v \in U\} && = \\
& \{v \in \mathbb{N} : \mathrm{ord}(v_{A_M^*}) \in U\} && = \\
& \{v \in \mathbb{N} : h(\mathrm{ord}(v_{A_M^*})) = 1\} && = \\
& \{v \in \mathbb{N} : u_P(v_{A_M^*}) = 1\} && = \\
& \{v \in \mathbb{N} : u_P(v_{A_M^*})\!\downarrow\} && = C_u
\end{aligned}
$$

∇

Exercise 13 Show that every true theory of arithmetic is ω-consistent.

Solution: Let Θ be a true theory of arithmetic. Assume that

$$\{(\neg[\alpha]_k^y) : k \in \mathbb{N}\} \subseteq \Theta$$

where α is such that $\mathrm{fv}_{\Sigma_{\mathbb{N}}}(\alpha) = \{y\}$. Then,

$$(\dagger) \quad \mathbb{N} \Vdash_{\Sigma_{\mathbb{N}}} (\neg[\alpha]_k^y), \text{ for every } k \in \mathbb{N},$$

since $\mathbb{N} \Vdash_{\Sigma_{\mathbb{N}}} \theta$ for every $\theta \in \Theta$, using the fact that Θ is true. Assume, by contradiction, that $(\exists y\, \alpha) \in \Theta$. Then, $\mathbb{N} \Vdash_{\Sigma_{\mathbb{N}}} (\exists y\, \alpha)$. That is,

$$\mathbb{N}\rho \Vdash_{\Sigma_{\mathbb{N}}} (\exists y\, \alpha)$$

for every assignment ρ in $\mathbb{N}$. Thus, for every ρ, there is an assignment σ such that $\sigma \equiv_y \rho$ and

$$(\dagger\dagger) \quad \mathbb{N}\sigma \Vdash_{\Sigma_N} \alpha.$$

Then, by the substitution lemma,

$$\mathbb{N}\sigma \Vdash_{\Sigma_N} \alpha \quad \text{iff} \quad \mathbb{N}\rho \Vdash_{\Sigma_N} [\alpha]\frac{y}{\sigma(y)}$$

and so, by ($\dagger\dagger$),

$$\mathbb{N}\rho \Vdash_{\Sigma_N} [\alpha]\frac{y}{\sigma(y)}.$$

Therefore,

$$\mathbb{N}\rho \nVdash_{\Sigma_N} (\neg[\alpha]\frac{y}{\sigma(y)}).$$

contradicting ($\dagger$). ∇

Exercise 14 Let $f : \mathbb{N} \to \mathbb{N}$ be a map, gr $f = \{(a, b) : f(a) = b\}$ and Θ an axiomatizable and true theory of arithmetic. Justify or refute the following assertions:

(a) If gr f is computably enumerable then f is computable.

(b) Let φ be a formula representing f in Θ. Then:

$$(a, b) \in \text{gr } f \quad \text{iff} \quad ([\varphi]_a^{x_1} \Leftrightarrow (x_2 \cong \bar{b})) \in \Theta.$$

(c) If f is representable in Θ then f is computable.

(d) If f is representable in $\text{Th}(\mathbb{N})$ then f is computable.

Solution:

(a) True. Indeed, let h be a computable enumeration of gr f. Then, f can be computed as follows:

$$f = \lambda a . (h(\mu k . a = (h(k))_1))_2.$$

(b) True. Indeed, the implication from left to right is a direct consequence of the fact that φ represents f in Θ. The converse also holds thanks to this fact together with the trueness of Θ and the totality of f, as established by *reductio ad absurdum* as follows. Assume

$$([\varphi]_a^{x_1} \Leftrightarrow (x_2 \cong \bar{b})) \in \Theta \quad (\dagger)$$

and $f(a) \neq b$. Since f is total there is $c \neq b$ such that $f(a) = c$ from whence, by representability, one obtains

$$([\varphi]_a^{x_1} \Leftrightarrow (x_2 \cong \bar{c})) \in \Theta. \quad (\ddagger)$$

From (†) and (‡) one obtains

$$((x_2 \cong \bar{b}) \Leftrightarrow (x_2 \cong \bar{c})) \in \Theta$$

from whence it follows $(\bar{b} \cong \bar{c}) \in \Theta$ and, thanks to the trueness of Θ, $b = c$, in contradiction with $b \neq c$.

(c) True. Indeed, let f be represented by φ in Θ and h the characteristic function of $g_{\mathbb{N}}(\Theta)$. That is,

$$h = \lambda k. \begin{cases} 1 & \text{if } k \in g_{\mathbb{N}}(\Theta) \\ \text{undefined} & \text{otherwise} \end{cases}$$

Function h is computable since $g_{\mathbb{N}}(\Theta)$ is computably enumerable. Clearly, $g_{\mathbb{N}}(\Theta)$ is computably enumerable because Θ is computably enumerable (thanks to the Gödelization theorem) and the latter is computably enumerable because it is axiomatizable (thanks to the semi-decidability of FOL). Furthermore,

$$\lambda a, b. ([\varphi]_{\bar{a}}^{x_1} \Leftrightarrow (x_2 \cong \bar{b})) : \mathbb{N}^2 \to L_{\Sigma_{\mathbb{N}}}$$

is also computable. Thus,

$$h' = \lambda a, b. h(g_{\mathbb{N}}([\varphi]_{\bar{a}}^{x_1} \Leftrightarrow (x_2 \cong \bar{b}))) : \mathbb{N}^2 \rightharpoonup \mathbb{N}$$

is computable. Observe that h' is the characteristic function of $\operatorname{gr} f$, thanks to (b) above. Therefore,

$$\operatorname{gr} f = \operatorname{dom} h'$$

is computably enumerable and, so, by (a) above, f is computable.

(d) False. Take as counterexample the characteristic map of $g_{\mathbb{N}}(\mathbf{P})$. Indeed, $\chi_{g_{\mathbb{N}}(\mathbf{P})}$ is representable in $\operatorname{Th}(\mathbb{N})$ (see below) but $\chi_{g_{\mathbb{N}}(\mathbf{P})}$ is not computable since, as a corollary of Church's theorem, theory $\mathbf{P}$ is not decidable. Concerning the representability of $\chi_{g_{\mathbb{N}}(\mathbf{P})}$ in $\operatorname{Th}(\mathbb{N})$, first observe that $\mathbf{P}$ is computably enumerable. Thus, thanks to the Gödelization theorem, $g_{\mathbb{N}}(\mathbf{P})$ is also computably enumerable. Therefore, by the projection theorem, there is a computable binary relation $R_{\mathbf{P}}$ such that $a \in g_{\mathbb{N}}(\mathbf{P})$ iff there is s such that $(a, s) \in R_{\mathbf{P}}$. Let $\varphi_{\mathbf{P}}$ represent $R_{\mathbf{P}}$ in $\operatorname{Th}(\mathbb{N})$. Then:

- If $a \in g_{\mathbb{N}}(\mathbf{P})$ then there is s such that $[\varphi_{\mathbf{P}}]_{\bar{a},\bar{s}}^{x_1,x_2} \in \operatorname{Th}(\mathbb{N})$. That is, there is s such that $[[\varphi_{\mathbf{P}}]_{\bar{a}}^{x_1}]_{\bar{s}}^{x_2} \in \operatorname{Th}(\mathbb{N})$. So,

$$(\exists x_2 [\varphi_{\mathbf{P}}]_{\bar{a}}^{x_1}) \in \operatorname{Th}(\mathbb{N}).$$

That is

$$[(\exists x_2 \varphi_{\mathbf{P}})]_{\bar{a}}^{x_1} \in \operatorname{Th}(\mathbb{N}).$$

- If $a \notin g_{\mathbb{N}}(\mathbf{P})$ then $(\neg[\varphi_{\mathbf{P}}]_{\overline{a},\overline{s}}^{x_1,x_2}) \in \mathrm{Th}(\mathbb{N})$ for every s. So, $[(\neg[\varphi_{\mathbf{P}}]_{\overline{a}}^{x_1})]_{\overline{s}}^{x_2} \in \mathrm{Th}(\mathbb{N})$ for every s. Thus, by ω-consistency of $\mathrm{Th}(\mathbb{N})$,

$$(\exists x_2 \, [\varphi_{\mathbf{P}}]_{\overline{a}}^{x_1}) \notin \mathrm{Th}(\mathbb{N})$$

and, so, by exhaustiveness of $\mathrm{Th}(\mathbb{N})$,

$$(\neg(\exists x_2 \, [\varphi_{\mathbf{P}}]_{\overline{a}}^{x_1})) \in \mathrm{Th}(\mathbb{N}).$$

That is,

$$(\neg[(\exists x_2 \, \varphi_{\mathbf{P}})]_{\overline{a}}^{x_1}) \in \mathrm{Th}(\mathbb{N}).$$

Therefore, $g_{\mathbb{N}}(\mathbf{P})$ is represented in $\mathrm{Th}(\mathbb{N})$ by $(\exists x_2 \, \varphi_{\mathbf{P}})$ and, so, its characteristic map is also representable in $\mathrm{Th}(\mathbb{N})$. ∇

Exercise 15 Let Θ_1 and Θ_2 theories of arithmetic such that: (i) $\Theta_1 \subseteq \Theta_2$; (ii) Θ_1 is suitable; (iii) Θ_2 is ω-consistent. Justify or refute the following assertion:

$$(\square_{\Theta_1} \alpha) \in \Theta_2 \text{ iff } \alpha \in \Theta_1.$$

Solution: First observe that $(\square_{\Theta_1} \alpha)$ is well defined since Θ_1 is assumed to be a suitable theory of arithmetic (that is, a computably enumerable theory of arithmetic in which computable maps are representable). Recall also that $(\square_{\Theta_1} \alpha)$ is an abbreviation of

$$[\exists x_2 \, \varphi_{\Theta_1}]_{\lceil \alpha \rceil}^{x_1}$$

where $\lceil \alpha \rceil$ is the numeral corresponding to the Gödel number of α and φ_{Θ_1} is a formula representing in Θ_1 a computable binary relation R_{Θ_1} such that a is the Gödel number of a theorem of Θ_1 iff there exists $s \in \mathbb{N}$ such that $(a, s) \in R_{\Theta_1}$. The existence of such a relation follows by the projection and the Gödelization theorems, since Θ_1 is computably enumerable. Furthermore, the assertion at stake is true. Indeed:

$(\rightarrow)$ By contraposition:

if	$\alpha \notin \Theta_1$,
then	$g_{\mathbb{N}}(\alpha) \notin g_{\mathbb{N}}(\Theta_1)$
	for any $s \in \mathbb{N}, (g_{\mathbb{N}}(\alpha), s) \notin R_{\Theta_1}$
	for any $s \in \mathbb{N}, (\neg[\varphi_{\Theta_1}]_{\lceil\alpha\rceil,\overline{s}}^{x_1,x_2}) \in \Theta_1$ (representability of R_{Θ_1} in Θ_1)
	for any $s \in \mathbb{N}, (\neg[\varphi_{\Theta_1}]_{\lceil\alpha\rceil,\overline{s}}^{x_1,x_2}) \in \Theta_2$ (since $\Theta_1 \subseteq \Theta_2$)
	for any $s \in \mathbb{N}, (\neg[[\varphi_{\Theta_1}]_{\lceil\alpha\rceil}^{x_1}]_{\overline{s}}^{x_2}) \in \Theta_2$
	$\{(\neg[[\varphi_{\Theta_1}]_{\lceil\alpha\rceil}^{x_1}]_{\overline{s}}^{x_2}) : s \in \mathbb{N}\} \subseteq \Theta_2$
	$(\exists x_2 \, [\varphi_{\Theta_1}]_{\lceil\alpha\rceil}^{x_1}) \notin \Theta_2$ (ω-consistency of Θ_2)
	$[(\exists x_2 \, \varphi_{\Theta_1})]_{\lceil\alpha\rceil}^{x_1} \notin \Theta_2$
	$(\square_{\Theta_1} \alpha) \notin \Theta_2$

($\leftarrow$) Directly:

if $\quad \alpha \in \Theta_1$,

then $\quad g_{\mathbb{N}}(\alpha) \in g_{\mathbb{N}}(\Theta_1)$

$\qquad$ there is $s \in \mathbb{N}$ s.t. $(g_{\mathbb{N}}(\alpha), s) \in R_{\Theta_1}$

$\qquad$ there is $s \in \mathbb{N}$ s.t. $[\varphi_{\Theta_1}]_{\lceil \alpha \rceil, s}^{x_1, x_2} \in \Theta_1$ $\qquad$ (representability of R_{Θ_1} in Θ_1)

$\qquad$ $(\exists x_2 \, [\varphi_{\Theta_1}]_{\lceil \alpha \rceil}^{x_1}) \in \Theta_1$ $\qquad$ (ExR1)

$\qquad$ $[(\exists x_2 \, \varphi_{\Theta_1})]_{\lceil \alpha \rceil}^{x_1} \in \Theta_2$ $\qquad$ (since $\Theta_1 \subseteq \Theta_2$)

$\qquad$ $(\square_{\Theta_1} \alpha) \in \Theta_2$

$$\nabla$$

Exercise 16 Let $\alpha, \beta \in cL_{\Sigma_{\mathbb{N}}}$ and Θ a suitable and true theory of arithmetic. Assume also that Θ fulfills the HBL2 e HBL3 conditions. Justify or refute the following assertions:

($\dagger$) $((\square_{\Theta}(\alpha \wedge \beta)) \Leftrightarrow ((\square_{\Theta}\alpha) \wedge (\square_{\Theta}\beta))) \in \Theta$.

($\ddagger$) $((\square_{\Theta}(\alpha \Rightarrow ((\square_{\Theta}\alpha) \Rightarrow \beta))) \Rightarrow ((\square_{\Theta}(\square_{\Theta}\alpha)) \Rightarrow (\square_{\Theta}(\square_{\Theta}\beta)))) \in \Theta$.

Solution:

($\dagger$) True. Indeed, recall that Θ also fulfills the HBL1 condition since it is suitable. Thus:

$$((\square_{\Theta}(\alpha \wedge \beta)) \Rightarrow (\square_{\Theta}\alpha)) \in \Theta$$

since

1	$((\alpha \wedge \beta) \Rightarrow \alpha)$	TAUT
2	$(\square_{\Theta}((\alpha \wedge \beta) \Rightarrow \alpha))$	HBL1 1
3	$((\square_{\Theta}((\alpha \wedge \beta) \Rightarrow \alpha)) \Rightarrow ((\square_{\Theta}(\alpha \wedge \beta)) \Rightarrow (\square_{\Theta}\alpha)))$	HBL2
4	$((\square_{\Theta}(\alpha \wedge \beta)) \Rightarrow (\square_{\Theta}\alpha))$	MP 2, 3

Similarly,

$$((\square_{\Theta}(\alpha \wedge \beta)) \Rightarrow (\square_{\Theta}\beta)) \in \Theta.$$

Therefore, tautologically,

$$((\square_{\Theta}(\alpha \wedge \beta)) \Rightarrow ((\square_{\Theta}\alpha) \wedge (\square_{\Theta}\beta))) \in \Theta.$$

Moreover,

$$(((\square_{\Theta}\alpha) \wedge (\square_{\Theta}\beta)) \Rightarrow (\square_{\Theta}(\alpha \wedge \beta))) \in \Theta$$

since

$$1 \quad (\alpha \Rightarrow (\beta \Rightarrow (\alpha \wedge \beta))) \qquad\qquad\qquad \text{TAUT}$$

$$2 \quad (\square_\Theta(\alpha \Rightarrow (\beta \Rightarrow (\alpha \wedge \beta)))) \qquad\qquad \text{HBL1 1}$$

$$3 \quad ((\square_\Theta(\alpha \Rightarrow (\beta \Rightarrow (\alpha \wedge \beta))))$$
$$\Rightarrow$$
$$((\square_\Theta\alpha) \Rightarrow (\square_\Theta(\beta \Rightarrow (\alpha \wedge \beta))))) \qquad \text{HBL2}$$

$$4 \quad ((\square_\Theta\alpha) \Rightarrow (\square_\Theta(\beta \Rightarrow (\alpha \wedge \beta)))) \qquad\qquad \text{MP } 2,3$$

$$5 \quad ((\square_\Theta(\beta \Rightarrow (\alpha \wedge \beta))) \Rightarrow ((\square_\Theta\beta) \Rightarrow (\square_\Theta(\alpha \wedge \beta)))) \qquad \text{HBL2}$$

$$6 \quad ((\square_\Theta\alpha) \Rightarrow ((\square_\Theta\beta) \Rightarrow (\square_\Theta(\alpha \wedge \beta)))) \qquad\qquad \text{TAUT } 4,5$$

$$7 \quad (((\square_\Theta\alpha) \Rightarrow ((\square_\Theta\beta) \Rightarrow (\square_\Theta(\alpha \wedge \beta))))$$
$$\Rightarrow$$
$$(((\square_\Theta\alpha) \wedge (\square_\Theta\beta)) \Rightarrow (\square_\Theta(\alpha \wedge \beta)))) \qquad \text{TAUT}$$

$$8 \quad (((\square_\Theta\alpha) \wedge (\square_\Theta\beta)) \Rightarrow (\square_\Theta(\alpha \wedge \beta))) \qquad\qquad \text{MP } 6,7$$

(‡) True. Indeed,

$$1 \quad ((\square_\Theta(\alpha \Rightarrow ((\square_\Theta\alpha) \Rightarrow \beta)))$$
$$\Rightarrow$$
$$((\square_\Theta\alpha) \Rightarrow (\square_\Theta((\square_\Theta\alpha) \Rightarrow \beta)))) \qquad \text{HBL2}$$

$$2 \quad ((\square_\Theta((\square_\Theta\alpha) \Rightarrow \beta)) \Rightarrow ((\square_\Theta(\square_\Theta\alpha)) \Rightarrow (\square_\Theta\beta))) \qquad \text{HBL2}$$

$$3 \quad ((\square_\Theta(\alpha \Rightarrow ((\square_\Theta\alpha) \Rightarrow \beta)))$$
$$\Rightarrow$$
$$((\square_\Theta\alpha) \Rightarrow ((\square_\Theta(\square_\Theta\alpha)) \Rightarrow (\square_\Theta\beta)))) \qquad \text{TAUT } 1,2$$

$$4 \quad ((\square_\Theta\alpha) \Rightarrow (\square_\Theta(\square_\Theta\alpha))) \qquad\qquad\qquad \text{HBL3}$$

$$5 \quad ((\square_\Theta(\alpha \Rightarrow ((\square_\Theta\alpha) \Rightarrow \beta))) \Rightarrow ((\square_\Theta\alpha) \Rightarrow (\square_\Theta\beta))) \qquad \text{TAUT } 3,4$$

$$6 \quad ((\square_\Theta(\alpha \Rightarrow ((\square_\Theta\alpha) \Rightarrow \beta)))$$
$$\Rightarrow$$
$$(\square_\Theta(\square_\Theta(\alpha \Rightarrow ((\square_\Theta\alpha) \Rightarrow \beta))))) \qquad \text{HBL3}$$

$$7 \quad (\square_\Theta((\square_\Theta(\alpha \Rightarrow ((\square_\Theta\alpha) \Rightarrow \beta))) \Rightarrow ((\square_\Theta\alpha) \Rightarrow (\square_\Theta\beta)))) \qquad \text{HBL1 5}$$

$$8 \quad ((\square_\Theta((\square_\Theta(\alpha \Rightarrow ((\square_\Theta\alpha) \Rightarrow \beta))) \Rightarrow ((\square_\Theta\alpha) \Rightarrow (\square_\Theta\beta))))$$
$$\Rightarrow$$
$$((\square_\Theta(\square_\Theta(\alpha \Rightarrow ((\square_\Theta\alpha) \Rightarrow \beta))))$$
$$\Rightarrow$$
$$(\square_\Theta((\square_\Theta\alpha) \Rightarrow (\square_\Theta\beta))))) \qquad \text{HBL2}$$

$$9 \quad ((\square_\Theta(\square_\Theta(\alpha \Rightarrow ((\square_\Theta\alpha) \Rightarrow \beta))))$$
$$\Rightarrow$$
$$(\square_\Theta((\square_\Theta\alpha) \Rightarrow (\square_\Theta\beta)))) \qquad \text{MP } 7,8$$

10   $((\Box_\Theta(\alpha \Rightarrow ((\Box_\Theta\alpha) \Rightarrow \beta)))$
      $\Rightarrow$
    $(\Box_\Theta((\Box_\Theta\alpha) \Rightarrow (\Box_\Theta\beta))))$ TAUT $6, 9$

11   $((\Box_\Theta((\Box_\Theta\alpha) \Rightarrow (\Box_\Theta\beta)))$
      $\Rightarrow$
    $((\Box_\Theta(\Box_\Theta\alpha)) \Rightarrow (\Box_\Theta(\Box_\Theta\beta))))$ HBL2

12   $((\Box_\Theta(\alpha \Rightarrow ((\Box_\Theta\alpha) \Rightarrow \beta)))$
      $\Rightarrow$
    $((\Box_\Theta(\Box_\Theta\alpha)) \Rightarrow (\Box_\Theta(\Box_\Theta\beta))))$ TAUT $10, 11$

∇

Bibliography

[1] J. Barwise. Axioms for abstract model theory. *Annals for Mathematical Logic*, 7:221–265, 1974.

[2] J. L. Bell and A. B. Slomson. *Models and Ultraproducts: An Introduction*. North-Holland Publishing Co., 1969.

[3] G. Birkhoff. *Lattice Theory*, volume 25 of *American Mathematical Society Colloquium Publications*. American Mathematical Society, Third edition, 1979.

[4] G. Boolos. *The Logic of Provability*. Cambridge University Press, 1993.

[5] N. Bourbaki. *Elements of the History of Mathematics*. Springer, 1994. Translated from the 1984 French original by J. Meldrum.

[6] D. S. Bridges. *Computability*, volume 146 of *Graduate Texts in Mathematics*. Springer, 1994.

[7] S. R. Buss. An introduction to proof theory. In *Handbook of Proof Theory*, volume 137 of *Studies in Logic and the Foundations of Mathematics*, pages 1–78. North-Holland, 1998.

[8] J. Carmo, A. Sernadas, C. Sernadas, F.M. Dionísio, and C. Caleiro. *Introdução à Programação em Mathematica – Segunda Edição*. IST Press, Lisboa, 2004.

[9] W. A. Carnielli, M. E. Coniglio, D. Gabbay, P. Gouveia, and C. Sernadas. *Analysis and Synthesis of Logics - How To Cut And Paste Reasoning Systems*, volume 35 of *Applied Logic*. Springer, 2008.

[10] A. Church. An unsolvable problem of elementary number theory. *American Journal of Mathematics*, 58(2):345–363, 1936.

[11] R. Cori and D. Lascar. *Mathematical Logic, Part I, Propositional Calculus, Boolean Algebras, Predicate Calculus*. Oxford University Press, 2000.

[12] R. Cori and D. Lascar. *Mathematical Logic, Part II, Recursion Theory, Gödel Theorems, Set Theory, Model Theory.* Oxford University Press, 2000.

[13] W. Craig. On axiomatizability within a system. *The Journal of Symbolic Logic*, 18:30–32, 1953.

[14] R. David, K. Nour, and C. Raffalli. *Introduction à la Logique.* Dunod, 2001.

[15] H. B. Enderton. *A Mathematical Introduction to Logic.* Academic Press, Burlington, MA, Second edition, 2001.

[16] R. L. Epstein and W. A. Carnielli. *Computability: Computable Functions, Logic and the Foundations of Mathematics.* Wadsworth, Second edition, 2000.

[17] M. Fitting. *First-order Logic and Automated Theorem Proving.* Graduate Texts in Computer Science. Springer, Second edition, 1996.

[18] G. Gentzen. *The Collected Papers of Gerhard Gentzen.* Edited by M. E. Szabo. Studies in Logic and the Foundations of Mathematics. North-Holland Publishing Co., 1969.

[19] G. Gierz, K. H. Hofmann, K. Keimel, J. D. Lawson, M. W. Mislove, and D. S. Scott. *A Compendium of Continuous Lattices.* Springer, 1980.

[20] K. Gödel. *Collected Works. Vol. I.* Oxford University Press, 1986. Edited by S. Feferman, J. W. Dawson, Jr., S. C. Kleene, G. H. Moore, R. M. Solovay and J. van Heijenoort.

[21] J. A. Goguen and R. M. Burstall. Institutions: Abstract model theory for specification and programming. *Journal of the Association for Computing Machinery*, 39(1):95–146, 1992.

[22] G. H. Hardy and E. M. Wright. *An Introduction to the Theory of Numbers.* Oxford University Press, Fifth edition, 1979.

[23] L. Henkin. The completeness of the first-order functional calculus. *The Journal of Symbolic Logic*, 14:159–166, 1949.

[24] L. Henkin, J. D. Monk, and A. Tarski. *Cylindric Algebras. Part I*, volume 64 of *Studies in Logic and the Foundations of Mathematics.* North-Holland, 1985. With an introductory chapter: General theory of algebras, Reprint of the 1971 original.

[25] R. Herken, editor. *The Universal Turing Machine: A Half-Century Survey*. Oxford Science Publications. Oxford University Press, 1988.

[26] D. Hilbert. *Grundlagen der Geometrie*, volume 6 of *Teubner-Archiv zur Mathematik. Supplement [Teubner Archive on Mathematics. Supplement]*. B. G. Teubner Verlagsgesellschaft mbH, Fourteenth edition, 1999. With supplementary material by P. Bernays, With contributions by M. Toepell, H. Kiechle, A. Kreuzer and H. Wefelscheid, Edited and with appendices by M. Toepell.

[27] D. Hilbert and W. Ackermann. *Grundzüge der Theoretischen Logik*. Springer, Third edition, 1949.

[28] D. Hilbert and W. Ackermann. *Principles of Mathematical Logic*. American Mathematical Society, 2003.

[29] W. Hodges. *A Shorter Model Theory*. Cambridge University Press, 1997.

[30] P. T. Johnstone. *Stone Spaces*, volume 3 of *Cambridge Studies in Advanced Mathematics*. Cambridge University Press, 1982.

[31] J. L. Kelley. *General Topology*. Springer, 1975. Reprint of the 1955 edition [Van Nostrand, Toronto, Ont.], Graduate Texts in Mathematics, No. 27.

[32] S. C. Kleene. General recursive functions of natural numbers. *Mathematische Annalen*, 112(1):727–742, 1936.

[33] S. Lang. *Algebra*, volume 211 of *Graduate Texts in Mathematics*. Springer, Third edition, 2002.

[34] S. Mac Lane. *Categories for the Working Mathematician*, volume 5 of *Graduate Texts in Mathematics*. Springer, Second edition, 1998.

[35] E. Mendelson. *Introduction to Mathematical Logic*. Chapman and Hall, Fourth edition, 1997.

[36] E. L. Post. Formal reductions of the general combinatorial decision problem. *American Journal of Mathematics*, 65:197–215, 1943.

[37] M. Presburger. On the completeness of a certain system of arithmetic of whole numbers in which addition occurs as the only operation. *History and Philosophy of Logic*, 12(2):225–233, 1991. Translated from the German and with commentaries by Dale Jacquette.

[38] H. Rogers, Jr. *Theory of Recursive Functions and Effective Computability*. MIT Press, second edition, 1987.

[39] V. V. Rybakov. *Admissibility of Logical Inference Rules*, volume 136 of *Studies in Logic and the Foundations of Mathematics*. North-Holland Publishing Co., 1997.

[40] A. Sernadas, C. Sernadas, and C. Caleiro. Fibring of logics as a categorial construction. *Journal of Logic and Computation*, 9(2):149–179, 1999.

[41] J. C. Shepherdson and H. E. Sturgis. Computability of recursive functions. *Journal of the Association for Computing Machinery*, 10:217–255, 1963.

[42] J. R. Shoenfield. *Mathematical Logic*. Association for Symbolic Logic, Urbana, IL, 2001. Reprint of the 1973 second printing.

[43] R. M. Smullyan. *First-order Logic*. Dover Publications Inc., 1995. Corrected reprint of the 1968 original.

[44] W. Szmielew. Elementary properties of Abelian groups. *Polska Akademia Nauk. Fundamenta Mathematicae*, 41:203–271, 1955.

[45] A. Tarski. The semantic conception of truth and the foundations of semantics. *Philos. and Phenomenol. Res.*, 4:341–376, 1944.

[46] A. Tarski. A lattice-theoretical fixpoint theorem and its applications. *Pacific Journal of Mathematics*, 5:285–309, 1955.

[47] A. Tarski. What is elementary geometry? In *The Axiomatic Method. With Special Reference to Geometry and Physics. Proceedings of an International Symposium held at the Univ. of Calif., Berkeley, Dec. 26, 1957-Jan. 4, 1958 (edited by L. Henkin, P. Suppes and A. Tarski)*, Studies in Logic and the Foundations of Mathematics, pages 16–29. North-Holland, 1959.

[48] A. Turing. On computable numbers, with an application to the Entscheidungsproblem. *Proceedings of the London Mathematical Society*, 2(42):230–265, 1936.

[49] A. Turing. *Pure Mathematics*. Collected Works of Alan Turing. North-Holland Publishing Co., 1992. Edited and with an introduction and postscript by J. L. Britton, With a preface by P. N. Furbank.

[50] L. van den Dries. Alfred Tarski's elimination theory for real closed fields. *The Journal of Symbolic Logic*, 53(1):7–19, 1988.

[51] J. van Heijenoort, editor. *From Frege to Gödel. A Source Book in Mathematical Logic, 1879–1931*. Harvard University Press, 2002. Reprint of the third printing of the 1967 original.

[52] D. A. Wolfram. *The Mathematica Book*. Wolfram Media, Fifth edition, 2003.

[53] Y. Yin. Quantifier elimination and real closed ordered fields with a predicate for the powers of two. Master's thesis, Carnegie-Mellon University, 2005. Supervised by J. Avigad.

[54] R. Zach. Hilbert's program then and now. In D. Jaquette, editor, *Handbook on the Philosophy of Science*, volume 5, pages 411–447. Elsevier, 2006.

Table of symbols

317

Subject Index